Video and Multimedia Transmissions over Cellular Networks

Video and Multimedia Transmissions over Cellular Networks

Analysis, Modelling and Optimization in Live 3G Mobile Communications

Editor Markus Rupp

University of Technology Vienna, Austria

A John Wiley and Sons, Ltd, Publication

This edition first published 2009
© 2009 John Wiley & Sons Ltd

Registered office
John Wiley & Sons Ltd, The Atrium, Southern Gate, Chichester, West Sussex, PO19 8SQ,
United Kingdom.

For details of our global editorial offices, for customer services and for information about how to apply
for permission to reuse the copyright material in this book please see our website at www.wiley.com.

Library of Congress Cataloging-in-Publication Data

Rupp, Markus, 1963-
    Video and multimedia transmissions over cellular networks : analysis, modelling, and optimization
in live 3G mobile communications / Markus Rupp.
        p. cm.
    Includes bibliographical references and index.
    ISBN 978-0-470-69933-1 (cloth)
1. Multimedia communications. 2. Wireless communication systems. 3. Mobile computing. 4. Video
telephone. I. Title.
    TK5105.15.R86 2009
    006.7–dc20   2009007440

A catalogue record for this book is available from the British Library.

ISBN 978-0-470-69933-1 (H/B)

Set in 10/12pt Times by Sunrise Setting Ltd, Torquay, UK.
Printed in Great Britain by CPI Antony Rowe, Chippenham, UK.

Contents

II   Analysis and Modelling of the Wireless Link   35

III Video Coding and Error Handling   97

List of Contributors

Wolfgang Karner, Olivia Nemethova, Michal Ries, Markus Rupp, Luca Superiori and Philipp Svoboda
Institute of Communications and Radio Frequency Engineering
Vienna University of Technology (TU Wien)
Gusshausstrasse 25/389
A-1040 Vienna
Austria

Fabio Ricciato, Claudio Weidmann and Peter Romirer-Maierhofer
Forschungszentrum Telekommunikation Wien (ftw.)
Tech Gate Vienna
Donau-City-Strae 1/3
A-1220 Vienna
Austria

Thomas Stockhammer
Nomor Research GmbH
Brecherspitzstraße 8
D-81541 Munich
Germany

Jiangtao Wen
Department of Computer Science
Tsinghua University
Beijing 100084
P.R. China

About the Contributors

Wolfgang Karner received his Dipl.-Ing. (MS) degree in Electrical Engineering with special subjects Communications and Information Technology from Vienna University of Technology, Austria, in 2003 and his Dr. techn. (PhD) degree in 2007, both with distinction. His research interests include performance evaluation, modelling, simulation and cross-layer optimization of wireless communication systems with the focus on radio resource optimization in UMTS (DCH, HSDPA, HSUPA) and the transmission of multimedia services. He is currently working for mobilkom austria AG. Email: w.karner@mobilkom.at

Olivia Nemethova received her BS and MS degrees from Slovak University of Technology in Bratislava in 1999 and 2001, respectively, both in Informatics and Telecommunications. She received her Dr. techn. (PhD) in Electrical Engineering from Vienna University of Technology with distinction in 2007. From 2001 until 2003 she was with Siemens as a systems engineer. She worked on UMTS standardization within 3GPP TSG RAN2 as a Siemens delegate. In parallel she worked within an International Property Rights management team responsible for evaluation of IPRs regarding RANs. In 2003 she joined the Institute of Communications and Radio-Frequency Engineering at Vienna University of Technology as a research and teaching assistant. Her current research interests include error resilient transmission of multimedia over wireless networks, video processing and mobile communications. Email: onemeth@nt.tuwien.ac.at

Peter Romirer-Maierhofer received his Dipl.-Ing. (FH) degree in Telecommunications Engineering from the Salzburg University of Applied Sciences, Austria, and an MS degree in Computer Systems Engineering from Halmstad University, Sweden, in 2005. He is a member of the DARWIN project on traffic analysis in GPRS/UMTS networks at the Telecommunications Research Center, Vienna. Currently, he is working towards his PhD degree in Computer Science in the field of anomaly detection in 3G cellular networks. Email: romirer@ftw.at

Fabio Ricciato received a Laurea degree in Electrical Engineering and PhD in Telecommunications from the University La Sapienza of Rome in 1999 and 2003, respectively. In 2004 he joined the Telecommunications Research Center Vienna (ftw.) as Senior Researcher and Project Manager for the METAWIN project. For the past five years he has been leading a series of research projects on traffic measurements in 3G networks. Currently he holds an Assistant Professorship at the University of Salento, Italy, while at the same time serving as Key Researcher and Scientific Manager for the Networking Area at ftw. He is the proposer and Chair of the COST Action IC0703 on Traffic Monitoring and Analysis (www.tma-cost.eu) launched in 2008. His research interests are in the networking area for wireless and backbone technologies, including QoS, traffic engineering, resilience, traffic measurement and analysis, and network security. Email: ricciato@ftw.at

Michal Ries received his BS and Dipl.-Ing. degrees in 2002 and 2004 at the Slovak University of Technology, Faculty of Electrical Engineering and Information Technology in Bratislava and his Dr. techn. degree in 2008 at the Institute of Communications and Radio-Frequency Engineering at Vienna University of Technology. Before he joined TU Vienna he was working for Siemens PSE as a systems engineer. In 2004 he joined the Institute of Communications and Radio-Frequency Engineering as a research assistant. His research interests include perceptual video and audiovisual quality evaluation, video and audiovisual metric design, monitoring of QoS in wireless networks and video streaming in wireless network optimization. Email: mries@nt.tuwien.ac.at

Markus Rupp received his Dipl.-Ing. degree in 1988 at the University of Saarbruecken, Germany, and his Dr.-Ing. degree in 1993 at the Technische Universitaet Darmstadt, Germany, where he worked with Eberhardt Haensler on designing new algorithms for acoustical and electrical echo compensation. From November 1993 until July 1995, he had a postdoctoral position at the University of Santa Barbara, California, with Sanjit Mitra and where he worked with Ali H. Sayed on a robustness description of adaptive filters with impact on neural networks and active noise control. From October 1995 until August 2001 he was a member of the Technical Staff in the Wireless Technology Research Department of Bell-Labs at Crawford Hill, NJ, where he worked on various topics related to adaptive equalization and rapid implementation for IS-136, 802.11 and UMTS. From October 2001 he has been a full professor for Digital Signal Processing in Mobile Communications at the Technical University of Vienna where he served as Dean from 2005 to 2007. He was associate editor of *IEEE Transactions on Signal Processing* from 2002 to 2005, and is currently associate editor of *JASP EURASIP Journal of Advances in Signal Processing*, *JES EURASIP Journal on Embedded Systems*, *Research Letters in Signal Processing*. He has been elected AdCom member of EURASIP since 2004, serving as president of EURASIP from 2009 to 2010. He has authored and co-authored more than 300 papers and patents on adaptive filtering, wireless communications and rapid prototyping, as well as automatic design methods. Email: mrupp@nt.tuwien.ac.at

Thomas Stockhammer received his Dipl.-Ing. and Dr.-Ing. degrees from the Munich University of Technology, Germany, and was visiting researcher at Rensselear Polytechnic Institute (RPI), Troy, NY, and at the University of San Diego, California (UCSD). He has published more than 120 conference and journal papers, is member of different technical program committees and holds about 50 patents. He regularly participates in and contributes to different standardization activities, for example JVT, ITU-T, IETF, 3GPP and DVB, and has co-authored more than 200 technical contributions. He was chairing the video ad hoc group of 3GPP SA4 and is now the chair of the DVB IPTV Application Layer FEC and Content Download System Task Force and also acts as rapporteur/editor of several standardization documents. He is also co-founder and CEO of Novel Mobile Radio (Nomor) Research, a company developing simulation and emulation platforms of future mobile networks such as HSxPA, WiMaX, MBMS and LTE as well as Mobile and IPTV-related matters. The company also provides consulting services in the respective areas. After his work as research assistant at the Munich University of Technology until 2004, he was working for two years as a research and development consultant for Siemens Mobile Devices, later BenQ mobile in Munich, Germany. Since June 2006, he has been consultant for Digital Fountain, Inc., in research and standardization matters for CDPs, IPTV and mobile multimedia communication. His research interests include video transmission, cross-layer and system design, forward error correction, content delivery protocols, rate-distortion optimization, information theory and mobile communications. Email: stockhammer@nomor.de

Luca Superiori received his BS and MS degrees in Electronic Engineering in 2002 and 2005, respectively, both from the University of Cagliari, Italy. In 2006 he joined the Institute of Communications and Radio-Frequency Engineering at Vienna University of Technology where he is currently working as Research Assistant in the field of video streaming over wireless networks. His current research interests are focused on error detection mechanisms for H.264/AVC streams, content-aware encoding for specific video contents, such as football, as well as cross-layer optimization of low-resolution video streams transmitted over 3G networks. Email: lsuper@nt.tuwien.ac.at

Philipp Svoboda received his Dipl.-Ing. (MS) degree in Electrical Engineering in 2004 and his Dr. techn. (PhD) with distinction in 2008 from the Vienna University of Technology, Austria. He has worked on the METAWIN (Measurement and Traffic Analysis in Wireless Networks) project focusing his research on the user behaviour in the GPRS and UMTS core network of mobilkom austria AG. At the moment he is involved in the Data Analysis and Reporting in Wireless Networks (DARWIN) project extending his work from METAWIN towards anomaly detection. In his spare time he investigates traffic models for online games and measures service performance in 3G networks. His current research interests include simple traffic generation, statistical analysis of IP level information and modelling of new services, specifically on all kinds of mobile networks. Email: psvoboda@nt.tuwien.ac.at

Claudio Weidmann received the Diploma degree in Electrical Engineering from ETH Zurich, Switzerland, in 1993 and the Docteur ès Sciences degree from EPF Lausanne, Switzerland, in 2000. From 1992 to 1996 he was involved in a start-up company, and from 1996 to 2001 he was a research assistant at the Audiovisual Communications Laboratory at Ecole Polytechnique Fédérale de Lauranne (EPFL). He spent 2002 as a post-doctoral researcher at IRISA–INRIA Rennes, France. From 2003, he has been a senior researcher at the Telecommunications Research Center Vienna (ftw.), Austria. In 2006 and 2007, he also spent time as an invited assistant professor at Université Paris–Sud XI, France, hosted by the Laboratoire des Signaux et Systèmes (LSS). His research interests include applied information theory and signal processing, in particular, distributed source coding and joint source-channel methods. Email: claudio.weidmann@ieee.org

Jiangtao (Gene) Wen received BS, MS and PhD degrees from Tsinghua University in 1992, 1994 and 1996, respectively, all in Electrical Engineering. From 1996 to 1998, he was a Staff Research Fellow at the University of California, Los Angeles (UCLA), where he conducted cutting-edge research on multimedia coding and communications that later became international standards. Later in his career, he served as the Principal Scientist of PacketVideo Corp., the CTO of Morphbius Technology Inc., a Director of Video Codec Technologies of Mobilygen Corp., the Senior Director of Technology at Ortiva Wireless Inc., and consulted for many Fortune 500 or start-up companies. Dr Wen is a world-renowned expert in multimedia communication over hostile networks, video coding and communications. He has authored many widely referenced papers in related fields. Products deploying technologies that Dr Wen developed are currently used by over 200 million customers worldwide. Dr Wen holds 12 granted US patents with numerous others pending. A Senior Member of IEEE, Dr Wen is an Associate Editor for IEEE Transactions CSVT. Email: jiangtao.wen@gmail.com

Foreword

Telecommunication as a Responsibility in Modern Society

When the predecessor company of mobilkom austria AG started its first research cooperation with the Institute of Communications and Radio-Frequency Engineering of the Technical University Vienna, in the late 1980s, a societal change was emerging that few people, if any, were aware of at that time: the era of mobile communications. No other technology ever invented by mankind was embraced so eagerly and as quickly as the mobile phone. In retrospect, after thousands of person years invested in research and development all over the world, the open questions sound very simple today: how to get a cellular system for voice calls working, how to deploy a digital system such as GSM, including positioning of antenna posts, 'downtilt' angles, speech quality as well as the number of base stations in general; the questions we had were very much down to earth. Looking at this book we realize how quickly this field has developed in just 20 years. Questions now relate to network optimization, video streaming and quality of multimedia services including a cornucopia of services.

In taking the next step the open question is, what will our research questions in the future be like? Here, we strongly believe that true advances will be not so much in physical, technological or information-theoretical aspects as in usability and usefulness for human beings. *Easytech* rather than *Hightech* is the keyword. Do not forget that the backdrop of the sweeping victory of the mobile phone was its satisfaction of a key personal need: the desire for communication, independent of time and place. With telecommunication we can serve our society in much desired needs and – astonishing to many people – we can do it while saving cost, preserving nature and improving the quality of life as well as lifestyle, all at the same time.

How is this possible, you may well ask. By analysing trends in our society we recognize that in future we will live in more and more different environments. At work we are in our work environment with very particular needs for communication with our colleagues and company partners; at home we are in our home environment, interacting with our family; and if we follow our favourite sports activity, we are again in a different environment. For every environment, we have different needs and, in particular, different communication needs. Similar to an animal living in a certain biotope, human beings are living in a so-called infotope that needs to provide them with all kinds of information and access required. Translated into technical requirements it follows that modern devices have to become more user centric. A single individual technical contribution becomes less important but the interplay of all requirements fulfilled to serve the customer will become the key factor.

What is true for the individual is also true for households, and medium as well as large companies in general. Today, geolocation makes it possible to optimally route fleets of traffic through Europe, saving cost and energy and lowering pollution. Modern buildings measure temperature, air condition, humidity, pollution and much more, and set optimal values for their inhabitants but at the same time save energy and cost.

Telecommunications will change the world as we know it today. In fact it has already had substantial impact. As Bob Geldof pointed out correctly: mobile telecommunications has not simply been a communication asset but has triggered a social revolution in Africa. We live in exciting telecommunications times and it will be even more exciting to live in 'the foreseeable future'.

Boris Nemsic
mobilkom austria AG

Preface

The idea for this book started some years ago when I endeavoured to work with mobilkom austria AG. Although my personal background was in signal processing, and my expertise in wireless that I had gained during my six years at Bell-Labs Research was more focused on the algorithmic side of wireless communications, I saw great potential in learning the view of the network provider.

We began working on several hot topics at the same time: resilient video streaming over the newly introduced Universal Mobile Telecommunications System (UMTS) network, measurements-based modelling of the UMTS link layer, and finally Quality of Service (QoS) in the core network and in particular for video services. In parallel the METAWIN project started at Forschungszentrum Telekommunikation Wien (ftw.) to monitor all relevant data in the core network in order to identify bottlenecks and to model the various services encountered. Over the years much insight was gained on all counts and eventually the PhD students were finishing, about to leave for new adventures. This was the time to keep them in the loop for a few extra months to ensure that their work was well prepared for a few book chapters, each providing an overview of many new and interesting as well as challenging topics on wireless communications.

The particular characteristic of the methods presented in this book is that everything is based on measurements in life networks, rather than on pure simulation results. This makes the findings in this book rather unique since most researchers rely on models these days. However, the real world is often difficult and different, often not even closely related, from the models we have. Take, for example, the Internet, invented in 1969, with many astonishing queuing theory results in the 1970s. Since then, traffic has changed considerably, the Internet has changed, and on top of everything, services have also changed. The simple queuing theoretical approaches do not reflect gaming or email traffic at all. It was thus time to measure traffic and to model it anew to reflect the real world of today.

How to Read this Book

Parts I and III of this book are devoted to people that come from different areas of expertise. Experts in video coding, for example, often have problems understanding the complicated network architecture of cellular wireless systems that is required to transmit their videos. For them and all those not so familiar with the network architecture, we devote the first part of this book to explaining in detail the core network architecture for UMTS and GPRS as well as the UMTS Terrestrial Radio Access Network and HSPA of Rel. 5 and Rel. 6.

On the other hand, experts in wireless network architectures often face the problem of understanding the basic principles of video coding. Until recently video streams were simply treated as IP traffic – either TCP or UDP. We now realize that even only a basic knowledge of video streams can significantly improve transmission quality. Thus, understanding video coding is becoming more and more important for the traffic engineer.

For those interested in the basic principles of video coding we devote the chapters in Part III where we explain the principles of modern video coding as it is being applied in current standards H. 263 and H. 264. Receiver quality is largely achieved by error concealment techniques. We thus present the state of the art in error concealment techniques and discuss recently developed methods. Error concealment methods can only work if errors are detected and, in particular, if their location is detected correctly. We therefore devote a chapter to error detection in video streams.

Part II with Chapters 2–4 presents link measurements at the User Equipment. While classical Gilbert Elliot models are not sufficient to explain the observed statistics, newer so-called recurrent models are introduced that allow a perfect fit of measurement data with a small set of describing parameters. Moreover, due to these advanced models, predicting the channel behaviour is now possible with high quality.

In Part IV all previous parts are combined to explain error resilient techniques for video transmission. Here, Chapter 7 explains the 3GPP video services in more detail. Finally, in Chapter 8, cross-layer designs allowing modern error-resilient techniques are explained. By exploiting the knowledge that a video stream is carried over from the core network, schedulers can take advantage and use different coding, for example for I and P packets of a video. Also, channel prediction methods as explained in Chapter 4 facilitate the work of the scheduler and considerably improve the transmission quality.

In Part V we focus on QoS. In order to provide such quality it is required to monitor the network. Such a monitoring technique was developed in the so-called METAWIN project at the Forschungszentrum Telekommunikation Wien (see www.ftw.at). It is reported on in Chapter 9, allowing statistics out of a live wireless core network to be extracted. In Chapter 10 various case studies are presented of what can be found in a wireless network once it is being monitored. Here, bottleneck detection and anomalies are discussed and simple statistical measures are presented to detect them. Finally, Chapter 11 deals with video quality, providing an overview of the state of the art techniques and reporting on recently developed techniques. While video quality for high resolution transmission such as television has become a standard, for example by ANSI, low resolution videos, common for wireless handheld devices, behave entirely differently. It was thus necessary to study the user expectations on mobile devices and design quality estimators that reflect human perception.

The last part of this book is devoted to traffic models. Chapters 12–15 report on typical traffic mixes, their measurements and statistics as they are found in today's Internet. Measurement points were the Gn and Gi interface as provided from the METAWIN project, the measurement period was from 2005 to 2007. Due to the introduction of UMTS in Austria, the relative service shares change considerably during this period. Although we are not allowed to reveal absolute volume numbers, a multitude of fascinating aspects can be extracted in such measurements. The last two chapters deal with models describing modern services that were not present at the introduction phase of the Internet, such as email, html, gaming and push to talk service. Here, original models from the 1990s are adopted and entirely new models are presented.

Due to our NDA with mobilkom austria AG, many absolute numbers, in particular, in terms of traffic volume, cannot be revealed. I apologize to the readers for this and ask for their understanding.

Markus Rupp
Vienna

Acknowledgements

This book is the outcome of many years of research and teaching in the field of signal processing and wireless communications. It was made possible by Ernst Bonek who gave me the singular opportunity to continue his long-term research collaboration with mobilkom austria and to whom I owe more than a simple word of gratitude.

Special thanks also to the reviewers of this book. The manuscript was critically read by many anonymous reviewers selected by the publisher. Their comments were very helpful in finding the right focus for this book. A particular thank you goes to all authors, most of whom have been companions during several years of research, many of them starting as PhD students. From them I learnt the most, but not every contributor appears as an author in this book. I am grateful to the many students that helped our understanding of wireless systems owing to their bachelor, master and diploma theses: J.C. Rodriguez, A. Al-Moghrabi, E. Dijort-Romagosa, B. Lopez-Garcia, T. Tebaldi, R. Puglia, A. Paier, M. Zavodsky, C.T. Castella, M. Braun, G.C. Forte, A. Dancheva, I. Cort-Todoli, C. Crespi de Arriba, I. Rodriguez-Losanda, M. Salvat, E. Recas de Buen, J. Gero and J. Colom-Ikuno.

This book and the research results that it provides would not have been possible without mobilkom austria AG and its continuous financial as well as strategic support in our joint research adventures. In particular, I would like to thank Dr. Werner Wiedermann, who is responsible for steering the research activities of mobilkom austria; he is, in spite of all the financial problems over the past years, doing a splendid job. There are also many other people at mobilkom austria AG to thank for their permanent willingness to discuss our problems with them. The list is certainly not complete and I apologize if I have left out some names by mistake: T. Ergoth, W. Weiler, A. Ciaffone, U. Rokita, M. Steinbauer, T. Baumgartner, G. Schuster and M. Preh. The views expressed in this book are those of the authors and do not necessarily reflect the views within mobilkom austria AG.

Such a book would never have been possible without the constant support by the many helpful people from John Wiley & Sons Ltd: S. Hinton, T. Ruonamaa, A. Smart, A. Smith, S. Tilley and G. Woodward. I cannot thank you enough.

List of Abbreviations

2G 2nd Generation (wireless)

3G 3rd Generation (wireless)

3GPP 3rd Generation Partnership Program

AAL ATM Adaptation Layer

AC Alternating Current

ACF Auto Correlation Function

ACK (positive) ACKnowledgement

ACR Absolute Category Rating

ADSL Asymmetric Digital Subscriber Line

AEC Adaptive temporal and spatial Error Concealment

AGCH Access Grant CHannel

ALC Asynchronous Layered Coding

AM Acknowledged Mode
Amplitude Modulation

AMB AMBiguous retransmission timeout

AMC Adaptive Modulation and Coding

AMPS Advanced Mobile Phone Service Telephony

AMR Adaptive Multi Rate

ANN Artificial Neural Network

APN Access Point Name

APP A Posteriori Probability

ARQ Automatic Repeat reQuest

AS Application Server

ASCII American Standard Code for Information Interchange

ASO Arbitrary Slice Ordering

ATM Asynchronous Transfer Mode

ATS Ahead-of-Time Streaming

AUC AUthentification Centre

AVC Advanced Video Coding

AVT Audio/Video Transport

AWGN Additive White Gaussian Noise

BCH Broadcast CHannel

BCCH Broadcast Control CHannel

BEC Binary Erasure Channel

BER Bit Error Ratio

BIAWGN Binary Input Additive White Gaussian Noise

BICM Bit–Interleaved Coded Modulation

BLER BLock Error Ratio

BMC Broadcast Multicast Control

BPSK Binary Phase Shift Keying

BR Bit Rate

BSC Binary Symmetric Channel
Base Station Controller

BSS Base Station Subsystem

BSSGP Base Station Subsystem GPRS Protocol

BST Base Station Transmitter

BTS Base Transceiver Station

CABAC Context Adaptive Binary Arithmetic Coding

CAVLC Context Adaptive Variable Length Coding

CBR Constant Bit Rate

CC Content Class

ccdf complementary cumulative distribution function

CCH Control CHannel

CCCH Common Control CHannel

cdf cumulative distribution function

CDMA Code Division Multiple Access

CE Contextual Error

CIF Common Intermediate Format

CN Core Network

COTS Commercial Off The Shelf

CPCH Common Packet CHannel

CPICH Common PIlot CHannel

CTCH Common Traffic CHannel

CQI Channel Quality Indication

CRC Cyclic Redundancy Check

CS Circuit Switched
Code Set

CSD Circuit Switched Data

CUDP Complete UDP

DC Direct Current

DCCH Dedicated Control CHannel

DCH Dedicated Channel

DCR Degradation Category Rating

DCT Discrete Cosine Transform

DiffServ Differentiated Services

DL Downlink

DNS Domain Name System

DoS Denial of Service

DP Data Partitioning

DPCH Dedicated Physical Channel

DPCCH Dedicated Physical Control CHannel

DPDCH Dedicated Physical Data CHannel

DPI Deep Packet Inspection

DSCH Downlink Shared CHannel

DTCH Dedicated Traffic CHannel

DTX Discontinuous Transmission

DVB-H Digital Video Broadcasting – Handheld

ecdf empirical cumulative distribution function

eccdf empirical complementary cumulative distribution function

ECSD Enhanced CSD

ECSQ Entropy Constrained Scalar Quantization

ECVQ Entropy Constrained Vector Quantization

EDF Early-Deadline First

EDGE Enhanced Data rates for GSM Evolution

eecdf estimated empirical cumulative distribution function

EEP Equal Error Protection

EFI Error–Protection Feedback Information

EFR Enhanced Full Rate

EGPRS Enhanced GPRS

EIF Error Indication Flag

EIR Equipment Identity Register

EM Expectation Maximization

epmf empirical probability mass function

ESI Encoding Symbol ID

ETSI European Telecommunications Standards Institute

EUDCH Enhanced Uplink Data CHannel

EZW Embedded Zerotree Wavelet

FACH Forward Access CHannel

FACCH Forward Access Common CHannel

FDD Frequency Division Duplex

FEC Forward Error Correction

FEW Force Even Watermarking

FF File Format

FFT Fast Fourier Transform

FIFO First In First Out

FIR Finite Impulse Response

FLC Fixed Length Code

FLUTE File Delivery over Unidirectional Transport

FLV FLash Video

FMO Flexible Macroblock Ordering

FPS First Person Shooter

FR Frame Rate

FRTX Fast Retransmit Retransmissions

FSMC Finite State Markov Channel

FSMM Finite State Markov Model

FTP File Transfer Protocol

FUNET Finnish University NETwork

GCRA General Cell Rate Algorithm

GEC Gilbert–Elliot Channel Model

GERAN GSM Edge Radio Access Network

GGSN Gateway GPRS Supporting Node

GMM GPRS Mobility Management

GMSC Gateway MSC

GOB Group Of Blocks

GOP Group Of Pictures

GPRS General Packet Radio Service

GSM Global System for Mobile communications

GTP GPRS Tunnelling Protocol

HARQ Hybrid Automatic Repeat reQuest

HLR Home Location Register

HMM Hidden Markov Model

HRD Hypothetical Reference Decoder

HSDLA High Speed DownLink Architecture

HSDPA High Speed Downlink Packet Access

HSDPCCH High Speed Dedicated Physical Control CHannel

HSPA High Speed Packet Access

HS-SCCH High Speed Control CHannel

HSUPA High Speed Uplink Packet Access

HTML HyperText Mark-up Language

HTTP Hypertext Transfer Protocol

HTTPS Hypertext Transfer Protocol Secure

HVS Human Visual System

ICMP Internet Control Messaging Protocol

ID Identification

IDR Insert subscriber Data Request

IEC International Electrotechnical Commission Interactive Error Control

IETF Internet Engineering Task Force

IL Internet Link

ILPC Inner Loop Power Control

IMAP Internet Message Access Protocol

IMEI International Mobile Equipment Identity

IMS IP Multimedia Subsystem

IMSI International Mobile Subscriber Identity

IMT International Mobile Telecommunications

IP Internet Protocol

IRPROP Improved Resilient Propogation

IS International Standard

ISD Independent Segment Decoding

ISDN Integrated Services Digital Network

ISO International Organization for Standardization

ISO/IEC International Organization for Standardization/International Electrotechnical Commission

ISMA Internet Streaming Media Alliance

ISP Internet Service Provider

ITU International Telecommunication Union

ITU-T International Telecommunication Union-Telecommunications Sector

JM Joint Model

JSCC Joint Source Channel Coding

JVT Joint Video Team

kNN k-Nearest Neighbour

KPI Key Performance Indicator

KQI Key Quality Indicator

KS Kolmogorov–Smirnoff

LAN Local Area Network

LER LLC Error Rate

LI Length Indicator

LLC Link Logical Control

LLR Log Likelihood Ratio

LRTO Loss-induced Retransmission Time Out

LTE Long Term Evolution

MAC Medium Access Controller

MAP Maximum A Posteriori

MB Macro Block

MBLC MacroBlock Level Concealment

MBMS Multimedia Broadcast/Multicast Service

MCCH MBMS point-to-multipoint Control CHannel

MCP Motion Compensated Prediction

MCS Modulation and Coding Scheme

MDC Multiple Description Coding

ME Mobile Equipment

MIB Management Information Base

MILP Mixed Integer Linear Programming

ML Maximum Likelihood

MLE Maximum Likelihood Estimator

MMOG Massively Multiplayer Online Game

MMS Multimedia Messaging Service

MMUSIC Multiparty Multimedia Session Control

MOS Mean Opinion Score

MPEG Moving Picture Experts Group

MS Mobile Station

MSC Mobile Switching Center

MSCH Multipoint Scheduling CHannel

MSE Mean Square Error

MSISDN Mobile Subscriber ISDN Number

MSRN Mobile Subscriber Roaming Number

MT Mobile Termination

MTCH Multipoint Traffic CHannel

MTSI Multimedia Telephony Service for IMS

MTU Maximum Transfer Unit

MV Motion Vector

M-SMMM Mixed Semi-Markov/Markov Model

NAL Network Abstraction Layer

NALU Network Abstraction Layer Unit

NDA Non Disclosure Agreement

NPC Non Playing Character

NRI NAL Reference Identification

NTSC National Television Systems Committee

OLPC Outer Loop Power Control

OMC Operation and Maintenance Centre

OR Out of Range

OSI Open System Interconnection

PAL Phase Alternation by Line

PAM Pulse Amplitude Modulation

PC Pair Comparison

PCA Principal Component Analysis

PCCH Paging Control CHannel

PCH Paging CHannel

PCU Packet Control Unit

PDA Personal Digital Assistant, palmtop

PDCP Packet Data Convergence Protocol

PDP Packet Data Protocol

PDSCH Physical Downlink Shared CHannel

PDU Protocol Data Unit

pDVD percentage of Degraded Video Duration

PESQ Perceptual Evaluation of Speech Quality

PI Position Indicator

pdf probability density function

PLMN Public Land Mobile Network

pmf probability mass function

PoC PTT over Cellular

POP Post Office Protocol

PPP Point to Point Protocol

PPS Picture Parameter Set

PS Packet Switched

PSC Packet Switched Conversational

PSH Push

PSNR Peak Signal to Noise Ratio

PSS Packet Switched Streaming

PSTN Public Switched Telephone Network

PTT Push To Talk

QAM Quadrature Amplitude Modulation

QCIF Quarter Common Intermediate Format

QoE Quality of Experience

QoS Quality of Service

QP Quantization Parameter

QPSK Quadrature Phase Shift Keying

QVGA Quarter Video Graphics Array

RA Routing Area

RACH Random Access CHannel

RAN Radio Access Network

RANAP Radio Access Network Application Part

RAU Routing Area Update

RB Radio Bearer

RBW Relation Based Watermarking

RD Rate Distortion

RDO Rate Distortion Optimization

RF Radio Frequency

RGB Red, Green and Blue

RIR Random Intra macroblock Refresh

RLC Radio Link Control

RLM Run Length Model

RMT Reliable Multicast Transmission

RNC Radio Network Controller

RoHC Robust Header Compression

ROI Region of Interests

RPC Remote Procedure Call

RRC Radio Resource Control

RRM Radio Resource Management

RS Redundant Slice

RSCP Received Signal Code Power

RTCP Real Time Control Protocol

RTO Retransmission Time Out

RTP Real-time Transport Protocol

RTS Real Time Strategy

RTSP Real Time Streaming Protocol

RTT Round Trip Time

SA Sub Aggregate

SACCH Slow Associated Control CHannel

SAD Sum of Absolute Difference

SAP Service Access Point

SC Synchronization Channel

SCCP Signalling Connection Control Part

SCH Synchronization CHannel

SD Straight Decoding

SDU Service Data Unit

SECAM SÉquentiel Couleur À Mémoire

SEI Supplemental Enhancement Information

SF Spreading Factor

SH Slice Header

SHCCH SHared Control CHannel

SI Spatial Information

SIF Standard Interchange Format

SIM Subscriber Identity Module

SIP Session Initiation Protocol

SIR Signal to Interference Ratio

SGSN Serving GPRS Support Node

SLC Slice Level Concealment

SM Synchronization Marker

SMB Sub Macro Block

SMTP Simple Mail Transfer Protocol

SNDCP Sub Network Dependent Convergence Protocol

SNIR Signal to Noise and Interference Ratio

SNMP Simple Network Management Protocol

SNR Signal to Noise Ratio

SPS Sequence Parameter Set

SRTO Spurious Retransmission Time Out

SVC Scalable Video Coding

TA Terminal Adapter

TAC Type Approval Code

TB Transport Block

TBS Timestamp Based Streaming

TBSS Transport Block Set Size

TCH Traffic CHannel

TCP Transmission Control Protocol

TDD Time Division Duplex

TDMA Time Division Multiple Access

TE Terminal Equipment

TEID Tunnel Endpoint IDentifier

TF Transport Format

TFCI Transport Format Combination Identifier

TFCS Transport Format Combination Set

TI Temporal Information

TM Transport Mode

TMA Traffic Monitoring and Analysis

TML Test Model Long-term

TPC Transmit Power Control

TTI Transmission Time Interval

UDP User Datagram Protocol

UE User Equipment

UEP Unequal Error Prediction

UID Unique Identification Listing

UL UpLink

UM Unacknowledged Mode

UMTS Universal Mobile Telecommunications System

URL Uniform Resource Locator

UTRA Universal Terrestrial Radio Access

USB Universal Serial Bus

USIM UMTS Subscriber Identity Module

UT99 Unreal Tournament '99 (online game)

UTRA Universal Terrestrial Radio Access

UTRAN UMTS Terrestrial Radio Access Network

VBR Variable Bit Rate

VCL Video Coding Layer

VGA Video Graphics Array

VCEG Video Coding Experts Group

VLC Variable Length Code

VLR Visitors Location Register

VoIP Voice over Internet Protocol

VQEG Video Quality Experts Group

WAN Wide Area Network

WAP Wireless Application Protocol

WCDMA Wideband CDMA

WLAN Wireless Local Area Network

WM WaterMarking

WoW World of Warcraft

WWW World Wide Web

XML eXtended Markup Language

Y-PSNR Peak Y-Signal to Noise Ratio

YUV composite video signal

Part I

Cellular Mobile Systems

Introduction

In this first part of the book a short introduction to wireless cellular systems is presented. This introduction summarizes the most important wireless cellular digital standards (in Europe) and presents the background knowledge required to understand the following parts. It particularly addresses people in video coding who are unfamiliar with the principles of cellular systems. They, like every newcomer in the field of cellular network architectures, feel overwhelmed with tons of abbreviations, making it very hard to read about this subject and even prohibiting access to this new field of knowledge. While standards seem to be entirely unreadable, even excellent introductory books such as those by H. Holma and A. Toskala are difficult for the layman in this field.

Focus is placed on the European standard Universal Mobile Telecommunications System (UMTS) and its UMTS Terrestrial Radio Access Network (UTRAN). It has been most successful even outside of Europe and can guide as a central theme to understand principles of core networks. It has been around since the Second Generation (2G) 'Global System for Mobile communications' (GSM), and its successor 'General Packet Radio Service' (GPRS). Here, the UTRAN protocol architecture and its physical layer data processing in the UTRAN Radio Interface are explained to some extent. An introduction to the Third Generation (3G) packet switched core network is provided in which the most important parts are explained, such as the Radio Network Controller (RNC), Serving GPRS Support Node (SGSN), Gateway GPRS Supporting Node (GGSN), the Home Location Register (HLR), the Visitors Location Register (VLR), the Operation and Maintenance Center (OMC), the Equipment Identity Register (EIR) and the AUthentication Center (AUC). An example of a typical data session in a 3G network explains the interplay of the protocols from the Radio Access Network Application Part (RANAP), the GPRS Tunnelling Protocol (GTP v0) as well as the GPRS Mobility Management (GMM).

Moreover, the small but important differences between 2G, 2.5G and 3G core network entities are explained, revealing some details of the GPRS protocol stack and the bearer speed in GPRS and 'Enhanced GPRS' (EDGE). Finally, the last section discusses the differences to High Speed Packet Access (HSPA) and nicely demonstrates the main novelties that HSPA offers.

For those readers interested in learning even more details, we allied the references with their corresponding '3rd Generation Partnership Program' (3GPP) standards. They are freely available under www.3gpp.org.

1

Introduction to Radio and Core Networks of UMTS

Philipp Svoboda and Wolfgang Karner

Mobile networks were first designed for one single service which was voice telephony. The very first mobile radio telephone system was introduced in 1918 by the German national railway *Deutsche Reichsbahn*, which offered their first-class passengers a radio-based telephone link in the Berlin area (Feyerabend *et al.* 1927, p. 872). However, the first large mobile network was established in 1958. It was the so-called *A-Netz*. The terminals were huge and their cost immense. Also, the number of subscribers was limited due to the simple implementation.

The successor, the so-called *B-Netz*, was introduced in 1972. An important new feature of the system allowed the subscribers to set up a call on their own. In the previous networks a central operator was involved in any call setup procedure.

The last analogue technology was the *C-Netz*, which began in 1986 and was shut down in 2000. The terminals were still quite expensive, but at *only* 6.5 kg they were real lightweights compared to the previous generations.

Mobile telephony as we know it today began in the early 1990s. In 1990 the second generation (2G) of mobile communication technology, namely the Global System for Mobile Communications (GSM), was introduced by the European Telecommunications Standards Institute (ETSI), supporting digital transmission of voice data. The number of mobile terminals started to ramp up very quickly and after 15 years they had already exceeded the number of fixed telephone systems in Austria.

Video and Multimedia Transmissions over Cellular Networks   Edited by Markus Rupp
© 2009 John Wiley & Sons, Ltd

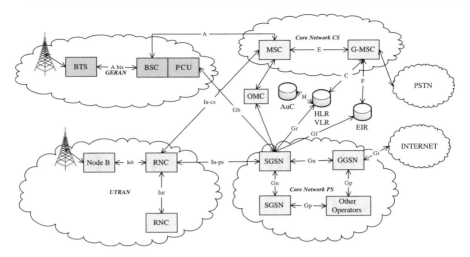

Figure 1.1 Network elements for 2.5G and 3G mobile networks.

Meanwhile, another technology began to emerge: the Internet. The number of Internet hosts also started to grow rapidly. This evolution also had an impact on mobile communication networks. End terminals could only process audio/voice data as input data. Therefore, users had to use a modem to transfer data traffic via GSM. This method of data transport is quite inefficient. To prevent such shortcomings the GSM group standardized a new technology for packet-switched traffic only: the General Packet Radio Service (GPRS). To minimize changes and costs, only minor parts of the GSM system were adopted. Consequently, GPRS is referred to as 2.5G. GPRS was introduced to the Austrian market in autumn 1999 by mobilkom austria AG. Thereafter, the Internet and the mobile networks went into a merging process, which today (2009) here in Austria-Europe, results in mobile flat rate contracts being cheaper than rates for fixed-line access.

At the end of the last century the standardization process of the third generation (3G) of mobile communication technologies, the Universal Mobile Telecommunications System (UMTS), was finalized (Holma and Toskala 2004). This new technology was a leap forward to the replacement of fixed Internet access technologies. While GPRS supports only data rates in the order of fixed analogue modems (for example, 10–60 kbit/s), UMTS, in Dedicated Channel (DCH) mode, can support up to 386 kbit/s and beyond. With the increase of data rate came a reduction of the Round Trip Time (RTT) from 1000 ms to 140 ms. These parameters are already close to the performance of an Asymmetric Digital Subscriber Line (ADSL). Just six years later UMTS was further improved by the introduction of High Speed Downlink Packet Access (HSDPA). Currently (2007), it enables user download rates of up to 7.2 Mbit/s per host. Introduced in 2002, UMTS increased the number of available services to the end terminal even further. The higher data rate, the possibility to use advanced Quality of Service (QoS) settings and to choose between Circuit Switched (CS) and Packet Switched (PS) bearers enabled new advanced services such as live video streaming, video telephony and so on.

Figure 1.1 depicts a top-level view of a 3G core network including a GPRS and a UMTS Radio Access Network (RAN). From this figure we learn that GPRS was *attached* to the existing GSM by adding the Packet Control Unit (PCU), while in UMTS the packet-switched data is processed in the same device as the voice and video calls. GPRS is the first mobile technology purely to target data traffic.

This part of the book is but a brief introduction to the large topic of mobile communication systems. Later, we will refer to the associated standards of Third Generation Partnership Program (3GPP), which some may find hard to read. For a comprehensive overview of the topic of mobile cellular communications, see Eberspächer and Vögel (1999), Taferner and Bonek (2002) and Holma and Toskala (2004, 2006). Furthermore, detailed information on parts of each system can be found in Heine (2001, 2002) (GPRS), Heine (2004, 2006) (UMTS) and Blomeier (2005, 2007) and Blomeier and Barenburg (2007) (HSDPA).

1.1 UMTS Network Architecture

The UMTS system is built according to the same well-known architecture that has been used by all major second generation systems in Europe. At a high level, the UMTS network consists of three parts (3GPP TS 23.002 2002). These are the Mobile Station (MS), in 3G now called User Equipment (UE) as the interface between the user and the radio part, the UMTS Terrestrial Radio Access Network (UTRAN) containing radio-related functionality and the Core Network (CN) responsible for the connection to external networks. This high-level system architecture, as well as the most important nodes and interfaces, is presented in Figure 1.2.

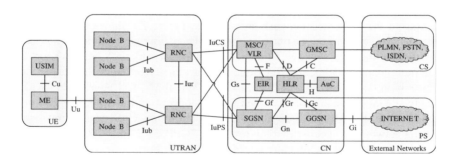

Figure 1.2 UMTS network architecture and interfaces.

The UE consists of the physical equipment used by a Public Land Mobile Network (PLMN) subscriber, which comprises the Mobile Equipment (ME) and the Subscriber Identity Module (SIM). It is called the UMTS Subscriber Identity Module (USIM) for Rel. 99 and following. The ME comprises the Mobile Termination (MT), which, depending on the application and services, may support various combinations of Terminal Adapter (TA) and Terminal Equipment (TE) functional groups to provide end-user applications and to terminate the upper layers.

Within the UTRAN, several Base Stations (NodeB) – each of them controlling several cells – are connected to one Radio Network Controller (RNC). The main task of the NodeB is the performance of physical layer processing including channel coding, interleaving, rate adaptation, spreading and so on. Furthermore, some Radio Resource Management (RRM) operations such as the Inner Loop Power Control (ILPC) as well as the fast Hybrid Automatic Repeat reQuest (HARQ), scheduling and priority handling for HSDPA have to be performed in the NodeB. The RRM tasks performed in the RNC are the load and congestion control of its own cells, admission control and code allocation for new radio links to be established in those cells as well as handover decisions and the Outer Loop Power Control (OLPC). The RNC performs the layer-two processing of the data to/from the radio interface and macrodiversity combining in case of soft handover.

While the UE and the UTRAN contain new specific protocols as well as a new radio interface (WCDMA), Rel. 99 UMTS CN was inherited from the GSM system and both UTRAN and GSM Edge Radio Access Network (GERAN) connect to the same core network. As presented in Figure 1.2, the core network consists of the circuit switched domain for the real time data and the packet switched domain for non-real-time packet data. In the CS domain the Mobile Switching Center (MSC) including the Visitor Location Register (VLR) connects to the RNCs. It switches the CS data transactions and stores the visiting user's profiles and location. The Gateway MSC (GMSC) connects UMTS to external networks such as, for example, the Public Switched Telephone Network (PSTN). In the Home Location Register (HLR) the user's service profiles and the current UE locations are stored and the Equipment Identity Register (EIR) is a database for identification of UEs via their International Mobile Equipment Identity (IMEI) numbers. The Serving GPRS Support Node (SGSN) is the equivalent to the MSC but for the PS domain. It is responsible for the user mobility and for security (authentication). With the Gateway GPRS Support Node (GGSN) the connection to external networks such as the Internet is realized.

According to the presented network architecture, Figure 1.3 shows the layered UMTS bearer architecture, where each bearer on a specific layer offers its individual services using those provided by the layers below. To realize a certain network QoS, a bearer service with clearly defined characteristics and functionality is to be set up from the source (left TE) to the destination (right TE) of a service, passing MT, RAN, CN edge node (SGSN) and CN gateway (GGSN). Details of the UMTS QoS concept and architecture can be found in reference 3GPP TS 23.107 2002 and the interaction and QoS negotiation with other neighbouring networks is specified within reference 3GPP TS 23.207 2005.

Note, the presented CN architecture refers to Rel. 99/4. Further details of the UMTS CN architecture and its evolution within Rel. 5, 6 and 7 can be found in Holma and Toskala (2004) as well as in the corresponding versions of reference 3GPP TS 23.002 2002.

1.2   UTRAN Architecture

In this section the overall architecture of UTRAN is described. The UTRAN consists of two basic elements – base stations NodeB and RNC. The functions of both elements are discussed as well as their interfaces and the corresponding protocol architecture of the user plane and the control plane. As a part of the International Mobile Telecommunications at 2000 MHz (IMT-2000) standards of the International Telecommunications Union (ITU),

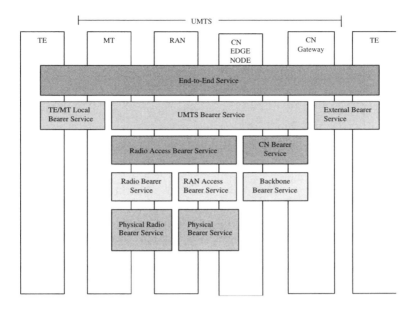

Figure 1.3 Architecture of UMTS bearer services.

UMTS is specified within the 3G Partnership Project (3GPP) where its main radio access technologies based on WCDMA are called Universal Terrestrial Radio Access (UTRA), Frequency Division Duplex (FDD) and Time Division Duplex (TDD).

3GPP specifies UMTS in several steps, from Rel. 99/4 offering theoretical bit rates of up to 2 Mbit/s, to Rel. 5 and 6 reaching higher bit rates beyond 10 Mbit/s with the introduction of HSDPA and High Speed Uplink Packet Access (HSUPA). Whereas 2G systems such as GSM were designed for voice communications, UMTS as a 3G communication system with its high data rates, low delay and high flexibility is designed for the delivery of multimedia services.

1.2.1 UTRAN Protocol Architecture

A general overview of the UMTS radio interface protocol architecture (3GPP TS 25.301 2005) is presented in Figure 1.4. The radio interface is in three protocol layers: L1 (physical layer), L2 (data link layer) and L3 (network layer), and L2 is further split into the following sublayers: Medium Access Control (MAC), Radio Link Control (RLC), Packet Data Convergence Protocol (PDCP) and Broadcast Multicast Control (BMC). Vertically, L3 and RLC are divided into control and user planes, where the control plane is used for all UMTS-specific control signalling including the Radio Resource Control (RRC) as the lowest L3 sublayer.

In Figure 1.4 the Service Access Points (SAPs) for peer-to-peer communication are marked with ellipses at the interface between sublayers. The service provided by Layer 2 is referred to as the Radio Bearer (RB). The control plane RBs, which are provided by RLC to RRC, are denoted as signalling RBs.

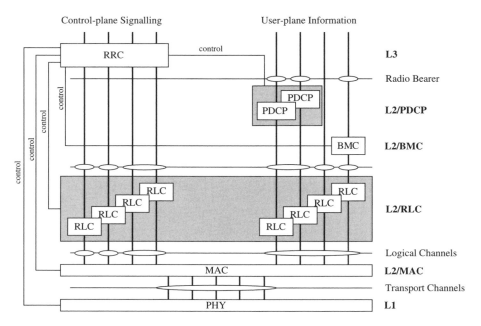

Figure 1.4 UMTS radio interface protocol architecture.

A fundamental part of the UTRAN architecture is the channel concept – the different functions of the channels and the channel mapping. The logical channels provide an interface for the data information exchange between the MAC protocol and the RLC protocol. There are two types of logical channel: control channels for the transfer of control plane information and traffic channels for the transfer of user plane information. Table 1.1 presents an overview of available logical channels in UTRAN.

Table 1.1 Logical channels.

Control Channels (CCHs):	Broadcast Control Channel (BCCH)
Traffic Channels (TCHs):	Paging Control Channel (PCCH)
	Dedicated Control Channel (DCCH)
	Common Control Channel (CCCH)
	Shared Control Channel (SHCCH)
	MBMS point-to-multipoint Control Channel (MCCH)
	MBMS point-to-multipoint Scheduling Channel (MSCH)
	Dedicated Traffic Channel (DTCH)
	Common Traffic Channel (CTCH)
	MBMS point-to-multipoint Traffic Channel (MTCH)

Whereas the logical channels are separated by the information they are transporting, the transport channels are separated by how the information is transmitted over the air interface (in a shared connection or via a dedicated link). The transport channels provide the bearers for the information exchange between the MAC protocol and the physical layer. In contrast to the logical channels, which can be bidirectional, all transport channels are unidirectional. A list of transport channels as well as the possible mapping to the logical channels is presented in Figure 1.5. The arrows show whether the mapping is for DownLink (DL) and UpLink (UL) or unidirectional only.

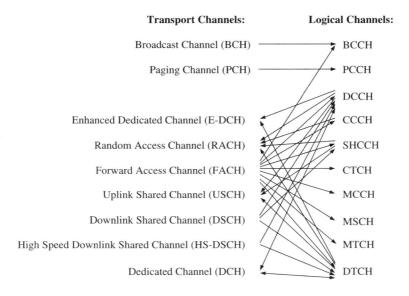

Figure 1.5 Mapping between transport channels and logical channels (seen from UE side).

The RRC (3GPP TS 25.331 2004) is the central and most important protocol within the UTRAN protocol stack as it controls most of the UE, NodeB and RNC protocols and configures the physical layer through the transfer of peer-to-peer RRC-signalling messages. The main tasks of the RRC are the establishment, maintenance and release of an RRC connection between the UE and UTRAN as well as paging, QoS control, UE measurement reporting and the OLPC. The PDCP (3GPP TS 25.323 2006) performs header compression and decompression of IP data streams, for example TCP/IP and RTP/UDP/IP headers,[1] at the transmitting and receiving entity, respectively.

Several RLC instances are placed in the control and user planes without differences. As a classical data link layer (L2) application the main function of the RLC (3GPP TS 25.322 2006) is the exchange of higher layer Protocol Data Units (PDUs) between RNC and UE. Further tasks of the RLC are the segmentation and de-segmentation of higher layer PDUs, overflow protection via discard of Service Data Units (SDUs), for example after a maximum number of retransmissions or timeout, error correction by retransmissions and in-sequence delivery in case of retransmissions. The RLC can work in three different modes. In RLC

---

[1] Internet Protocol (IP), Transport Control Protocol (TCP), Real-time Transport Protocol (RTP), User Datagram Protocol (UDP).

Transparent Mode (TM) the RLC-protocol simply conveys higher layer SDUs to the peer RLC-entity without error detection and correction mechanisms, ciphering or in-sequence delivery. The RLC Unacknowledged Mode (UM) enables in-sequence delivery and error detection but no retransmissions and thus error correction mechanisms need to be taken care of by higher layers. The RLC Acknowledged Mode (AM) guarantees the error-free transmission (by means of retransmissions) and in-sequence delivery of upper layer PDUs to the peer entity.

In the MAC-layer (3GPP TS 25.321 2006) the logical channels are mapped to the transport channels and an appropriate Transport Format (TF) is selected from the Transport Format Combination Set (TFCS) for each transport channel, depending on the instantaneous source rate. Further functions of the MAC protocol are priority handling between data flows of one UE and also between UEs by means of dynamic scheduling (for the Forward Access CHannel (FACH) and the DSCH), service multiplexing for RACH/FACH/Corrosion Packet Channel (CPCH) and the DCH, ciphering in case of RLC TM and dynamic transport channel type switching. The processing of the layer-three packets within the UTRAN protocol stack can also be seen in Figure 1.6, where the data flow for non-transparent RLC and non-transparent MAC is shown.

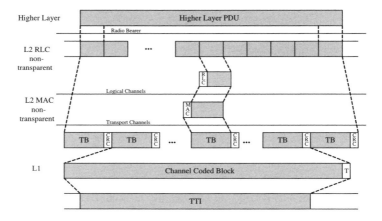

Figure 1.6   Schematic illustration of a packetization example for the transmission over UMTS.

One higher layer PDU (for example, IP packet) coming from the user plane in layer three will be delivered via the corresponding radio bearers to layer two where it can be processed either by PDCP to perform header compression (the packet can be handled by the BMC), or it can be delivered directly to the RLC layer. There, the higher layer PDUs may be segmented into smaller packets. In case of a bearer with a data rate below or equal to 384 kbit/s, usually a 320-bit (40-byte) payload within the RLC packets is used (3GPP TS 25.993 2006). For non-transparent RLC (RLC AM/UM), a header will be added to the packets, then forming RLC PDUs. The header size for the RLC AM is 16 bit, whereas the RLC UM packet header contains 8 bits. These RLC PDUs are then transported via the logical channels to the MAC layer where a MAC header is added if transport channel multiplexing (non-transparent MAC) is used in the system. After that, the Transport Blocks (TB = MAC PDU) are sent via the transport channels to the physical layer.

1.2.2   Physical Layer Data Processing in the UTRAN Radio Interface

The first process in the physical layer after delivering the so-called TBs via the transport channels is a Cyclic Redundancy Check (CRC) of the packet data (3GPP TS 25.201 2005; 3GPP TS 25.302 2003). Then, after attaching the CRC bit to the TBs (3GPP TS 25.212 2006), these are segmented or concatenated in order to fit to the block size of the channel coding (3GPP TS 25.944 2001). For packet oriented applications, usually turbo coding is used with a coding rate of 1/3, which can further be punctured to match the rate with the physical resources.

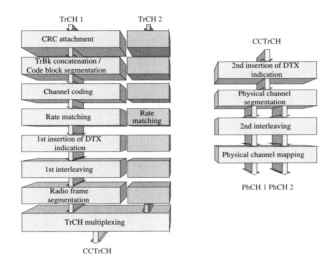

Figure 1.7 UTRAN physical layer procedures.

Figure 1.7 shows a sequential illustration of the physical layer processes such as rate matching, first Discontinuous Transmission (DTX) insertion indication, first interleaving (over one coded block), radio frame segmentation and then multiplexing of the various transport channels. The Coded Composite Transport Channel (CCTrCH) is then, after a second insertion of DTX indication, segmented into the appropriate physical channels. After a second interleaving (over one radio frame) the data bits are mapped onto the correct physical channel. In order to illustrate the physical layer procedures and their sequential processing, the data flow of an example for a bearer with 64 kbit/s user data rate is presented in Figure 1.8.

In this example user data from a DTCH enters the physical layer in the form of one TB of 1280-bit size. It is shown that this 1280-bit TB is transmitted within two radio frames (10-ms/radio frame) which gives the required 64 kbit/s. The first processing step in the physical layer is the adding of a CRC information to the TB followed by the coding of the resulting data block via turbo code with rate 1/3. Rate matching has to be performed

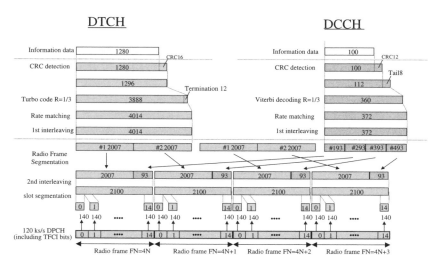

Figure 1.8 Data flow in the UTRAN physical layer for 64 kbit/s reference channel (3GPP TS 34.121 2004).

so that the coded data bit fits into the associated radio frames. After a first interleaving and radio frame segmentation, the bit of the DTCH get multiplexed with the information data of the DCCH which transmits TBs of 100 bits over four radio frames (40 ms) and thus reaches a data rate of 2.5 kbit/s. After the transport channel multiplexing, a second interleaving over one radio frame (10 ms) is performed. The resulting 2100-bit data blocks are transmitted within one radio frame, reaching a bit rate of 210 kbit/s at that point. Every 2100-bit data block is then segmented into 15 140-bit blocks each for fitting into the 15 slots per radio frame. Considering Quadrature Phase Shift Keying (QPSK)[2] modulation of the Dedicated Physical Channel (DPCH), together with physical layer control data, results in 120 kbaud/s which become 3.84 Mchip/s due to the spreading operation with a Spreading Factor (SF) of 32 (3GPP TS 25.213 2003). In Figure 1.9 the mapping of transport channels onto corresponding physical channels is shown. In the case of the DCH, the data from DTCH and DCCH is mapped onto the Dedicated Physical Data Channel (DPDCH) and multiplexed with the Dedicated Physical Control Channel (DPCCH) which contains the Transmit Power Control (TPC), Transport Format Combination Indicator (TFCI) and Pilot bit. In the UMTS DL the DPDCH and the DPCCH are time multiplexed and modulated via QAM whereas in the UL the DPCCH and the DPDCH are modulated according to two orthogonal PAM schemes separately in order to prevent an interference of the transmitting UL signal with audio equipment as in GSM. An exemplary illustration of the slot structures in UL and DL is presented in Figure 1.10.

---

[2]Despite the fact that QPSK and Binary Phase Shift Keying (BPSK) are mentioned throughout the documents in 3GPP, a modulation scheme AM in the form of Quadrature Amplitude Modulation (QAM) and Pulse Amplitude Modulation (PAM) with root raised cosine transmit pulse-shaping filter (roll-off factor $\alpha = 0.22$ (3GPP TS 25.104 2007)) is used while the terms QPSK or BPSK just indicate the symbol constellation (3GPP TS 25.213 2003).

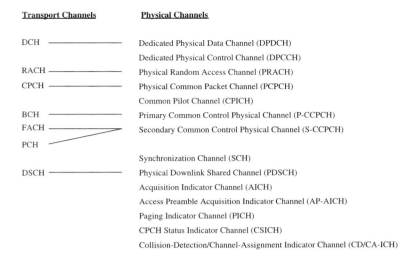

Transport Channels

Physical Channels

DCH ——————————— Dedicated Physical Data Channel (DPDCH)

Dedicated Physical Control Channel (DPCCH)

RACH ——————————— Physical Random Access Channel (PRACH)

CPCH ——————————— Physical Common Packet Channel (PCPCH)

Common Pilot Channel (CPICH)

BCH ——————————— Primary Common Control Physical Channel (P-CCPCH)

FACH ——————————— Secondary Common Control Physical Channel (S-CCPCH)

PCH

Synchronization Channel (SCH)

DSCH ——————————— Physical Downlink Shared Channel (PDSCH)

Acquisition Indicator Channel (AICH)

Access Preamble Acquisition Indicator Channel (AP-AICH)

Paging Indicator Channel (PICH)

CPCH Status Indicator Channel (CSICH)

Collision-Detection/Channel-Assignment Indicator Channel (CD/CA-ICH)

Figure 1.9 Mapping of transport channels onto physical channels (3GPP TS 25.211 2002).

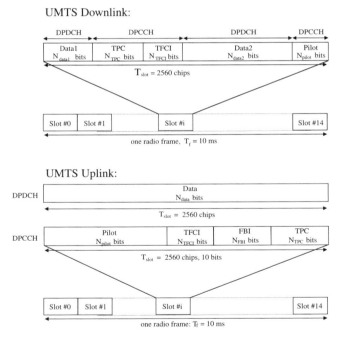

Figure 1.10 Downlink and uplink slot structure example (3GPP TS 25.211 2002).

1.3   UMTS PS-core Network Architecture

The initial design goal of GSM was to support voice services which are on the same level as the Integrated Services Digital Network (ISDN) combined with mobility. In the late 1990s the user focus started to shift from pure voice to voice-and-data traffic. The GPRS standard was agreed upon as a basis to build for data-only services. Although this was a first step to mobile packet-switched networks, it was only a placeholder for a new technology which could serve PS and CS services by default. UMTS was designed to serve the needs of both the CS and the PS domains. To minimize the cost of the core network the structure and functions of the components are very similar to the GSM/GPRS units. In fact, from UMTS Rel. 5 on, the UMTS units can also serve GPRS and GSM RANs. Figure 1.11 depicts the key elements of the 3G mobile network.

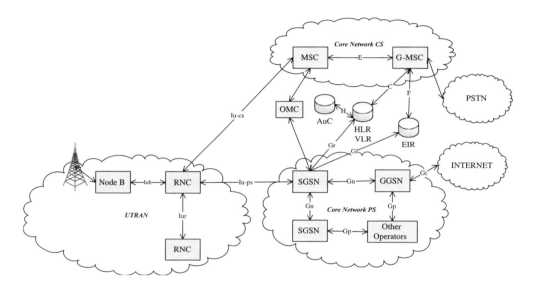

Figure 1.11 Network elements of a 3G cellular mobile network.

Radio Network Controller (RNC)   The RNC covers all radio resource management tasks. The NodeB itself has quite a simple function set; therefore, the RNC has to manage the scrambling code tree and the transmit power for each active radio link. The RRC protocol is established between the UE and the RNC to support the manipulation of the radio link between the NodeB and the UE. Towards the core network the RNC terminates the GPRS Tunnel Protocol tunnels. This is different from the GPRS network where the SGSN was the endpoint of the GPRS Tunnelling Protocol (GTP) tunnel. The NodeB has no caching for data packets; therefore, the RNC also has to process the flow control algorithms.

Serving GPRS Support Node (SGSN)   The SGSN is the switching centre for the data traffic. In the downlink direction the SGSN is connected to several RNCs using Iu-Ps protocols. A certain SGSN serves a group of RNCs and therefore covers a given geographical area. The number of necessary SGSNs is given simply by the processing power that is needed to serve the given traffic in the area. The common tasks of an SGSN are: session management including attach, detach and mobility management, ciphering, cell updates, paging, compression and so on. The billing in mobile cellular networks is volume based, therefore the SGSN generates billing tickets per user and sends these tickets to a central database. The protocols used by the SGSN are the Sub Network Dependent Convergence Protocol (SNDCP), the Logical Link Control (LLC), the Base Station Subsystem GPRS Protocol (BSSGP) and GTP.

Gateway GPRS Supporting Node (GGSN)   The GGSN is the boarder node between the core network of the mobile operator and the external packet data network. GPRS supports different Packet Data Protocols (PDPs) such as IP, Point-to-Point Protocol (PPP) and X.25. The GGSN must be able to handle all these PDPs. The type of PDP can be chosen by the mobile subscriber by creating a PDP-context. The PDP-context creation request marks the start of a data session. It holds the information about the Access Point Name (APN) and the settings the user requests from the mobile network. The GGSN is connected to the external network via the Gi. There is a firewall in common between the GGSN and the external network, protecting the mobile core infrastructure from attacks. If the user accesses an IP network, the GGSN will convert the user datagram from the mobile network to IP packets and replace the GTP identifier, which is the ID for a specific user within the mobile network, with an external IP-address. The GGSN also takes care of the QoS profiles for each PDP-context. One user can have several PDP-contexts, each with a different QoS profile. This can be used to access different services with different QoS settings. For more details, see reference 3GPP TS 23.060 2006.

The Home Location Register (HLR)   The HLR is the heart of the GSM network (3GPP TS 11.131 1995). It is a database holding management data for each user of the mobile operator. The HLR holds all permanent user data such as Mobile Subscriber ISDN Number (MSISDN), available services, QoS, international ID, the IMSI and further temporal data such as the location area where the ME was last seen, the actual VLR or the Mobile Subscriber Roaming Number (MSRN). The HLR can be accessed by the MSC via the C interface and by the VLR via the D interface. The HLR itself is closely connected to the Authentification Centre (AUC). The AUC takes care of the generation of security-related data that the HLR needs to authenticate users. The HLR has to hold at least one entry per subscriber and to fulfil real-time requests from the MSC units. To solve this issue a HLR unit normally consists of several discrete units managing the huge load of data and requests. This can also be seen in the International Mobile Subscriber Identity (IMSI). The IMSI is structured as shown in Figure 1.12.

The first three digits are fixed by the country of the operator. The next two digits identify the operator itself. The following HLR part identifies the HLR in which the user data is stored. Finally, the last eight digits are the unique identifier at this HLR for the searched user. At the terminal side this information is stored in the SIM card.

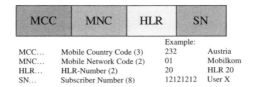

Figure 1.12 Structure of the IMSI.

The Visitors Location Register (VLR)   The VLR is a database holding MS specific data allocated to one or several MSC unit(s). One could think of the visitors of one MSC unit, either in their home operators network or in roaming mode. As the user population will change over time, this database, in contrast to the HLR, is highly dynamic. The first request of a MSC will target the VLR; this instance takes the load from the central HLR unit and can be seen as a kind of cache instance. The VLR makes more sense when thinking about an instance directly implemented in the MSC enabling local caching of user information over a period of time. More details can be found in references 3GPP TS 29.016 2005 and 3GPP TS 11.132 1995.

The Operation and Maintenance Center (OMC)   The OMC uses the O-interface to monitor and control all the network components. The protocols deployed on this interface are SS7 and X.25. Typical tasks are status reports, generation of billing tickets, user billing and security screening.

The Equipment Identity Register (EIR)   The EIR holds user equipment information in the form of IMEIs (3GPP TS 22.016 2008). The idea of the EIR was to blacklist stolen or malfunctioning devices to ensure that they are not able to enter the network of a mobile subscriber. Although the intention itself was good, it suffers from the fact that the mobile providers do not update this list regularly and that the IMEI is easily re-programmable on most devices.

The AUthentication Center (AUC)   The AUC holds the authentication key Ki, which is also stored at the SIM-card. By using this shared secret a new key, called Kc, can be derived to secure the radio link of the mobile network. Although treated separately from the HLR in this introduction, the AUC normally is a part of the HLR because the relation between these units is very close.

1.4 A Data Session in a 3G Network

In a mobile cellular network several steps are necessary in order to set up a data connection for IP transmissions. Figure 1.13 presents the most important steps to establish a connection. The PDP-context must not be activated prior to a GPRS Mobility Management (GMM) activation. The mobility management transfers GPRS related subscription data of the subscriber. This subscription data is needed to clarify which PDP-contexts may be established by the user interface. After the GPRS attachment the subscriber can activate a PDP-context at any time.

The session management for PDP-context activation takes place only between the UE and the SGSN. The SGSN communicates the PDP-context activation data to the GGSN via a GTP tunnel.

After a successful PDP-context activation procedure the subscriber can now transmit user data on the IP layer to an external packet data network. In the case of roaming, the setup procedure has to initiate an intra-SGSN handover first. The QoS profiles may be modified to the needs of the new SGSN (3GPP TS 4.008 2000).

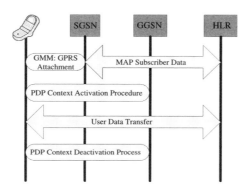

Figure 1.13 Session management procedures.

The PDP-context is similar to a dial-up session which is known from the fixed wired networks. For each context the subscriber is assigned a unique IP address. In the following chapters the IP address information at the Gn interface is sometimes used to represent a user session. A context represents a user session with volume, duration and frequency of use.

1.4.1 The UMTS (PS-core) Protocol Stack

In UMTS the lower layers rely on the Asynchronous Transfer Mode (ATM). This allows for a simple integration of high-speed optical fibre systems as a physical layer. The ATM protocol connects RNC, SGSN, GGSN and GMSC. For user data the ATM tunnels even reach up to the NodeB. The CS domain uses the ATM Adaptation Layer v2 (AAL2) version of ATM which is connection oriented. AAL2 supports the transmission and multiplexing of many real-time data streams, offering a low delay, small jitter and less data rate fluctuation. The PS domain uses the AAL5. It is connection-less and implements more or less a best-effort approach, suitable for non-real-time services such as Internet traffic. The AAL5 layer does not support connection management. Therefore, the PS domain in UMTS needs an additional layer using the GTP. The GTP protocol builds up a user-specific data tunnel between the GGSN and the RNC. At the RNC the GTP protocol is converted over to the PDCP. The PDCP supports a more efficient coding of the headers, which is suitable for the radio link where resources are expensive.

1.4.2   The Protocols

Figure 1.14 presents the UMTS protocol stack for all three domains. From this figure we learn that there is a second splitting into 'Access Stratum' and 'Non Access Stratum'. The idea is to separate the services in the upper layers. The breakdown allows the UMTS network to migrate to different parts on its own. For example, the change to an all-IP network in the 'Access Stratum' will have no effect on the 'Non Access Stratum'. As long as the interface stays the same, both systems can still interact.

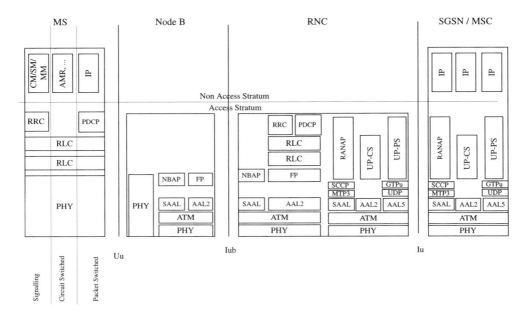

Figure 1.14 UMTS protocol stack.

Radio Access Network Application Part (RANAP)

The Radio Access Network Application Part (RANAP) handles the signalling between the UTRAN and the core network, via the Iu interface. It is responsible for tasks such as booking ATM lines, changing radio setup and so on; see reference 3GPP TS 29.108 2006. All control procedures needed by the UTRAN can be executed by using instances from the three elementary classes:

- general control service;

- notification service;

- dedicated control service.

All necessary functions can be constructed by using these three elementary classes. Examples for these procedures are

- Iu release;

- overload control;

- RAB assignment.

Signalling Connection Control Part (SCCP)

The Signalling Connection Control Part (SCCP) delivers an abstraction between UMTS-related layers and the used transport layers (3GPP TS 29.800 2006). It allows different transport systems (ATM, IP) to be used. The main functions are:

- connection-less and connection-oriented extension to MTP;

- address translation;

- full layer 3 Open Systems Interconnection (OSI) compatibility;

- below SS7 protocol.

GPRS Tunnelling Protocol (GTP v0)

The GPRS Tunnelling Protocol (GTP v0) is the main protocol in the core network. It allows the end users in a GPRS or UMTS network to move between different cells while having continuing connects to the Internet. This is achieved by transmitting the subscriber's data from the current sub-network to the GGSN. It is used for connections between RNC, SGSN and GGSN. The data payload is attached to the GTP headers (8 byte). It can handle signalling and data traffic (3GPP TS 9.060 2003; 3GPP TS 29.060 2008). The header of the GTP v0 protocol is shown in Figure 1.15. The GTP-C(ontrol) is used to transport control information. It transmits GPRS mobility management messages between GGSN and SGSN nodes. Logically GTP-C is attached to the GTP-U(ser) tunnel – physically it is separated. The main functions are:

- Create/Update/Change PDP Context;

- Echo Request/Response;

- RAN Information.

The GTP-U(ser) is used to transport user data. It basically hides terminal mobility from the IP layer of the user supporting the reordering of Transport-PDUs. Note, a Transport-PDU is the encapsulation of data communicated by the transport layer via the network layer. The used Tunnel Endpoint Identifier (TEID) is always unique. The main functions are:

- data transmission;

- tunnel setup/release/error;

- echo request/response.

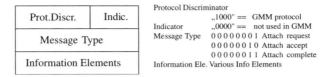

Prot.Discr.	Indic.	Protocol Discriminator
		„1000" == GMM protocol
		Indicator „0000" == not used in GMM
Message Type		Message Type 0 0 0 0 0 0 0 1 Attach request
		0 0 0 0 0 0 1 0 Attach accept
		0 0 0 0 0 0 1 1 Attach complete
Information Elements		Information Ele. Various Info Elements

Figure 1.15 GTP header and description.

GPRS Mobility Management (GMM)

This protocol is defined in reference 3GPP TS 23.060 2006. It offers in UMTS the same functions as in GPRS: managing the mobility of the terminals. This protocol was designed to reduce the number of terminals in active state consuming radio resources. Therefore, three states were defined:

- idle;

- ready;

- standby.

The transition between these states is initialized by well-defined events. A normal mobile sending data will be in the ready state. After a time period (set timeout) of not sending data the mobile will drop to standby. The state indicates that the mobile is expected to become active again. If there is no data transmission up to a second timeout the mobile will finally drop to the IDLE state. The algorithm is known by the RNC and the mobile terminal. Therefore, we do not need any signalling to initialize the state transitions. Figure 1.16 presents the state transmission diagram.

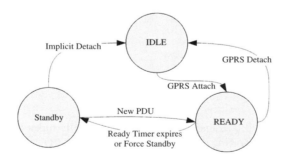

Figure 1.16 States of the GMM protocol.

However, if a mobile terminal tries to change its state it has to send a signal to the higher instances in the core network. GMM offers this functionality. It can handle basic procedures in the attachment process, such as 'attach', 'accept', 'request' and 'complete'.

1.4.3 Bearer Speed in UMTS

The bearer speed in a wireless mobile network is a term for the net data rate that is available to the UE. User data in UMTS may be transferred using two different implementations: DCH or High Speed Packet Access (HSPA) (3GPP TS 25.213 2006; 3GPP TS 25.308 2007). In case a very low amount of user data has to be transmitted, a random or common channel can also serve for data transmission. However, normal Internet applications will initiate data transfers triggering a DCH or HSPA channel assignment.

The DCH channel has different bearer speeds depending on the chosen spreading factor. For a fixed transmit power, a larger spreading factor allows more reliable transmission at the cost of a lower user data rate. Therefore, users with a higher distance to the base station will only achieve a lower data rate. In addition to this, as part of the network optimization process, the RNC monitors the actual data rate the user needs and adjusts it, via the SF, accordingly. Table 1.2 shows the available options for the DCH from our live network.

Table 1.2 DCH data rates for different spreading codes.

User Data Rate	Interface Data Rate	Spreading Factor
12.2 kbit/s	30 kbit/s	128
32 kbit/s	60 kbit/s	64
64 kbit/s	120 kbit/s	32
128 kbit/s	240 kbit/s	16
384 kbit/s	480 kbit/s	8

HSPA extends the radio interface of the UMTS network. A data symbol on the radio interface can transmit up to 4 bits of data, while standard UMTS symbols transmit only 2 bits of data. The data rate assignment in HSDPA differs from DCH. The physical channel is set to a fixed spreading factor of 16, which equals a data rate of 14.4 Mbit/s. This is a strong improvement over the 384 kbit/s in the DCH. However, 14.4 Mbit/s is the total rate of the entire HSDPA cell. All users have to share this resource. HSDPA uses a slot length of 2 ms; within each slot 15 different code channels are transmitted. A scheduler in the NodeB assigns code channels to the specific users according to the UE capabilities and the data rate need. A UE capable of class five can decode five code channels within one time slot, which equals a user data rate of 3.6 Mbit/s.

1.5 Differences between 2.5G and 3G Core Network Entities

The GSM standard was introduced to build a telephone system that could carry the services found in ISDN and combine it with mobility all over the world. At the air interface GSM initially used 935–960 MHz in the downlink and 890–915 MHz in the uplink. Later upgrades, also called GSM 1800, introduced more frequencies at around 1700–1900 MHz.

GPRS was initially standardized in GSM phase 2+. Today, in 2007, it is hosted by the 3GPP. The integration of GPRS into GSM was introduced in a very smooth way; the physical channels stayed unchanged and most of the infrastructure was reused. The only two new nodes introduced were the GGSN and the SGSN. In 2.5G GPRS implemented packet-oriented data services to the GSM network. GPRS directly supports packet-oriented protocols such as IP or X.25. It is therefore possible to communicate directly with the Internet – no modem is needed. The billing is implemented volume based and not per time interval. To extend the data rate the GPRS specification allows the use of all eight time slots by a single user. In practice, most mobile equipments feature only one common receiver/sender unit and therefore will only support up to four time slots in downlink and two in uplink. In GPRS the connection between the core network and the mobile equipment is only permanent for the logical layer – the physical resources in the cell will be scheduled according to the actual load in the cell and the user data in the buffer of the SGSN.

GPRS offers packet-switched IP-based services to users in GSM environments. The IP routing is available through the entire network, beginning at the UE and ending at the GGSN. In contrast to GSM, where each active user occupies exactly one time slot, GPRS users can use up to eight time slots in parallel in order to boost their data rate.

1.5.1   GPRS Channels

The GSM relies on a Time Division Multiple Access (TDMA) transmission system. The TDMA technique uses a fixed time grid to serve different users at the same frequency slot. All the active mobile stations are synchronized and each of them is assigned a certain time slot that it can use to transmit its data. Figure 1.17 depicts a simple example of the GSM time frames. Bins with the same number belong to the same physical channel.

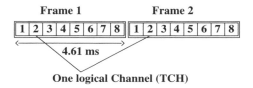

Figure 1.17 GPRS physical channel.

Logical Channels

The Traffic CHannels (TCHs) are used to transport user data, for example the output of the Adaptive Multi Rate (AMR) coder. The data rate of a full rate TCH is 22.8 kbit/s. It consumes a full slot in every frame. Therefore, signalling data has to be sent using a different time slot. There exists also a different implementation of the voice codec, which needs less data rate and leaves some room for strong encryption. The Control CHannels (CCHs), in contrast to the traffic channels, consist of three different channel types, each group featuring four different channels. These channels have a low bit rate. In fact the signalling traffic normally consumes only one of the eight time slots. The different channels are multiplexed to this time slot. In other words a channel X that occupies only one time slot every tenth frame can be multiplexed for several mobile equipments. To structure this multiplexing, GSM knows hyper, super and multiframes.

The Broadcast Control CHannel (BCCH) acts like a lighting house for a GSM cell. It broadcasts all the important information that mobile equipment needs to attach to the cell. The data transmitted includes cell ID, schedule of the signalling channels and information about the cell neighbours. The BCCH features sub-channels for frequency correction, Frequency Correction CHannel (FCCH), and time synchronization, Synchronization CHannel (SC). Also, paging is realized via the BCCH, therefore, everything powered on mobiles will monitor this channel. The channel is unidirectional only.

The Dedicated Control CHannel (DCCH) is a bidirectional signalling channel that can be used by the mobile to interact with the cell, for example register. Its schedule is broadcast via the BCCH. The TCH is allocated using the standalone DCCH (SDCCH). Combined with a booked TCH, the mobile station uses the slow associated DCCH (SACCH) to exchange system data such as the channel quality and the receiving power strength. Should the need for signalling data rate not be fulfilled by the SACCH, the mobile can book an additional signalling channel called the Forward Access Common CHannel (FACCH). This channel steals time slots from the TCH channel and uses them for signalling information. Such situations typically arise when there are handovers between different Base Transceiver Station (BTS) units.

The Common Control CHannel (CCCH) carries all the call management information. If a BTS has a call for an MS, it broadcasts this information using the Paging CHannel (PCH). A mobile station that wants to react to this paging uses the Random Access CHannel (RACH) to send its information to the BTS. The access is obtained using the ALOHA protocol, which is an OSI layer 2 protocol for LAN networks. A successful connection will receive a free TCH via the Access Grant CHannel (AGCH).

1.5.2 GPRS Core Network Architecture

The network elements of the GPRS core network are very similar to the elements found in UMTS. Figure 1.18 displays all interfaces and core nodes of a GPRS network. In order to make the integration of GPRS smooth, the Base Station Subsystem (BSS) extended the Base Station Controller (BSC) with the PCU which handles the new packet-switched signalling procedures. It converts data received via the Gb interface from the SGSN in order that it can be processed by the BSC. The second component of the BSS is the Base Station Transmitter (BST), which is only a relay station transmitting the information via the air interface. This element was also present in GSM.

The GGSN and SSGN nodes have already been introduced in the UMTS Section. A more detailed description of the nodes can be found in the standard (3GPP TS 23.060 2006).

The GGSN, SGSN, HLR and VLR nodes have the same functions as in UMTS. Some of the interfaces connecting the nodes have names that differ from the UMTS scheme, as the protocol stack is not identical for these interfaces; for example Gb replaces Iu.

1.5.3 The GPRS Protocol Stack

The protocol stack of GPRS is split into transmission and signalling planes. Figure 1.19 depicts the transmission plane for GPRS (3GPP TS 29.060 2008). As in UMTS the GTP protocol routes user-packets within a tunnel from the GGSN to the actual position of the UE.

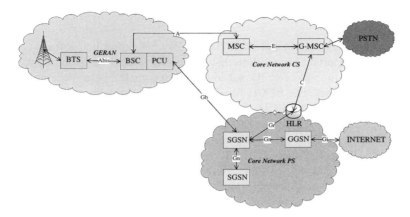

Figure 1.18 GPRS core nodes and interfaces (3GPP TS 23.060 2006).

The GTP builds up a tunnel between the GGSN and the MS. Below the GTP, UDP is used to transport the information between the different GPRS core nodes, for example SGSN and GGSN. At the SGSN the GTP protocol is replaced by the SNDCP protocol to adapt the data flow to different implementations of the PCUs. The CS design of the BSS had no features to provide reliable data transport. Therefore, the LLC layer was introduced featuring different kinds of ARQ and FEQ modes, granting reliable transmission of the data packet units. Finally, BSSGP is used for QoS aware routing between the SGSN and the target BSS. At the BSS the MAC manages the media access and maps the LLC frames to physical channels. The RLC layer provides a reliable connection over the radio interface.

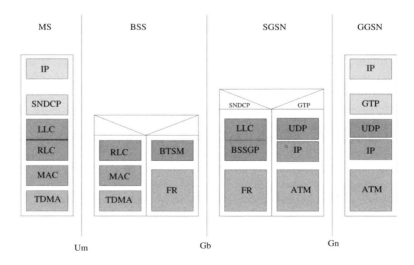

Figure 1.19 GPRS stack: transmission plane.

The SNDCP is a transparent network layer protocol for IP data (3GPP TS 44.065 2006). The protocol offers two important features: header and data compression. Therefore, it can improve performance as it reduces the amount of data transferred.

The RLC protocol layer transfers PDUs from the Logical Link Layer (LLC) protocol. The LLC offers a logical link from the SGSN to the UE over Gb and Um interfaces. It covers flow control and ciphering for the logical link (3GPP TS 43.064 2006; 3GPP TS 44.064 2007). Finally, the MAC protocol takes care of the physical properties of the radio channel.

1.5.4 Bearer Speed in GPRS and EDGE

The physical link in GPRS is a TDMA implementation. It offers eight slots in the uplink and the downlink directions, respectively. A normal GSM voice call uses one time slot, GPRS UEs can allocate up to eight time slots. Each added slot upgrades the data rate available to the user. The number of free slots is a function of the cell load and has an upper limit which is bound by the capabilities of the UE.

In addition to this, three new Code Sets (CSs) were introduced with the start of GPRS. The code sets offer different strengths of data protection. Higher data protection secures the transmission and the signal is more resistant to noise and interference. However, a stronger code needs more parity bit, thereby reducing the user data rate. The assignment of the code is limited due to the Signal-to-Noise Ratio (SNR) at the UE; the fastest code set, CS-4, can only be activated close to the base station transmitter. The maximum possible data rate for GPRS is 160 kbit/s. Table 1.3 shows the data rates for one time slot and different code sets in GPRS and Enhanced GPRS (EGPRS) also called Enhanced Data rates for GSM Evolution (EDGE).

Table 1.3 GPRS and EDGE data rates for one time-slot and different CSs.

Code Set	User kbit/s	Interface kbit/s	User kbit/s	Interface kbit/s
CS-1	8.0	9.0	22.4	27.0
CS-2	12.0	13.4	29.6	40.2
CS-3	14.4	15.6	44.8	46.8
CS-4	20.0	21.4	59.2	64.6

The EDGE service uses 3 bits per symbol at the air interface, in contrast to the one bit per symbol of GPRS. Therefore, EDGE pushes the data rate by a factor of three to a maximum of approximately 473.6 kbit/s.

1.6 HSDPA: an Evolutionary Step

The goal of HSDPA was the introduction of higher bit rates for the UE, hence keeping the changes to the architecture to a minimum. HSDPA was introduced by the 3GPP in Rel. 5. It is an extension to UMTS Rel. 99. The features introduced in HSDPA are:

- shorter radio frames (2 ms instead of 10 ms);

- introduction of Channel Quality Indication (CQI) as feedback means from UEs to NodeB;

- new up- and downlink channels (HS-PDSCH, HS-DPCCH and HS-SCCH);

- 16 QAM modulation type additionally to 4 QAM;

- Adaptive Modulation and Coding (AMC);

- Hybrid-ARQ (HARQ);

- MAC scheduling functionality within NodeB.

We will now give a short introduction to the main changes of architecture compared to UMTS, as well as to the new features of HSDPA.

1.6.1   Architecture of HSDPA

The migration from GPRS towards UMTS was a paradigm change in the connection of UEs and NodeBs. In GSM and GPRS a mobile is connected to one base station at a time; in UMTS the UEs support so-called soft handover modes. In this mode the UE is connected to several (up to six) base stations. This feature reduces the risk of call drops which often occur in hard-handover scenarios. However, as all NodeBs have to offer the same data stream to the UE, the next hierarchical entity, in this case the RNC, has to manage all the packets and radio link parameters. Therefore, the local NodeB cannot adapt the data rates accordingly to the actual channel conditions. As this was considered necessary for higher data rates, the idea of soft handover was withdrawn and NodeB was given a local scheduler to allow for adaptive modulation and coding and fast scheduling. Figure 1.20 depicts the changes that took place, starting with Release 5.

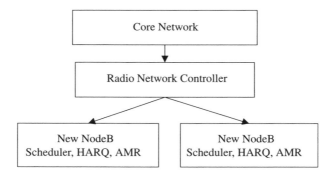

Figure 1.20 Changes in architecture from Rel. 99 towards HSDPA.

The MAC-hs

The new functions of HSDPA are implemented into a new logical layer called MAC-hs. The MAC-hs is a new entity. It transmits data over the HS-DSCH channel, a new set of channels introduced in HSDPA. It also manages the physical resources allocated and can be configured from higher layers; see references 3GPP TS 25.308 2007; 3GPP TS 25.321 2006. The function set of the MAC-hs is depicted in Figure 1.21.

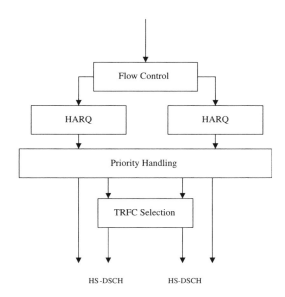

Figure 1.21 Structure of the MAC-hs entity.

1.6.2   Difference between UMTS and HSDPA

UMTS Rel. 99 allows for up to 384 kbit/s while HSDPA targets for much higher data rates of up to 14.4 Mbit/s. It achieves this by the implementation of new coding schemes and modulation techniques combined with scheduling techniques directly in the NodeBs. In other words the SF is no longer variable and there is no more fast power control available. These two elements of Rel. 99 are replaced by Adaptive Modulation and Coding (AMC), Fast retransmission strategy (called HARQ) and scheduling algorithms (3GPP TS 25.213 2006; 3GPP-25.848 2003). These new functions are described in the following sections.

Scheduling Algorithms

The place of the scheduling systems in Rel. 99 is inside the RNC. In HSDPA the function has been moved into the NodeBs, which allows for faster scheduling as there is no more 'reaction' delay present. The scheduler in HSDPA also has an additional task. Besides selecting the correct modulation and coding scheme and the HARQ process, it now schedules the transmission for all users. In Rel. 99 the scheduler was implemented on a per user base only. The implemented scheduler may follow different strategies such as:

- equal throughput per user (Round-Robin);

- balance cell and user throughput (Proportional Fair);

- maximum cell throughput (Maximum C/I).

Hybrid ARQ: A Fast Retransmission Strategy

The retransmission logic moved from the RNC entity into the NodeB. There exist two different error control and recovery methods to guarantee error-free transmissions to and from the UE, namely Forward Error Correction (FEC) and Automatic Repeat reQuest (ARQ); see references 3GPP TS 25.214 2008; 3GPP TS 25.302 2007; 3GPP TS 25.331 2008.

FEC introduces a set of redundant bit information added to the payload of each protocol and derived following some code scheme. This information allows the receiver to detect and recover errors as occurring from channel impairments. FEC information is added to each packet regardless of the actual channel state. Therefore, no feedback channel is needed. However in good channel conditions available data rate is wasted by the redundant information.

The ARQ error correction scheme improves on the main disadvantage of the FEC scheme. Error correction information is only requested on erroneous received packets. An ARQ system offers functions for error detection, acknowledgment, time-out and retransmission request. Typically, the basic functions are implemented using either a selective retransmission or a stop-and-wait procedure.

The disadvantage of these two methods is the delay that occurs in the case of a packet error. This can be overcome by combining the ARQ and the FEC methods in a so-called HARQ mode. The FEC is set to cover the most frequent error patterns and therefore will reduce the number of retransmissions. The ARQ part covers less frequent error patterns, allowing the number of bits added by the FEC to be reduced. There are different types of HARQ method available and the performance in total depends on the channel conditions, receiver equipment and other related parameters. Considering the complexity of a UMTS radio implementation, choosing the 'correct' or 'best' retransmission strategy is a wide field for ongoing research.

Adaptive Modulation and Coding (AMC)

The original implementation of UMTS-Rel. 99 offered one fixed modulation scheme. The adaptation to the actual radio channel is then performed using a power control algorithm. The instantaneous data rate is set by choosing an appropriate spreading factor offering the necessary gain for the given signal to interference situation.

In HSDPA the method was changed. Instead of relying on a fast power control, the SF was fixed and the modulation now follows the channel conditions, both modulation and coding format adapting in accordance with variations in the channel conditions. This system is called AMC, or link adaptation. Compared to standard power control, such methods deliver higher data rates. In HSDPA the AMR scheme assigns higher order modulation with higher code rates, such as 16 QAM.

1.6.3 Transport and Control Channels

The implementation of HSDPA into the physical layer of UMTS required major changes. At layer two of the transport network new entities were created to allow for fast MAC handling. One of these changes is the definition of new high-speed physical channels, namely: the High Speed Physical Downlink Shared CHannel (HS-PDSCH), the High Speed Dedicated Physical Control CHannel (HS-DPCCH) and the High Speed Shared Control CHannel (HS-SCCH). The main features of the physical channels are now presented.

High Speed Downlink Shared Channel (HS-DSCH)

The HS-DSCH is the new transport channel for user data introduced in HSDPA. To achieve the full advantage of moving the scheduling from the RNC to NodeB the round-trip time had to be reduced. This was accomplished by reducing the Transmission Time Interval (TTI) from 10 ms (Rel. 99) to 2 ms; see reference 3GPP TS 25.211 2007. The higher data rate is possible due to a higher order modulation scheme, namely 16 QAM, and a dynamic error protection. The combination of these two allows for higher peak data rates.

As discussed, the new transport channel has a fixed SF equal to 16. Based on this every code slot can offer up to 15 parallel codes to transmit user data, each of them unique by its specific channelization code. Each of these codes represents an HS-DSCH channel. A single HS-DSCH with 16 QAM achieves a data rate of 960 kbit/s. All 15 codes in parallel allow for the peak data rate of HSDPA equal to 14.4 Mbit/s. The assignment of codes is assigned on the per slot base by the scheduler in the NodeB. A single UE can have several codes in parallel within the same TTI. Users can be served simultaneously within one slot. These features allow for a better utilization of the available data rate.

High Speed Shared Control Channel (HS-SCCH)

The new transport channel, the HS-DSCH, no longer belongs exclusively to a single user. Therefore, the NodeB must now transmit control information associated to the HS-PDSCH, indicating to the user terminal which schedule will take place in the upcoming TTI. This is the task of the HS-SCCH. Aside from the obvious task already mentioned, the channel contains signalling and control information such as modulation scheme, HARQ information and transport format. There must be one HS-SCCH per each user active on the HS-DSCH; see references 3GPP TS 25.211 2007; 3GPP TS 25.212 2006; 3GPP TS 25.321 2006. The rate of the HS-SCCH is fixed to 60 kbit/s and an SF of 128. This results in 40 bit/slot and 120 bit/subframe. The duration of such a frame is three slots equal to 2 ms and it consists of two parts. The first part contains the time-sensitive information, such as the codes to de-spread and the modulation in the next TTI. The terminal needs this data to start the decoding process of the HS-DSCH. The other part holds the CRC and the HARQ information, which is no longer time critical.

Uplink High Speed Dedicated Physical Control Channel (HS-DPCCH)

The uplink channel for HSDPA carries feedback signalling related to the correlated downlink HS-DSCH channel. This information consists of the HARQ part and the CQI part. Each subframe is of length 2 ms or three slots. The payload of the packet is 10 bits, which are encoded to 30 bits by the error protection. The SF is set to 256; see references 3GPP TS 25.211 2007; 3GPP TS 25.214 2008.

The first 10 bits of encoded information carry the Ack/Nack messages for the HARQ scheme of the NodeB. They indicate the receiver if the last transmission was successful or if a retransmission has to take place. The second part of the subframe, another 20 bits of encoded information, carry the CQI value. This value is a type of quality index reporting on the channel conditions at the UE side. The CQI indicates which estimated block size, modulation type and number of parallel codes could have been received correctly in the downlink direction, thus indicating the quality of the link back to the scheduling system in the NodeB. The CQI value is based on the quality of the CPICH (Ec/No) channel broadcast in the cell averaged over 2 ms. It is calculated at the UE side and can have values between zero and 30, where larger is better. With this value the UE reports an estimate for the maximum setting of estimated block size, modulation type and number of parallel codes considering a probability for a TB error of less than 10%. Values above 15 allow for 16 QAM modulations while values below only allow for 4 QAM modulations. The CQI estimation process is a difficult task for the UE as there exists no standardized mapping by the 3GPP.

References

Blomeier, S. (2005) *HSDPA – Design Details and System Engineering,* INACON.

Blomeier, S. (2007) *HSUPA – Design Details and System Engineering,* INACON.

Blomeier, S. and Barenburg, S. (2007) *HSPA+ – Design Details and System Engineering,* INACON.

Eberspächer, J. and Vögel, H. J. (1999) *GSM–Global System for Mobile Communication,* 2nd edn, Teubner.

Feyerabend, E. Heidecker, H. Breisig, F. and Kruckow, A. (1927) *Handwörterbuch des elektrischen Fernmeldewesens,* Springer, Berlin, **2**, 871–874.

Heine, G. (2001) *GPRS / UMTS Rel.5 – Signaling and Protocol Analysis (The Core Network),* INACON.

Heine, G. (2002) *GPRS – Signaling and Protocol Analysis (RAN and Mobile Station),* INACON.

Heine, G. (2004) *UMTS – Signaling and Protocol Analysis (UTRAN and User Equipment),* INACON.

Heine, G. (2006) *UMTS – Rel. 4, 5 and 6 Core Network Architecture and Signaling (BICC, IMS & SIP),* INACON.

Holma, H. and Toskala, A. (2004) *WCDMA for UMTS, Radio Access For Third Generation Mobile Communications,* John Wiley & Sons, Ltd.

Holma, H. and Toskala, A. (2006) *HSDPA/HSUPA for UMTS,* John Wiley & Sons, Ltd.

Taferner, M. and Bonek, E. (2002) *Wireless Internet Access over GSM and UMTS,* Springer, Berlin.

3GPP TS 4.008 2000 (2000) *Mobile Radio Interface Layer 3 Specification (R99),* v.8.0.0, Jun.

3GPP TS 9.060 2003 (2003) *GPRS Tunnelling Protocol across the Gn and Gp Interface (GPRS),* v.7.10.0, Jan.

3GPP TS 11.131 1995 (1995) *Home Location Register (HLR), Specification,* v.3.2.1, Jan.

3GPP TS 11.132 1995 (1995) *Visitor Location Register (VLR), Specification,* v.3.2.1, Jan.

3GPP TS 22.016 2008 (2008) *International Mobile Equipment Identities (IMEI),* v.7.1.0, Dec.

3GPP TS 23.002 2002 (2002) *Network Architecture,* v.3.6.0, Sep.

3GPP TS 23.060 2006 (2006) *General Packet Radio Service (GPRS); Service Description; Stage 2,* v.6.15.0, Dec.

3GPP TS 23.107 2002 (2002) *Quality of Service (QoS) Concept and Architecture,* v.3.9.0, Sep.

3GPP TS 23.207 2005 (2005) *End-to-end Quality of Service (QoS) Concept and Architecture*, v.6.6.0, Sep.

3GPP TS 25.104 2007 (2007) *BS Radio Transmission and Reception (FDD)*, v.4.9.0, Mar.

3GPP TS 25.201 2005 (2005) *Physical Layer — General Description*, v.6.2.0, Jun.

3GPP TS 25.211 2002 (2002) *Physical Channels and Mapping of Transport Channels onto Physical Channels (FDD)*, v.3.12.0, Sep.

3GPP TS 25.211 2007 (2007) *Physical Channels and Mapping of Transport Channels onto Physical Channels (FDD)*, v.6.9.0, Dec.

3GPP TS 25.212 2006 (2006) *Multiplexing and Channel Coding (FDD)*, v.6.9.0, Oct.

3GPP TS 25.213 2003 (2003) *Spreading and Modulation (FDD)*, v.4.4.0, Dec.

3GPP TS 25.213 2006 (2006) *Spreading and Modulation (FDD)*, v.7.1.0, Mar.

3GPP TS 25.214 2008 (2008) *Physical Layer Procedures (FDD)*, v.7.10.0, Dec.

3GPP TS 25.301 2005 (2005) *Radio Interface Protocol Architecture*, v.6.4.0, Sep.

3GPP TS 25.302 2003 (2003) *Services Provided by the Physical Layer*, v.4.8.0, Sep.

3GPP TS 25.302 2006 (2006) *Services Provided by the Physical Layer*, v.7.5.0, Oct.

3GPP TS 25.308 2007 (2007) *High Speed Downlink Packet Access (HSDPA); Overall Description; Stage 2*, v.6.4.0, Dec.

3GPP TS 25.331 2004 (2004) *Radio Resource Control (RRC) Protocol Specification*, v.3.21.0, Dec.

3GPP TS 25.321 2006 (2006) *Medium Access Control (MAC) Protocol Specification*, v.6.10.0, Sep.

3GPP TS 25.322 2006 (2006) *Radio Link Control (RLC) Protocol Specification*, v.6.9.0, Oct.

3GPP TS 25.323 2006 (2006) *Packet Data Convergence Protocol (PDCP) Specification*, v.6.7.0, Sep.

3GPP TS 25.331 2008 (2008) *Radio Resource Control (RRC); Protocol Specification*, v.6.20.0, Dec.

3GPP TS 25.848 2003 (2003) *Physical Layer Aspects of UTRA High Speed Downlink Packet Access*, v.4.0.0, Apr. 2003.

3GPP TR 25.944 2001 (2001) *Channel Coding and Multiplexing Examples*, v.4.1.0, Jun.

3GPP TR 25.993 2006 (2006) *Typical Examples of Radio Access Bearers (RABs) and Radio Bearers (RBs) Supported by Universal Terrestrial Radio Access (UTRA)*, v.4.2.0, Sep.

3GPP TS 29.016 2005 (2005) *General Packet Radio Service (GPRS); Serving GPRS Support Node SGSN - Visitors Location Register (VLR); Gs Interface Network Service Specification,* v.6.0.0, Jan.

3GPP TS 29.060 2008 (2008) *GPRS Tunnelling Protocol across the Gn and Gp Interface (UMTS)*, v.6.19.0, Sep.

3GPP TS 29.108 2006 (2006) *Application of the Radio Access Network Application Part (RANAP) on the E-interface*, v.7.1.0, Dec.

3GPP TS 29.800 2006 (2006) *Signalling System No. 7 (SS7) Security Gateway; Architecture, Functional Description and Protocol Details*, v.7.0.0, Mar.

3GPP TS 34.121 2004 (2004) *Terminal Conformance Specification; Radio Transmission and Reception (FDD)*, v.5.4.0, Jun.

3GPP TS 43.064 2006 (2006) *Overall Description of the GPRS Radio Interface; Stage 2*, v.6.11.0, Jul.

3GPP TS 44.064 2007 (2007) *Mobile Station – Serving GPRS Support Node; Logical Link Control (LLC) Layer Specification*, v.6.2.0, Mar.

3GPP TS 44.065 2006 (2006) *Mobile Station – Serving GPRS Support Node; Subnetwork Dependent Convergence Protocol (SNDCP)*, v.6.6.0, Jun.

Part II

Analysis and Modelling of the Wireless Link

Introduction

In Part II the focus is on modelling of the wireless link. While many sophisticated methods for channel modelling are now available, the accurate prediction of something as simple as the BLock Error Ratio (BLER) is still complicated owing to a multitude of physical phenomena describing the wave propagation effects and many nested control processes in the transmission scheme. On the one hand, available channel models reflecting the physical phenomena are rather complicated, on the other hand UMTS transmissions are not dominated by the channels alone. Moreover, the complex interaction with Inner Loop Power Control (ILPC) and Outer Loop Power Control (OLPC) as well as the interleaving and the turbo coding and finally the scheduling seem to make it impossible to derive simple but accurate models to describe the observed BLER.

In the following three chapters Wolfgang Karner describes how he measured BLER (Chapter 2) in various wireless networks in Austria and, from his measurements, how he modelled the transmission behaviour at link level in an astonishingly accurate way (Chapter 3). The measurements were performed in the wireless networks of three different service providers with a multitude of different cell phones as well as in various scenarios between static and speedy movements. Surprisingly, the measurement results are such that only two modes are important to discriminate: 'static' and 'non-static'. The modelling of two such modes was successfully performed with so-called recurrent state models, an extension to the standard Markovian models. Wolfgang Karner shows in his modelling approaches that it is possible to describe the occurrence of burst errors and their statistics very accurately and only with a small set of parameters that are obtained from the measurements.

Finally, these accurate models allow the behaviour of the wireless links to be predicted directly from observations (Chapter 4). Once some simplifying assumptions are made, the prediction is rather low in terms of complexity but surprisingly accurate. Based on such modelling, sophisticated scheduling techniques are derived and will be explained in detail in Chapter 8 in Part III of this book.

The here-reported work by Wolfgang Karner can be found in more detail in his thesis 'Link Error Analysis and Modeling for Cross-Layer Design in UMTS Mobile Communication Networks' (http://publik.tuwien.ac.at/files/pub-et_13114.pdf) as well as in the journal article by W. Karner, O. Nemethova, P. Svoboda and M. Rupp, 'Link Error Analysis and Modeling for Video Streaming Cross-Layer Design in Mobile Communication Networks', *ETRI Journal*, **29**(5), 569–595, October 2007. Reproduced by permission of © Electronics and Telecommunication Research Institute. Note that there are many other interesting PhD theses available under the EURASIP open library (www.eurasip.org).

2

Measurement-based Analysis of UMTS Link Characteristics[1]

Wolfgang Karner

The following link error analysis and the resulting modelling was performed based on measured link layer error traces. The measurements were realized in the live Universal Mobile Telecommunications System (UMTS) networks of three different operators in the city centre of Vienna, Austria.

Due to similar adjustments of the relevant radio parameters in the different networks, the measured results out of the three networks (Karner *et al.* 2005) led to the same conclusions. Thus, we are here focusing on the measured error traces of only one operator's live network utilizing the UMTS radio network elements from Ericsson (Ericsson TEMS) in the following.

Moreover, the measurements for this work have been performed with several different mobile stations and radio bearers as well as in various mobility scenarios. Throughout this part of the book the following structure for presenting the results from these measurements is maintained: as long as the resulting conclusions from the measurements with the different mobiles/bearers/scenarios are the same, the measured results from only one mobile/bearer/scenario are presented with the specific name of the mobile/bearer/scenario given as reference. In case more information or further conclusions become available from the comparison of the results measured with different mobiles/bearers/scenarios, these results are presented in parallel.

[1]W. Karner, O. Nemethove, P. Svoboda and M. Rupp, "Link Error Analysis and Modeling for Video Streaming Cross-Layer Design in Mobile Communication Networks", *ETRI Journal*, **29**(5) (2007), pp. 569–595. Reproduced by permission of © Electronics and Telecommunication Research Institute.

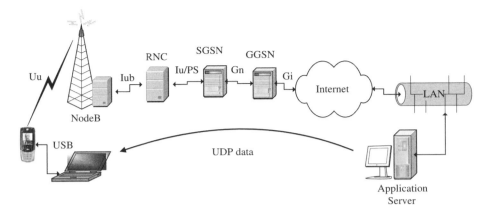

Figure 2.1 Scheme of the measurement setup in the live network.

2.1   Measurement Setup

2.1.1   General Setup

In Figure 2.1 a schematic illustration of the measurement setup for the measurements in the live networks is given. For the measurements a User Datagram Protocol (UDP) data stream with bit rates of 360 kbit/s, 120 kbit/s and 60 kbit/s (which, including the UDP/IP overhead, are 372 kbit/s, 125.6 kbit/s and 62.8 kbit/s, respectively) was sent from a PC over the UMTS network to a notebook using a UMTS terminal as a modem via a Universal Serial Bus (USB) connection.

Additionally, measurements were performed in a reference network, which is a separate network for acceptance testing of the same operator and with equal parameter settings as in the corresponding live UMTS network. As illustrated in Figure 2.2, in the reference network the radio link (Uu interface) is replaced by a cable connection (attenuation ≈ 60 dB) between the NodeB antenna connector and a Radio Frequency (RF) shielding box Willtek, 4920 (Willtek) where the mobile is enclosed to avoid interference, multipath propagation and fading effects.

In order to be capable of tracing the internal measurements of the mobiles, WCDMA 'TEMS' mobiles[2] were used as terminals in connection with 'TEMS Investigation' software as offered by Ericsson (Ericsson TEMS). In this document we refer to the used mobiles as 'mobile 1' to 'mobile 4' as listed in Table 2.1.

Throughout this work the following structure for presenting the results from different mobile terminals is maintained: as long as the resulting conclusions from the measurements with the different mobile terminals are the same, the measured results from only one are presented with the specific mobile name given as reference. In case more information or further conclusions can be given from the comparison of the results measured with different mobiles, the results are presented in parallel.

---

[2]Modified UMTS mobiles for TEMS data logging are offered by Ericsson (Ericsson TEMS). The mobiles are modified in a way that they provide the internal measurements via the interface to the notebook.

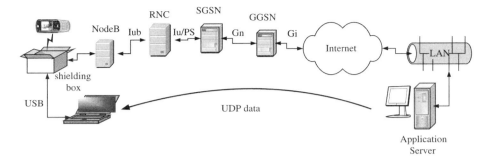

Figure 2.2 Scheme of the measurement setup in the reference network.

Table 2.1 Mobile equipment used for the measurements.

'mobile 1'	Motorola A835
'mobile 2'	Motorola E1000
'mobile 3'	Sony Ericsson Z1010
'mobile 4'	Sony Ericsson V800

After parsing the export files of the 'TEMS Investigation' software tool, various parameters of the UMTS system can be analysed. One of these parameters is the Cyclic Redundancy Check (CRC) information of the received Transport Blocks (TBs) which is used for the analysis of the UMTS link error characteristics as presented later in this document.

In the considered UMTS networks turbo coding with a coding rate of 1/3 was applied for the Dedicated Traffic Channel (DTCH) with a TB size of 336 bits, consisting of 320 bits for the RLC payload and 16 bits for the Radio Link Control (RLC) Acknowledged Mode (AM) header. Although using the RLC AM with its error detection and feedback mechanism, the link error analysis was performed without consideration of retransmissions, thus offering independence of selecting RLC AM or RLC Unacknowledged Mode (UM) and the specific RLC AM parameter adjustments such as discard timer or the maximum number of retransmissions (3GPP TS25.322 2006). The settings of Spreading Factor (SF), Transmission Time Interval (TTI) and the number of TBs that are jointly coded and transmitted per TTI is shown in a schematic illustration in Figure 2.3 for the three UMTS Down Link (DL) radio bearers that have been used during the measurement campaign.

The focus throughout this work is mainly on the 384 kbit/s bearer as it represents the most demanding of the available bearers. As the measurements from different networks and different mobile stations are already mentioned, likewise the measurement results for the other bearers are presented just in case new conclusions arise. Otherwise, only the results for the 384 kbit/s bearer are shown in this work and the conclusions for the other bearers can be derived similarly.

Another very important parameter of the UTRAN for evaluating the error characteristics of the DCH is the BLock Error Ratio (BLER) quality target value for the Outer Loop Power Control (OLPC) mechanism. This target value was set to 1% in the considered networks.

384 kbit/s bearer:

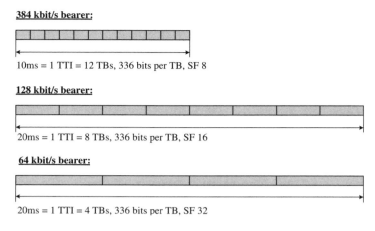

10ms = 1 TTI = 12 TBs, 336 bits per TB, SF 8

128 kbit/s bearer:

20ms = 1 TTI = 8 TBs, 336 bits per TB, SF 16

64 kbit/s bearer:

20ms = 1 TTI = 4 TBs, 336 bits per TB, SF 32

Figure 2.3 Illustration of bearer parameters.

Table 2.2 List of measurement scenarios with different mobility characteristics.

Reference network – without movement (within shielding box)	'reference'
Live network – without movement	'static'
Live network – with movement ('dynamic')	'small-scale movements'
	'walking indoor'
	'tramway'
	'car-city'
	'car-highway'

As a consequence, the OLPC tries to adjust the Signal to Interference Ratio (SIR) target for the Inner Loop Power Control (ILPC) mechanism in a way that the required link quality (1% TB error probability in this case) is satisfied. It will be shown further on in this chapter that this quality target is missed significantly in all the scenarios.

2.1.2   Mobility Scenarios

For the analysis of the UMTS Dedicated Channel (DCH) link error characteristics, we have considered several scenarios with different mobility characteristics which in this work we refer to as 'static', 'small-scale movements', 'walking indoor', 'tramway', 'car-city', 'car-highway' and 'reference' as listed in Table 2.2.

   The measurements for the **'static'** case were performed in an office room in the city centre of Vienna, Austria, with the UMTS terminal lying on the table in a typical Viennese office environment. Due to little movement of persons or other objects around the mobile station, there were few variations in the channel. The **'small-scale movements'** measurements were performed by a person sitting at the table and randomly tilting and moving the UMTS mobile

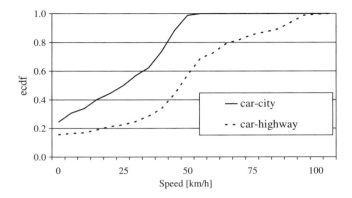

Figure 2.4 Comparison of speed in the scenarios 'car-city' and 'car-highway'.

with his hands. In the '**walking indoor**' scenario, as the label indicates, the measurement results were obtained while walking around inside the building.

The rest are outdoor scenarios with the measurements performed in a tramway going round the city centre of Vienna ('**tramway**') and going by car either on a street in Vienna ('**car-city**') with moderate speed of up to 50 km/h or on a highway with higher speeds of up to 100 km/h ('**car-highway**'). The speed distribution in the form of empirical cumulative density function (ecdf) for the measurements in the last two scenarios is presented in Figure 2.4.

As mentioned, for the '**reference**' scenario the mobile was enclosed within a shielding box to avoid interference, multipath propagation and fading effects, with the direct signal from the NodeB antenna connector fed into the box via a planar antenna coupler at the bottom inside the box. In order to obtain sufficient statistics of the measured link error characteristics, several data traces were recorded in each of the mentioned scenarios, all with a length of about one hour.

Note that the measured link error statistics within all the scenarios heavily depend on the actual system and service coverage in the network – if the mobile station moves towards or even over the cell edge and thus the link transmit power at the base station reaches its limit, the target SIR value at the receiver can no longer be met and the link error probability increases. Therefore, measured network performance statistics in general are to be seen as a snapshot at the actual state of the UMTS radio network deployment at the time when the measurements have been performed. This is especially the case for the scenarios with movement. It is observed in Figures 2.6 and 2.8 that the Common Pilot Channel (CPICH) E_c/I_0 (chip energy to noise and interference ratio) and CPICH Received Signal Code Power (RSCP) take on very low values during the measurements in the 'tramway' and 'car-highway' scenarios, measured with 'mobile 1'. Typically, a CPICH E_c/I_0 level of about -12 dBm is defined as the cell edge (Holma and Toskala 2004).

On the other hand, it is shown in Figures 2.5 and 2.7 that in the 'static' and the 'reference' scenarios, the CPICH levels are indicated to be well inside network coverage. Note, 'static 1' to 'static 3' are measurements in 'static' scenarios at different locations, on different days and at different times of day, thus representing different propagation and network load conditions.

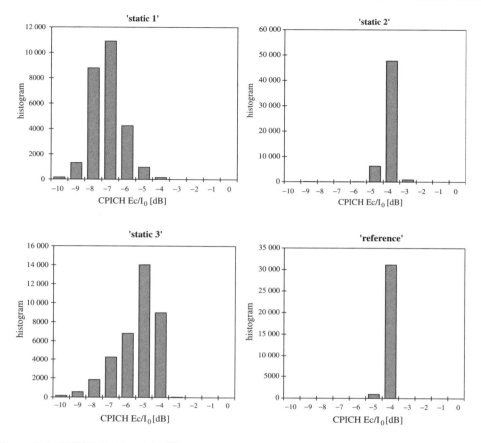

Figure 2.5 CPICH E_c/I_0 with different propagation and network load conditions in the live network ('static') and in the reference network ('reference'), 'mobile 1'.

Below, all these measurements in the 'static' scenarios will be shown to result in the same link error characteristics despite having different absolute values of CPICH E_c/I_0 and CPICH RSCP. Furthermore, measurements in heavily loaded cells have shown that the cell load has no impact on the DCH error characteristics – meaning that the admission control mechanism in the system works well. Owing to the arguments mentioned, we conclude that the measured results from the 'static' scenarios are broadly independent of the current state of network deployment and network load.

However, Figures 2.7 and 2.8 show great differences between the variance of the measured CPICH RSCP values of the 'static' and 'reference' scenarios and the movement case that is due to the differences in the fading properties. It will be shown in the next section that these differences in the CPICH RSCP variance caused by different fading effects lead to different link error characteristics owing to the non-optimality of the power control algorithm.

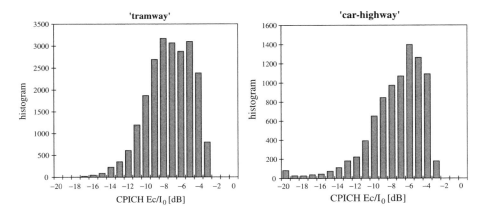

Figure 2.6 CPICH E_c/I_0 for 'tramway' and 'car-highway' scenarios, 'mobile 1'.

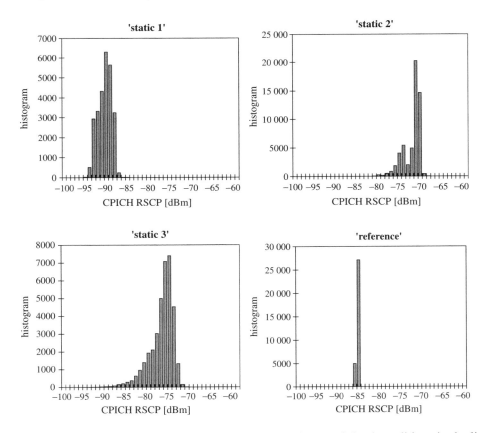

Figure 2.7 CPICH RSCP with different propagation and network load conditions in the live network ('static') and in the reference network ('reference'), 'mobile 1'.

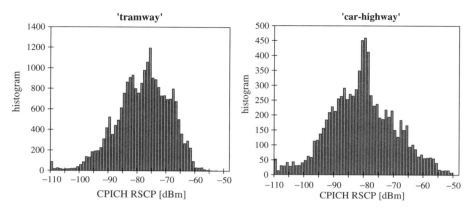

Figure 2.8 CPICH RSCP for 'tramway' and 'car-highway' scenarios, 'mobile 1'.

2.2   Link Error Analysis

When transmitting data over wireless mobile communication systems, most of the errors originate from the radio link, which will also become the main or even the only error source in the transmission chain as long as the frequency spectrum is a scarce resource and there is interference limitation in the system. Since the highest error probability is found in the radio link, we focus error analysis primarily on the wireless access part of the network.

As presented in Chapter 1 of this book, the UMTS standard specifies an error check in the radio link in the form of a CRC per TB in the physical layer (3GPP TS 25.302 2003). This information is perfect for our link error analysis, as all the important error sources within the radio link (fading, interference, multipath propagation, power control, as well as channel coding and interleaving effects) are included in the resulting link error statistics.

2.2.1   Link Error Probability

The first considered parameter for the analysis of the link error characteristics is the estimated total TB error probability[3] $\overline{P}_e(\mathrm{TB})$ measured over the complete trace. The values are listed in Table 2.3 for the different scenarios, measured with 'mobile 1' and a 384 kbit/s packet-switched (PS) bearer. Very similar results were obtained with other mobiles and bearers.

In all the scenarios with mobility, the estimated TB error probability is around 2×10^{-2}, whereas in the static case the estimated error probability is smaller by one order of magnitude. Due to these results, we not only conclude that the Transmit Power Control (TPC) algorithm of the UMTS is not capable of adjusting the required BLER target value (1%), but we also recognize that there are basically only two different link error characteristics with respect to estimated total link error probability: static and dynamic, regardless of the type of movement.

---

[3] As there is no proof of ergodicity for such transmissions, error ratios cannot be seen equivalently as error probabilities and thus, in this work, generally the terms 'estimated error probability' or 'error ratio' are used for measured values. Despite these arguments, sometimes the term 'error probability' is also used for measured results in order to prevent confusion with expected future probabilities, which then of course also refers to an 'estimated error probability' or 'error ratio'.

Table 2.3 Estimated probabilities of TB and TTI errors (384 kbit/s bearer).

Scenario	$\overline{P}_e(\text{TB})$	$\overline{P}_e(\text{TTI})$
static	2.66×10^{-3}	4.72×10^{-3}
small-scale movements	2.22×10^{-2}	2.34×10^{-2}
walking indoor	1.98×10^{-2}	2.44×10^{-2}
tramway	1.44×10^{-2}	1.70×10^{-2}
car-city	2.06×10^{-2}	2.48×10^{-2}
car-highway	2.34×10^{-2}	2.63×10^{-2}

An interesting and new conclusion, contradicting common research opinion, is that, for example, the small-scale movements performed by a person just sitting and moving the mobile with his hands result in the same error probability as when going by car.

The present work is based on measurements in live UMTS networks. Unfortunately, the analysis of the received signal within one TTI was not possible with the used measurement equipment and, therefore, the different underlying fading effects could not be analysed. Furthermore, it was beyond the scope of this research project to perform an analysis with a system level simulator, which would have to include propagation effects, channel coding and interleaving as well as a feedback channel and the inner closed loop power control and quality-based outer closed loop power control mechanisms. However, the very important conclusion can be drawn that the resulting channel (fading and interference effects in connection with power control, channel coding, interleaving and so on) behaves in the same way in all the scenarios considered, with movement, regardless of which kind of movement.

It may be assumed that the similarity in error characteristics, for example, of the 'small-scale movements' scenario and the scenario 'car-highway', is due to the spatial small-scale fading in the indoor scenario caused by multipath propagation. Together with the small-scale movements (changing the position of the mobile station by hands) the mobile is affected in time by fading which seems to be equivalent to the large-scale fading which occurs when going by car up to 100 km/h. Another reason for such behaviour can be attributed to the antenna characteristics of the mobile station in connection with the small-scale movements (tilting and turning the mobile station), again resulting in a time-variant channel. The resulting fading of received code power (or equally the changing of SIR) in time, together with the non-optimality of the power control mechanism leads to the observed high error probability in small-scale movement scenarios.

Moreover, the differences in the error characteristics between the 'static' case, where there is only a little movement of objects or persons around the mobile station, and the 'dynamic' case with the time variant channel, are, it is conjectured, caused by the dominating power control algorithm in the 'static' case and the fading effects in the 'dynamic' scenarios. When comparing $\overline{P}_e(\text{TB})$ in Table 2.3 with the estimated total TTI error probability $\overline{P}_e(\text{TTI})$ (from the same measurements), we note that the results are almost equal, especially in the scenarios with mobility. With 12 TBs per TTI in the case of the 384 kbit/s bearer and the assumption of a memoryless Binary Symmetric Channel (BSC) one would expect a much higher estimated total TTI error probability (for example, $\overline{P}_e(\text{TTI}) = 1 - (1 - 1.44 \times 10^{-2})^{12} = 0.159$ in the 'tramway' scenario) and thus we conclude that there is a high correlation between the error

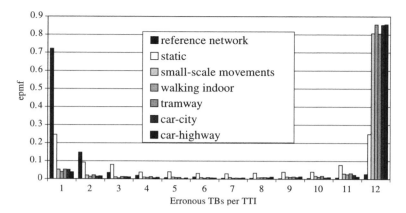

Figure 2.9 Number of erroneous TBs within erroneously received TTIs (384 kbit/s bearer, 'mobile 1').

states of the received TBs within one TTI leading to bursty TB error behaviour of the UMTS DCH which is further analysed below.

2.2.2   Number of erroneous TBs in TTIs

An analysis of the number of erroneous TBs within erroneously received TTIs presented in Figure 2.9 for the 384 kbit/s bearer and measured with 'mobile 1' leads to the same conclusions – especially for the scenarios with mobility, where the ratio of having all (12) TBs erroneously received within one erroneous TTI is between 0.8 and 0.9. On the other hand, the ratio of having only one out of 12 TBs received erroneously within one TTI increases, the less movement disturbs the transmission. We observe ratios of up to 0.7 with the mobile in the shielding box of the reference network where there is no interference, no multipath propagation and no fading in the propagation environment. This leads to the additional conclusion that the small variations of transmit power caused by the UMTS DCH TPC only, result in a very small TB error probability within one TTI (that is, within one jointly turbo-coded and interleaved data block), whereas the deep fades of the scenarios with movement heavily influence the channel and thus cause a high ratio of TB errors within an erroneous TTI. This conclusion is supported by comparison with a Bernoulli experiment with a low (0.05) error probability for the reference network and a high (0.98) value for the case with movement (see Figure 2.10). The static case can be regarded as a mixture of both cases. These results are the basis of the refinement of the link error models, especially in 'dynamic' scenarios, as presented in Chapter 3.2.1.

2.2.3   TTI-burstlength, TTI-gaplength

For further analysis of the correlation properties of the link errors outside one TTI (one jointly turbo-coded and interleaved data block), we build the statistics of the TTI-gaplengths and the TTI-burstlengths, first providing their definitions.

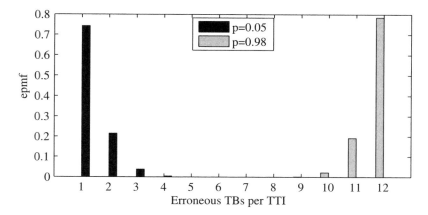

Figure 2.10 Theoretical analysis of the number of erroneous TBs within one TTI: binomial distributions with 0.05 and 0.98 TB error probability.

Definition 2.2.1 *The **TTI-gaplength** is the number of subsequently received error-free TTIs, while the number of subsequent erroneous TTIs is called **TTI-burstlength**.*

Definition 2.2.2 *According to references ITU-T Rec.M.60, 3008 and ITU-T Rec.Q.9, 0222, an **error burst** is defined as a group of successive units (bits, packets or TBs and TTIs in our case) in which two successive erroneous units are always separated by less than a given number L_c of correct units. For our work L_c is equal to zero for bursts.*

From the presentation of the TTI-gap- and TTI-burstlengths in Figure 2.11, we observe that the sequences of error-free TTIs are up to 700 and there are up to ten TTIs subsequently erroneously received. Again we recognize that the statistics for the scenarios with movement are similar while the 'static' scenario shows a considerably different distribution of the TTI-gaplengths and much shorter TTI-burstlengths of up to two only. The reason for this difference between 'static' and 'dynamic' scenarios is that in the scenarios with movement the influence of fading-effects dominates in the link error characteristics while in the 'static' scenario the impact of the OLPC algorithm becomes significant.

The fact that the TTI-gap- and TTI-burstlengths of the scenarios with only slow movement are similar to those measured in the scenarios with fast movement of up to 100 km/h also becomes interesting when analysing the expected coherence time of the channel. With a speed of 100 km/h and the assumption of a coherence length of half a wavelength, the coherence time of the channel due to small-scale fading is shorter than one TTI (10 ms), whereas with walking speed the coherence time would be of the length of 10 to 20 TTIs. Therefore, it may be concluded that in the case of high speeds it is the large-scale fading with an assumed coherence length of about 20 m which results in the same error characteristics as the small-scale fading in the case of slow movement.

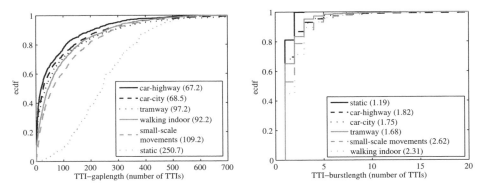

Figure 2.11 Statistics of TTI-gaplengths and TTI-burstlengths (mean values in the legends), 384 kbit/s bearer, 'mobile 1'.

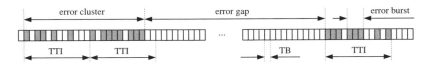

Figure 2.12 Schematic illustration of TB error bursts, gaps and clusters.

2.2.4   TB Error Bursts, TB Error Clusters

Due to the high ratio of having all TBs within one TTI erroneously received for the scenarios with movement as shown in Figure 2.9, the gap and burst analysis in terms of TTIs would be sufficient, whereas in the 'static' case the granularity of the analysis has to be refined to TB level in order to catch correlation properties of TB errors within one TTI.

Thus, we define TB error gaps and TB error bursts with the corresponding gaplength and burstlength in Definition 2.2.3, again defining a TB error burst as in Definition 2.2.2. Additionally, we define groups of erroneously received TBs as error clusters in Definition 2.2.4.

Definition 2.2.3 *The **gaplength** is the number of subsequently received error-free TBs, while the number of subsequent erroneous TBs is called **burstlength**.*

Definition 2.2.4 *An **error cluster** is a group of erroneously received TBs, if, for L_c equal to the number of TBs per TTI, the TB error bursts are separated by at most L_c error-free TBs.*

In Figure 2.12 we present a schematic illustration of the TB error bursts, gaps and error clusters.

Figure 2.13 shows the results of the analysis of the gaplengths and burstlengths (in number of TBs) for the 'static' scenario measured with a 384 kbit/s bearer and 'mobile 1' at three different locations as well as at different dates and times of day ('static 1' to 'static 3') in order to have different propagation, interference and network load situations (Karner and Rupp 2005). We observe that there is only a small difference between the statistics.

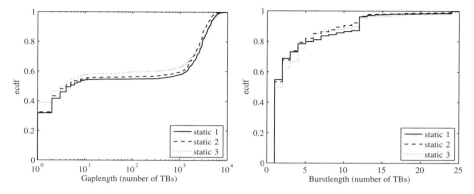

Figure 2.13 Statistics of TB gaplengths and burstlengths, 384 kbit/s bearer, 'static' scenario, 'mobile 1'.

Furthermore, in measurements with additional DCH and also High Speed Downlink Packet Access (HSDPA) users, for creation of intra- and inter-cell interference we discovered that additional interference in the DL does not change the measured statistics as long as the UMTS call admission control is working properly to keep the required link power off the power limit.

Note in Figure 2.13 that there are short gaps of at most 12 TBs which are the gaps within one or two successive TTIs. The statistics also include long gaps from 200 to 7000 TBs. The burstlengths are between one and 25 with two higher steps at 12 and 24 TBs caused by the ratio of having all TBs erroneously received within one TTI (there are 12 TBs within one TTI in the case of the 384 kbit/s bearer) which is still not negligible even in the 'static' case.

Note that the turbo coding as well as the interleaver algorithms of the UMTS radio access affect the link error characteristics, especially within one radio frame, TTI or coding block. However, the measurement-based analysis of the positions of the erroneous TBs within one TTI has shown that all possible TB locations exhibit almost the same estimated error probability (presented in Figure 2.14). Similar results are observed from the analysis of the output of turbo code simulations in Navratil (2001).

The uniform distribution of the positions of the erroneous TBs within one TTI, together with the high ratio of having not all the TBs within one TTI received with error, results in short error bursts (separated by short gaps) within one and even two successive TTIs.

Thus, in order to analyse the error properties within one or more successive TTIs – meaning the characteristics of and within an error cluster – it is also interesting to present the statistics of the clusterlength and the number of error bursts in a cluster as shown in Figure 2.15 for $L_c = 12$. There we show three 'static' measurements which have a mean clusterlength of 10.84, 11.19, 14.98, a mean number of erroneous TBs per error cluster of 7.97, 7.74, 9.81, and a mean number of bursts per cluster of 2.19, 2.23, 2.38, respectively. We observe that the clusterlength is at most 25 in most cases and we do have one to eight error bursts within one such error cluster. Since the influence of the quality based power control algorithm on the link error characteristics becomes significant in the 'static' scenarios, the error clusters are separated by long error gaps as shown in the following.

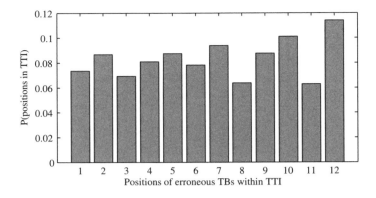

Figure 2.14 Estimated probabilities of TB error positions, 'static' scenario, 'mobile 1', 384 kbit/s bearer.

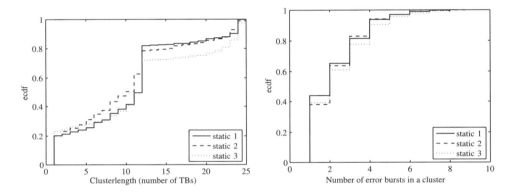

Figure 2.15 Statistics of TB clusterlengths and number of error bursts per cluster, 384 kbit/s bearer, 'static' scenario, 'mobile 1'.

2.2.5   The Influence of TPC on Link Error Characteristics

In order to use radio resources in cellular wireless communication systems efficiently, the minimum transmit power for each link has to be found while meeting the quality target of the link. In UMTS this task is performed by the OLPC algorithm (3GPP TS-25.101 2006) – a slow (100 Hz) closed loop control mechanism which sets the target SIR for the fast ILPC in order to converge to a required link quality given by the network (1% in our work). The ILPC then regulates the link power level with a rate of 1500 Hz to meet the target SIR.

Within the 3rd Generation Partnership Project (3GPP) the OLPC is not entirely specified. A common way of implementing the control procedure is to use CRC information feedback of the transmitted TBs together with a sawtooth algorithm as described in Holma and Toskala (2004) and Sampath, Kumar and Holtzman (1997). If a TB was received in error, the target SIR value is increased by $K \cdot \Delta$, whereas in the case of error-free transmission, it is reduced

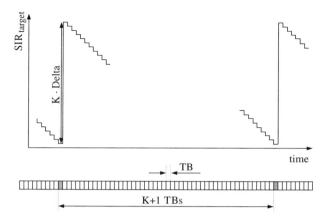

Figure 2.16 Quality based power control (ideal).

by Δ, where Δ is the step size in dB. If K is chosen to be $1/\text{BLER}_{\text{target}} - 1$, the algorithm tries to keep the BLER always less than or equal to $\text{BLER}_{\text{target}} = 1/(K + 1)$. Thus, by selecting a proper value K a quality level for the wireless link can be set.

For the work presented here there was no information available about the TPC algorithms of the three different manufacturers' equipment. Despite this fact, the analysis of the measured SIR signal and SIR target values led to the conclusion that a TPC algorithm, like the one described, is applied in their systems. This TPC algorithm is also the only one commonly referred to in the literature. Of course, this is an assumption and minor differences in the actual implemented TPC algorithms to the assumed ones are possible.

Figure 2.16 shows a schematic illustration of this OLPC algorithm for when an error occurs in the link with a probability of one if the $\text{SIR}_{\text{target}}$ and thus also the SIR at the receiver falls below a certain threshold. Obviously, in this special case there is an erroneous TB after K error-free TBs.

As we have seen in the UMTS DCH, error clusters are generated due to joint interleaving and coding over one TTI and also due to the stochastic nature of the channel. The power control algorithm which is still trying to meet the quality target for the link increases the target SIR by $K \cdot \Delta$ times the number of errors in the cluster. This leads to a multiple of K error-free TBs (long gap) between two error clusters.

From these arguments we conclude that, especially in the static case, the link error characteristics are a result of the quality-based closed outer loop power control mechanism of the UMTS DCH. Compared to the static case, in the scenarios with movement, fading effects dominate as reasons for radio-link errors but, nevertheless, the TPC still affects the link error characteristics in a way that the error process becomes recurrent.

Furthermore, we conclude that the predictability of the link errors is due to the fact that the quality-based control mechanism introduces recurrence to the error behaviour. Because of this, there is, for example, a predictive ending of an error free run and, therefore, there is memory in the error process and the error gaps are non-geometrically distributed. In a mobile communication system without a quality-based power control mechanism, there

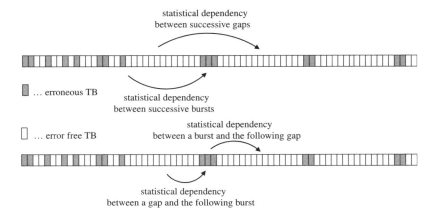

Figure 2.17 Schematic illustration of correlation between error gaps and bursts.

would, for example, be a high probability of having a very long error-free run with a non-predictive end in the case of a very good channel.

In the static case, stationarity of the UMTS DCH error process is assumed, as the statistics (for example, of the gap and burstlengths) are constant over time due to call admission control and also due to the closed loop power control mechanism which adjusts the same SIR and also the same link error characteristics for different received power levels and therefore for different cells and propagation scenarios. On the other hand, in scenarios with movement, as we move from cell to cell, the results depend on the current status of network deployment. However, it is argued here that there are no significant differences even between different scenarios with movement and we show that the statistics are also constant over one for this research relevant time period. Therefore, stationarity is assumed also for scenarios with movement.

Generally, the stochastic process of received power levels in a mobile communication network is a non-ergodic process as means are not fixed when performing measurements in live networks under mobility conditions. On the other hand, the error process of the UMTS DCH may be seen as an ergodic process owing to the fact that the closed loop power control mechanism (consisting of the inner loop power control and the quality-based outer loop power control mechanism) introduces recursiveness to the error process. Together with noise and fading, the process is also aperiodic, which would support the assumption that it has ergodicity.

2.2.6   Statistical Dependency between Successive Gaps/Bursts

Having several error bursts separated by short gaps forming an error cluster and the occurrence of long gaps following error clusters means not only having correlation between successive TBs but also having statistical dependencies between neighbouring gaps and bursts as indicated in the schematic illustration in Figure 2.17.

We validate this conclusion in Figure 2.18 where we find that the estimated probability of having short ($\leq 12\,$TBs) and long ($> 12\,$TBs) gaplengths does depend on previous

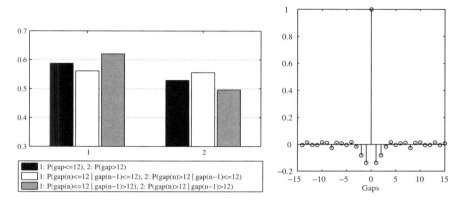

Figure 2.18 Statistical dependency between successive gaps (left) and autocorrelation of gaplengths (right), 384 kbit/s bearer, 'static 2' scenario, 'mobile 1'.

gaplengths. Furthermore, we observe from the autocorrelation function of the sequence of gaplengths (Figure 2.18, right) that the memory in the channel outreaches two successive gaps. Similar results were obtained for bursts and also for the dependencies between gaps and bursts.

2.2.7   Block Error Ratio (BLER)

For many applications as well as for the performance analysis of channel coding, another important measure of the link error characteristics is the statistics of the estimated error weight probability $P(m, n)$, the estimated probability of having exactly m errors occurring in a block of n (bits, packets or TBs in our case). This error weight probability is also well known as a performance indicator of the wireless link in mobile communication systems such as UMTS. There, it is called BLER, which is the number m of erroneously received TBs out of n TBs. Here, we have selected $n = 2400$ TBs to meet the trade-off between providing accurate results also for BLER $< 1\%$ and offering BLER values in fine time granularity. Moreover, $n = 2400$ TBs $= 200$ TTIs is adjusted for the performance measures in the considered UMTS networks for the 384 kbit/s bearer.

The measurements were applied in all the scenarios with a duration of ≈ 1 h each. This led to the fact that $\geq 3 \times 10^6$ TBs were available in each of the measured traces. Thus, at least 1000 blocks of $n = 2400$ TBs were considered for the analysis of the BLER.

Figure 2.19 shows the measurement results for the 'static' scenario and also for scenarios with movement ('tramway' and 'small-scale movements'), measured with 'mobile 2' and 'mobile 3' and with all the available bearers for the DCH in DL. We observe that in the 'static' scenario the TPC of the UMTS is capable of keeping the BLER around the target value of 1%, while in the scenarios with movement, the radio link with a very high probability experiences high BLER owing to fading effects, insufficient precision and lack of speed of the TPC algorithm. Again, we notice the similarity of the results for the scenarios with movement regardless of which kind of movement. For example, the small-scale movements performed by a person sitting at a table and tilting and turning the mobile station with his hands produce

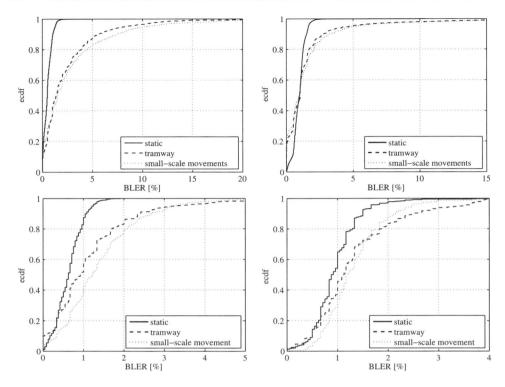

Figure 2.19 Comparison of DL BLER (%) for the 384 kbit/s bearer measured with 'mobile 2' (upper left) and 'mobile 3' (upper right) and for the 128 kbit/s (lower left) and 64 kbit/s (lower right) bearers measured with 'mobile 2'.

the same link error characteristics as when that person goes by tramway. Furthermore, the empirical cumulative distribution functions (cdfs) of the BLER in Figure 2.19 support the conclusion for having a high ratio of all TBs within one TTI being erroneously received (error bursts) as this is visible through the steps in the cdfs.

2.3   Dynamic Bearer Type Switching

High flexibility is one of the main advantages offered by UMTS. 3GPP specifications allow the physical resources allocated to the users dynamically to optimize the radio resource utilization in UMTS to be changed. This means that the data rate in UMTS can be changed every 10 ms, either by changing the transport channel type or by varying the dynamic or semi-static parameters of the Transport Format (TF) (Holma and Toskala 2004; 3GPP TS 25.302 2003). Network operators can make use of this optimization feature by properly adjusting the dynamic bearer type switching in their networks. Such switching of the channel characteristics optimizes the use of the radio resources and facilitates the provision of the

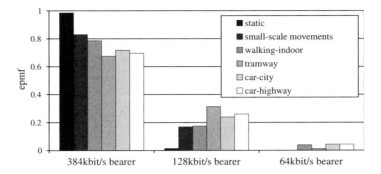

Figure 2.20 Bearer usage in TTIs with dynamic bearer type switching, 'mobile 1'.

required quality of service for the user. It can be triggered by admission control, congestion control, soft handover, required throughput or the radio channel quality (power threshold).

The focus of this work is on the dynamic switching between the three different DCH DL radio access bearers for the PS domain with 64 kbit/s, 128 kbit/s and 384 kbit/s, which are characterized by different values of TTI, TBSS (Transport Block Set Size) and SF, as shown in Figure 2.3. The analysis focuses on the dynamic bearer type switching due to fading and coverage reasons, where the decision of up or down switching is based on the link quality only. Therefore, it is assumed that all the bearers are admitted during the measurements and thus have a congestion-free network. This means that there were enough resources available in all the cells, so that the favourite bearer (384 kbit/s) could be assigned. Furthermore, a UDP data traffic with a constant bit rate of 372 kbit/s is sufficient throughput to be assigned a 384 kbit/s bearer due to throughput constraints.

For the performance evaluation of applications in system level simulators, models are needed which are capable of representing the error characteristics of the underlying channel properly. In Chapter 3 of this book such models are developed (Karner *et al.* 2005), assuming the absence of dynamic bearer switching. Due to the fact that each bearer shows different characteristics, a better representation of the link is obtained by including bearer switching in the error model which is performed, for example, in Umbert and Diaz (2004) based on simulations only. The approach in this work is to develop a model for dynamic bearer type switching based on measurements in live UMTS networks within typical mobility scenarios for using UMTS services. This model for dynamic bearer type switching can then be combined with the error model of Chapter 3 and the combination can be used as a representation of the lower layers in a system level simulation tool.

2.3.1   Measurement-based Analysis of Dynamic Bearer Type Switching

In Figure 2.20 the measured usage probabilities of the various bearers for 'mobile 1' are shown (based on the number of TTIs in each bearer – expressed in times, the values are different). We observe that in all the scenarios with mobility, the estimated probability of being in the 384 kbit/s bearer is around 75%, whereas in the static case it is almost 100%. Of course, the usage probabilities of the various bearers depend on the network coverage situation and also on the quality of the receiver in the mobile.

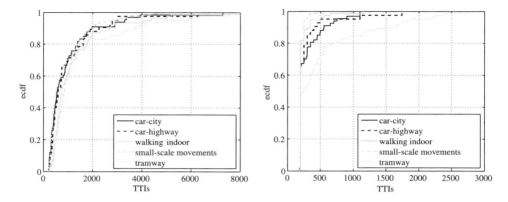

Figure 2.21   Measured runlengths of 384 kbit/s bearer (left) and 128 kbit/s bearer (upswitching) (right) in different scenarios, 'mobile 1'.

The analysis of dynamic bearer type switching is based on the runlength distribution of each of the bearers, measured in the number of subsequent TTIs in the case of 'mobile 1' and in seconds between the 'Transport Channel Reconfiguration' messages for 'mobile 3'. For 'mobile 3' the runlengths are provided in seconds instead of TTIs, due to the fact that with 'mobile 3' it is not possible to analyse the runlengths in the accuracy of TTIs but only by the 'Transport Channel Reconfiguration' messages.

In Figure 2.21 (left) the cdfs of the runlengths are shown for the 384 kbit/s bearer and for 'mobile 1' in various scenarios. We observe that the minimum runlength is 200 TTIs (= 2 s in the case of 384 kbit/s bearer) and that 80–90% of the runlengths are shorter than 2000 TTIs (= 20 s). Furthermore, we recognize that less than 5% of the runlengths are greater than 6000 TTIs (60 s) and the longest period to stay in the 384 kbit/s bearer is about 80 s.

The statistics of the runlengths of the 128 kbit/s bearer with subsequent upswitching to the 384 kbit/s bearer is shown in Figure 2.21 (right). Again, these results are for different mobility scenarios and are measured with 'mobile 1'. The runlengths in the 128 kbit/s bearer are all greater than 200 TTIs, which is equal to 4 s in that case. We observe a jump in the empirical cdfs up to 50% or even 80% in the 'walking indoor' scenario. The reason for having such a high step lies in the hysteresis time of the switching algorithm (4 s) and the decorrelation time (coherence time) of the channel, which is obviously shorter than 4 s. This means that the channel quality would allow an upswitching to the 384 kbit/s bearer earlier than 4 s after a previous downswitching, but this is prevented by the switching algorithm (hysteresis). Therefore, an optimized hysteresis parameter can accelerate the switching to the 384 kbit/s bearer in up to 80% of the observed cases and thus is capable of improving the throughput over the wireless link.

For the 128 kbit/s bearer with subsequent downswitching and in the case of the 64 kbit/s bearer, the sum[4] of the statistics of the different mobility scenarios is provided in Figure 2.22, measured with 'mobile 1'. Figure 2.22 (left) shows that in contrast to the runlength statistics

---

[4] As there is less usage of the 128 kbit/s and the 64 kbit/s bearer compared to the 384 kbit/s bearer, the sum of all mobility measurements is used for the smaller bearers to reach sufficient statistics.

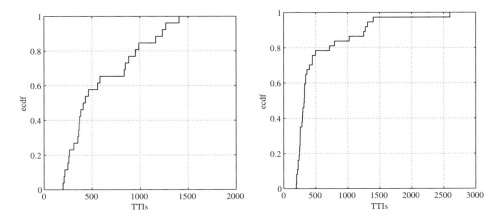

Figure 2.22 Measured runlengths of 128 kbit/s bearer (downswitching) (left) and 64 kbit/s bearer (right), sum of different scenarios, 'mobile 1'.

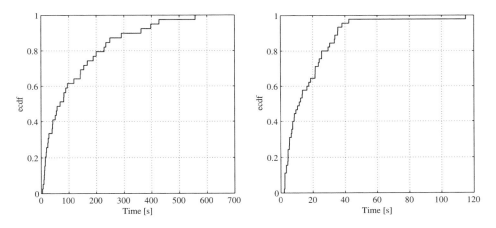

Figure 2.23 Measured runlengths of 384 kbit/s bearer (left) and 128 kbit/s bearer (right), sum of different scenarios, 'mobile 3'.

of the 128 kbit/s bearer with subsequent upswitching, there is no high step at 200 TTIs in the runlength statistics of the 128 kbit/s bearer when being followed by a downswitching to the 64 kbit/s bearer. That difference in the runlength distributions is the reason for splitting the 128 kbit/s bearer into two parts for analysis and modelling. The measured runlength distribution for 'mobile 3' with the 384 kbit/s and the 128 kbit/s bearer is shown in Figure 2.23, again as the sum of the measurements of all different mobility scenarios. From Figure 2.23 (left) we observe that with 'mobile 3', the runlengths of the 384 kbit/s bearer are greater than 2 s, and there are runlengths with a duration of up to 600 s, which is much longer than the measured runlengths of 'mobile 1' in the same bearer. The reason for the difference in the runlength distribution between 'mobile 1' and 'mobile 3' is that 'mobile 3' is less sensitive to fading effects and achieves a better service coverage in the network due to a better receiver.

References

Ericsson TEMS: http://www.ericsson.com/solutions/tems/

Holma, H. and Toskala, A. (2004) *WCDMA for UMTS, Radio Access For Third Generation Mobile Communications,* John Wiley & Sons, Ltd.

ITU-T Rec. M.60, 3008.

ITU-T Rec. Q.9, 0222.

Karner, W. and Rupp, M. (2005) Measurement based Analysis and Modelling of UMTS DCH Error Characteristics for Static Scenarios. In *Proceedings of the Eighth International Symposium on DSP and Communications Systems 2005 (DSPCS'2005),* Sunshine Coast, Australia, Dec.

Karner, W., Svoboda, P. and Rupp, M. (2005) A UMTS DL DCH Error Model based on Measurements in Live Networks. In *Proceedings of the 12th International Conference on Telecommunications 2005 (ICT 2005),* Capetown, South Africa, May.

Navratil, F. (2001) Fehlerkorrektur im physical Layer des UMTS. Master's thesis (in German), Institute of Communication and Radio-Frequency Engineering, Vienna University of Technology, Austria, Nov.

3GPP TS 25.101 2006, User Equipment (UE) Radio Transmission and Reception (FDD), v.6.12.0, Jun.

3GPP TS 25.302 2003, Services Provided by the Physical Layer, v.4.8.0, Sep.

3GPP TS 25.322 2006, Radio Link Control (RLC) Protocol Specification, v.6.9.0, Oct.

Papoulis, A. and Unnikrishna, P.S. (2002) *Probability, Random Variables, and Stochastic Processes,* McGraw-Hill.

Sampath, A., Kumar, P.S. and Holtzman, J.M. (1997) On Setting Reverse Link Target SIR in a CDMA System. In *Proceedings of the 47th IEEE Vehicular Technology Conference,* vol. 2, pp. 929–933.

Umbert, A. and Diaz, P. (2004) On the Importance of Error Memory in UMTS Radio Channel Emulation using Hidden Markov Models (HMM). In *Proceedings of the 15th IEEE International Symposium on Personal, Indoor and Mobile Radio Communications (PIMRC 2004),* vol. 4, pp. 2998–3002, Sep.

Willtek 4920: http://www.willtek.com/

3

Modelling of Link Layer Characteristics

Wolfgang Karner

Nowadays, simulations have become an important tool for the analysis of communication systems. In order to keep the computing effort manageable, different abstraction levels of the system are introduced in the simulator where models represent lower layer functionalities and their characteristics. With increasing complexity in coding, protocols and applications, increasingly large amounts of processed data are required for sufficient statistics in such analyses. Thus, simulations with stochastic link models are becoming necessary instead of, for example, using simulations with measured traces.

This work focuses on link error models for the radio access part of the Universal Mobile Telecommunications System (UMTS) network as it represents the most relevant error source in the communication chain. The presented models are situated in the link layer of UMTS Terrestrial Radio Access Network (UTRAN) and thus incorporate complete physical layer functionality and characteristics such as propagation effects, interference, power control, transmitter and receiver properties, channel coding and interleaving. On the other hand this modelling approach offers the possibility to evaluate the performance of link layer algorithms such as retransmission mechanisms and scheduling as well as to analyse all higher layer protocols and applications.

3.1 Modelling Erroneous Channels – A Literature Survey

The simplest way to model an erroneous channel is by a communication link in which errors occur with probability p. Considering the transmission of bits, this channel is called the Binary Symmetric Channel (BSC) (Papoulis and Unnikrishna 2002; Shannon 1948) as in case of an error the '0' bit is changed to '1' and vice versa, as depicted in Figure 3.1. This channel

Video and Multimedia Transmissions over Cellular Networks   Edited by Markus Rupp
© 2009 John Wiley & Sons, Ltd

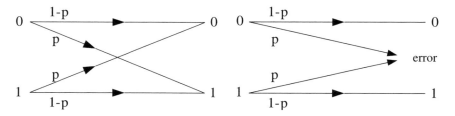

Figure 3.1 Binary Symmetric Channel (BSC) and Binary Erasure Channel (BEC).

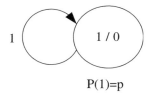

$P(1)=p$

Figure 3.2 Schematic illustration of the memoryless error source model.

model is frequently used in coding and information theory as it is one of the simplest channels to analyse.

Similarly to the BSC, the Binary Erasure Channel (BEC) (Elias 1954) is often used in information theory analyses. In the BEC, instead of 'flipping' the bits as in the BSC, either the transmitted bit is received correctly or an 'erasure' message is sent to the receiver in the event of an error.

The corresponding memoryless error source model is illustrated in Figure 3.2 comprising one state only, in which either an erroneous unit (1) with probability $P(1) = p$ or an error-free unit (0) with probability $P(0) = 1 - p$ is produced at each discrete time instant. Of course, the probability to return to the state is one.

Like comparing BSC with BEC, handling the error information from the error source model is different and depends on the application, for example whether bit 'flipping' is applied in the channel or whether packets are marked as erroneous or discarded. Thus, in the following, different error source models are presented without discussing the error handling in detail.

Since the work of Gilbert (1960), researchers have understood that a memoryless BSC or BEC is not capable of describing the bursty nature of real communication channels. Gilbert introduced an error source model where the error probability of the current unit depends on the error state of the last unit – meaning that there is correlation between two successive error events and thus there is memory and burstiness in the channel. Gilbert's model is a first-order two-state Markov model with one 'good' and one 'bad' state, where no errors occur in the 'good' state but in the 'bad' state the error probability takes some value larger than zero. In Figure 3.3 Gilbert's model is presented.

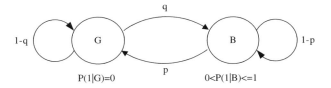

Figure 3.3 Schematic illustration of Gilbert's model.

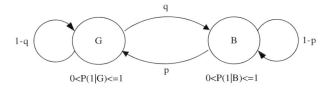

Figure 3.4 Schematic illustration of the Gilbert–Elliott model.

The probability to be in the bad state is $P(B) = q/(p+q)$ and thus the resulting total error probability is

$$\overline{P}_e = P(1|B)P(B) + P(1|G)P(G) = P(1|B) \cdot \frac{q}{p+q}.$$

Gilbert's model is a renewal model, meaning that the current burstlength or gaplength is independent of the previous burstlength or gaplength, and it is a Hidden Markov Model (HMM) as the current state of the model (G or B) cannot be determined from the model output. Furthermore, the model parameters $\{p, q, P(1|B)\}$ are not directly observable from training data and thus have to be estimated via trigram statistics or curve-fitting as proposed in Gilbert (1960).

Due to the straightforward parameter estimation, a simplified version of Gilbert's model has often been used in subsequent studies, where the error probability in the 'bad' state is one. Then the model changes from an HMM to a first-order two-state Markov chain. Thus, the two parameters $\{p, q\}$ for this 'simplified Gilbert' model can be calculated directly from the measured error trace by using the mean error burstlength ($p = 1/\bar{L}_{\text{burst}}$) and the mean gaplength ($q = 1/\bar{L}_{\text{gap}} = \overline{P}_e/(\bar{L}_{\text{burst}} \cdot (1 - \overline{P}_e))$) or the total error probability ($\overline{P}_e = q\bar{L}_{\text{burst}}/(1 + q\bar{L}_{\text{burst}})$).

Enhancements to Gilbert's model can be found in the work of Elliott (1963) where errors can also occur in the 'good' state, as shown in Figure 3.4, if rarely.

This model, also known as the Gilbert–Elliott Channel (GEC), overcomes the limitation of Gilbert's model in allowing only for geometric distributions of burstlengths. Besides being an HMM, the model is also non-renewal, that is, for example, the current burstlength is not statistically independent of the previous burstlength. This of course brings opportunities for channel modelling but also complicates the parameter estimation. The parameters for non-renewal HMMs such as the GEC have to be estimated as, for example, using the Baum–Welsh algorithm (Rabiner and Juang 1986).

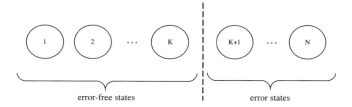

Figure 3.5 Partitioned Markov chain.

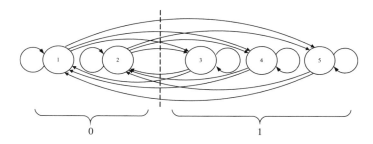

Figure 3.6 Partitioned Markov chain with two error-free states and three error states.

Later on in the 1960s, Berger and Mandelbrot (1963), Sussman (1963), Elliott (1965) and Mandelbrot (1965) proposed to use renewal processes to model the error characteristics of communication links, with the suggestion in Berger and Mandelbrot (1963) to use independent Pareto ($f(t|a) = a \cdot t^{-a}$, $t \geq 1$, $0 < a < 1$) distributions for the intervals between successive errors.

Further enhancements to Gilbert's model were published by Fritchman (1967), proposing partitioned Markov chains with several error-free and error states as shown in Figure 3.5. With the restriction of forbidden transitions between error states and also between error-free states, the model parameters can be estimated via separate fitting of polygeometric distributions to the ccdf of gaplengths and to the ccdf of burstlengths. The polygeometric cdfs are given by

$$F(x|K)_{\text{polygeometric,gaplength}} = 1 - \sum_{i=1}^{K} \mu_i \lambda_i^x, \qquad (3.1)$$

$$F(x|K, N)_{\text{polygeometric,burstlength}} = 1 - \sum_{i=K+1}^{N} \mu_i \lambda_i^x, \qquad (3.2)$$

with the constraints of $0 < \mu_i < 1$ and $0 < \lambda_i < 1$ (μ_i and λ_i correspond to the probabilities of transition to the state and within the state, respectively). K is the number of error-free states and N is the total number of states.

Even for this model configuration, as presented in Figure 3.6 for two error-free states and three error states, there is still statistical dependency between the current gaplength and the

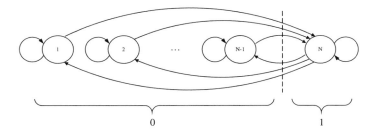

Figure 3.7 The well-known 'Fritchman' model configuration.

Table 3.1 Comparison of frequently used classical models.

Model	Properties
Two-state Markov chain ('simplified Gilbert' model)	renewal, geometric distributed burst- and gaplengths
'Gilbert' model	renewal, HMM, geom. burstlengths, approx. polygeom. gaplengths
GEC (Gilbert–Elliott Channel)	non-renewal, HMM, approx. polygeom. burst- and gaplengths
'Fritchman' model	renewal, HMM, geom. burstlengths, polygeom. gaplengths

previous burstlength and vice versa, and also between the current gaplength (burstlength) and the previous gaplength (burstlength). Therefore, for a full definition of the model, these dependencies have also to be considered. However, with the additional restriction of having constant proportions of allowed transition probabilities from one state to the successive states the model becomes renewal. For example, in the case of the 2/3 configuration of Figure 3.6 this requires $p_{13} : p_{14} : p_{15} = p_{23} : p_{24} : p_{25}$ and $p_{31} : p_{32} = p_{41} : p_{42} = p_{51} : p_{52}$.

The well-known 'Fritchman' model is a special case of the partitioned Markov chains and has one error state only as presented in Figure 3.7. In this configuration the error-free run (gaplength) distribution uniquely specifies the model and the model parameters can be found via curve-fitting (Fritchman 1967).

Note, each state of the 'Fritchman' model represents a memoryless error model and, therefore, Fritchman's model is limited to polygeometric distributions of the gaplengths as well as to burstlengths according to a single geometric distribution.

Table 3.1 presents a comparison of the main properties of the above-mentioned 'well-known' and frequently used classical models: the two-state Markov chain ('simplified Gilbert' model), the 'Gilbert' model, the GEC and the 'Fritchman' model.

These previous models (see Kanal and Sastry (1978), for a survey of proposed models until 1978) have attracted a great deal of attention from researchers, who have used the models or tried to improve the models' characteristics and usability. In Chouinard *et al.* (1988) and Pimentel and Blake (1998) methods for the parameter estimation of Gilbert's and Fritchman's models are presented.

Further models: The limitation of the 'Fritchman' model to polygeometric distributions led J.A. Adoul, B.D. Fritchman and L.N. Kanal to the proposal of slowly spreading Markov chains for error modelling in Adoul *et al.* (1972). In Rabiner (1989), Sivaprakasam and Shanmugan (1995), Turin and Sondhi (1993) and Turin (1999), HMMs in general are addressed. New models to describe special error characteristics of (wireless) communication links have also been developed to date, including a three-state model for Poisson distributed burstlengths (Aldridge and Ghanbari 1995), a two-state model with segmented exponential distributed burst- and gaplengths (Nguyen and Noble 1996), two- and three-state Markov models (Tralli and Zorzi 2005) and a Finite State Markov Channel (FSMC) (Wang and Moayeri 1995). Other recently published bipartite models (Willig 2002) are similar to Fritchman's model but they partition the burst- and gaplength statistics in subintervals. A bit error model for IEEE 802.11b using chaotic maps is presented in Koepke *et al.* (2003). In Konrad *et al.* (2003) a two-state process is used in which the gaps and bursts are generated from measured distributions. A Run Length Model (RLM) is presented in McDougall *et al.* (2004a), generating the error trace with a two-state renewal process with a mixture of geometric distributions which is then shown to be equal to a four-state Markov model. A three-state RLM is presented in McDougall *et al.* (2004b). Another RLM which uses half-normal distributions for burst- and gaplengths is shown in Poikonen (2006) and aggregated Markov processes are presented in Poikonen (2007), both for modelling the packet errors in Digital Video Broadcasting – Handheld (DVB-H).

3.2   Link Error Models for the UMTS DCH

Recent developments show that in most cases new models have been proposed to describe special properties of the links instead of taking general HMMs. The reason is only that in contrast to general models, which need a high number of parameters and complex parameter estimation methods, the new models are required to offer good usability by a small number of parameters which can easily be determined and adapted to changing system configurations. Furthermore, the goal in designing a new model is to represent the specific details of the measured error characteristics (for example, the error predictability) with high accuracy, whereas a small computational effort to generate an error trace has become less significant. Often (as is the case in this work), the specific link error properties observed in the measurement-based analysis strongly suggest a particular approach for designing a link error model. The well-known modelling approaches such as the famous 'Gilbert' model, the GEC or Fritchman's partitioned Markov chains are considered here but we demonstrate below that these models are not appropriate for describing the error statistics of the UMTS DCH.

As the link error characteristics of the UMTS Dedicated Channel (DCH) may change when adjusting different parameters in the UTRAN (such as different Transport Format (TF), Transmission Time Interval (TTI)-length, Spreading Factor (SF), and so on) the aim is to present a universal UMTS DCH modelling approach resulting from the specific link error characteristics of the UMTS DCH seen in the analysis in Chapter 2. Here in Part II the focus is particularly on the characteristics of the 384 kbit/s UMTS DCH Packet-Switched (PS) bearer in Down Link (DL) which currently represents the DCH PS bearer with the highest throughput available in the considered networks.

Despite the fact that the analysis in this book part is performed for the UMTS DCH, the basic idea behind the modelling approach, meaning to consider the specific properties of the underlying network, may as well be applied for other new or future mobile communication systems.

When modelling the UMTS DCH link error characteristics, only the two cases, 'static' and 'dynamic', have to be considered, as already observed. Due to the fact that the ratio of having all TBs erroneously received within one erroneous TTI in the scenarios with movement is very high (≈ 0.9), we propose to model the link error characteristics based on TTI granularity in the 'dynamic' case. On the other hand, in the 'static' scenario, the measured ratio of having all Transport Blocks (TBs) erroneously received within one erroneous TTI is approximately 25%. Furthermore, the ratio of having only one erroneous TB within an erroneous TTI is also 25%. Therefore, in the 'static' case, modelling in TB granularity is necessary.

3.2.1 Link Error Modelling – 'Dynamic' Case

We have shown by the existence of error bursts – up to ten TTIs long – that the simplest way of link error modelling via a memoryless BSC is not possible. One slightly more sophisticated modelling approach is to extend the single-state error model of the memoryless BSC (Bernoulli trials) to a two-state Markov chain, where in one state error-free TTIs are generated and in the other state erroneous TTIs are generated, as in Gilbert's model (1960) with an error probability of one in the bad state ('simplified Gilbert' model). This two-state Markov chain is uniquely specified by two parameters, namely the packet loss rate and the average TTI-burstlength (or the average TTI-gaplength) which can be directly determined from the measured error trace as the model is not an HMM.

As shown in Figure 3.8 (right), the geometric distribution of TTI-burstlengths, as generated by this simplified Gilbert model (two-state Markov chain), fits quite well to the measured empirical cdf of the lengths of TTI bursts (here shown for the 'car-city' scenario). On the other hand, when comparing the simulated statistics (geometric distribution) of TTI-gaplengths to the measured statistics (Figure 3.8, left), it can be observed that the two-state Markov chain is not capable of representing the correct TTI-gaplength distribution.

Enhancing the model from the simplified version of Gilbert's model (two-state Markov chain) to Gilbert's hidden Markov model with three parameters ($p = 0.384$, $q = 0.00139$ and $P(1|B) = 0.756$ for the 'car-city' scenario, estimated by using trigram-statistics as proposed in Gilbert (1960)) shows only a few improvements. Further improvements within the same modelling strategy were obtained via Fritchman's partitioned Markov chains with more than three states ($\gg$ four parameters). As we want to keep the number of model parameters as small as possible and because a Weibull distribution with only two parameters perfectly fits the measured TTI-gaplength statistics as shown in Figure 3.8, a change in the strategy towards other renewal processes such as semi-Markov models or RLMs is appropriate and well justified. The two-parameter Weibull (Murthy *et al.* 2004) pdf and cdf are given by

$$f(x|a, b) = ba^{-b}x^{b-1}e^{-(x/a)^b}, \tag{3.3}$$

$$F(x|a, b) = 1 - e^{-(x/a)^b}, \tag{3.4}$$

where a and b are scale and shape parameters, respectively.

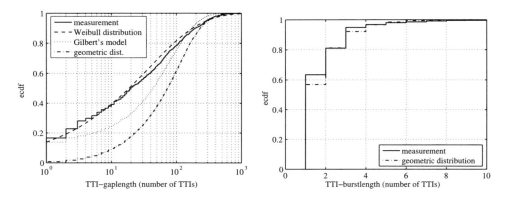

Figure 3.8   Comparison of empirical cdfs of TTI-gaplengths and TTI-burstlengths; measurements versus two-state Markov chain, Gilbert's model and Weibull distribution; 384 kbit/s bearer, 'car-city' scenario, 'mobile 1'.

The Weibull distribution was selected for the following reasons. Waloddi Weibull proposed this distribution in 1939 (Weibull 1939) for modelling the breaking strengths of materials. Current usage also includes reliability and lifetime modelling. When modelling the breaking strengths of materials, the model has in a way taken into account that the weakest link in the chain always breaks first. This idea can be applied analogically when modelling the link error characteristics. Also, in wireless channels the error occurs at the time instant of smallest SIR. In particular, when having a quality-based closed loop power control algorithm such as that for the UMTS DCH, the TPC algorithm continuously reduces the transmit power level until the weakest point breaks, implying that there is an error in the transmission.

Moreover, the Weibull distribution is very flexible in that extremely different distributions can be modelled when adjusting the parameters. For example, shape parameter $b = 1$ results in an exponential distribution, shape parameter $b = 2$ gives a Rayleigh distribution and shape parameter $b = 3$ generates a distribution similar to a Normal distribution (Gauss). This flexibility is also very welcome when modelling the link error characteristics.

For the estimation of the parameters for the Weibull distributions in this work, the following methods were considered: least squares parameter estimation, maximum likelihood parameter estimation and additional manual fine tuning by graphical parameter estimation methods such as, for example, Weibull plots (Murthy *et al.* 2004).

To generate the TTI-gaplengths via a Weibull distribution and the TTI-burstlengths in a discrete sense with a Markov chain (geometric distribution), a Mixed Semi-Markov/Markov Model (M-SMMM) with two states is proposed as depicted in Figure 3.9.

The state in which the error-free TTIs are generated is formed according to a renewal process with Weibull-distributed gaplengths, whereas erroneous TTIs are produced one by one in the discrete state. After generating an error gap, a single erroneous TTI is generated which may again be followed by a state change according to the transition probability $p_{ec} = 1 - p_{ee}$. Therefore, three parameters are required to define the model: two parameters for the Weibull distribution (scale parameter $a = 43.4$, shape parameter $b = 0.593$) and $p_{ee} = 0.43$

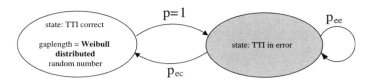

Figure 3.9 Mixed Semi-Markov/Markov model (M-SMMM) for the UMTS DCH 'dynamic' case.

(values for the 'car-city' scenario, 'mobile 1'), which are determined by curve fitting to the TTI-gaplength statistics and direct calculation from the mean TTI-burstlength, respectively.

The ratio of receiving all TBs in error within one erroneous TTI in the 'dynamic' scenario is around 85% (see Figure 2.9). Therefore, in the remaining 15%, refinements of the model are possible by considering the number of TB errors within the erroneous TTIs via the measured ratio of having x TBs erroneously received within one TTI (Karner *et al.* 2005). In Figure 2.9 the additional 11 required parameters are presented. Instead, we could use the fact that these values follow a binomial distribution (only one additional parameter required) as shown in Figure 2.10.

3.2.2 Link Error Modelling – 'Static' Case

For modelling the link error characteristics of the UMTS DCH in the 'static' scenario, a model based on TTI granularity is not sufficient owing to the high probability of receiving less than all TBs in error within one erroneous TTI. Thus, the aim of a first modelling approach is to generate a sequence of TBs with the correct statistics of gaplengths and burstlengths.

With the memoryless channel and the 'simplified Gilbert' model (classical two-state Markov chain), the produced gaplengths follow a geometric distribution (Papoulis and Unnikrishna 2002). As was shown in Chapter 2, Figure 2.13, there are two main areas with occurrence of gaplengths. Obviously, these statistics of gaplengths cannot be met via a single geometric distribution; therefore, the memoryless channel and the 'simplified Gilbert' model are inadequate tools in this case.

When using Fritchman's partitioned Markov chains with two error-free states (four parameters) and following the proposal of Chouinard *et al.* (1988), that is, estimating the model parameters by separate curve fitting for different parts of the cdf, the measured distribution of gaplengths is represented as demonstrated in Figure 3.10.

One of the error-free states is responsible for the distribution of the short gaps whereas the second state builds the form of the distribution for the long gaps. It can be shown that fitting of the short gaps can be improved by adding additional error-free states to the model but further states do not add accuracy to the fitting of the distribution for the long gaps. An explanation for this can be found in Figure 3.11, where the empirical cdfs of the measured data and a cdf of a geometric distribution are shown in linear scale. Note that in Figure 3.11 only the cdfs of the long gaplengths (> 12 TBs) are plotted. The curve of the measured data is convex between zero and the inflection point (mode). The geometric distribution on the other hand is concave over all its support. Therefore, fitting a geometric distribution (as formed by one state of Fritchman's partitioned Markov chains) to the measured distribution

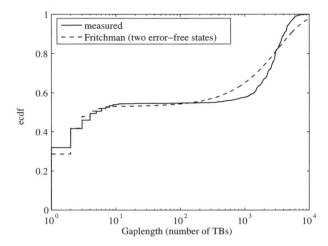

Figure 3.10   Gaplength distribution, measured versus Fritchman's model (two error-free states), 384 kbit/s bearer, 'static' scenario, 'mobile 1'.

is not possible. Furthermore, it becomes clear that the best fit is already obtained with only one geometric distribution, that is, only one error-free state exists in Fritchman's model for the long gaps. Fritchman's model offers the possibility of having more than one error-free state and, thus, having polygeometric distributions for the gaplengths, but every additional geometric distribution in a linear combination with only positive weighting factors adds concavity to the curve, and, thus, increases the fitting error. Therefore, all models producing polygeometric distributions of gaplengths such as Gilbert's model, Fritchman's model (with any number of states N) or the GEC (Gilbert–Elliott Channel) are not useful for modelling UMTS DCH link error characteristics in 'static' scenarios.

Moreover, as the best fit of Fritchman's model to the measured distribution of the long gaplengths is reached with one error state and due to the memoryless property (constant failure rate) of the geometric distribution, Fritchman's model is also incapable of describing the predictive nature of the measured link error characteristics (see Chapter 4 for details). The geometric distributions of gaplengths provided by Gilbert's model and by the GEC lead to the same conclusions.

As shown in Figure 3.11, a Weibull distribution with scale parameter $a = 3350$ and shape parameter $b = 2.018$ perfectly fits the measured distribution of the long gaps. Furthermore, a Weibull distribution ($a = 1.2$, $b = 0.7$) can be used to describe the distribution of the short gaps as depicted in Figure 3.12 (left) and another Weibull distribution ($a = 1.1$, $b = 0.55$) with two additional steps at 12 and 24 TBs represents the statistics of burstlengths properly for the 384 kbit/s bearer (see Figure 3.12 (right)).

Following the explained method of describing adequate statistics for gap- and burstlengths, we arrive at a two-state model (two-state alternating Weibull renewal process) as shown in Figure 3.13, where in one state correct TBs and in the other state erroneous TBs are generated. After each calculation of either burst- or gaplength, the state is changed.

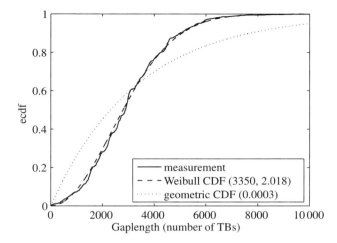

Figure 3.11 Distribution of long gaps, measured data versus Weibull and geometric distribution, 384 kbit/s bearer, 'static' scenario, 'mobile 1'.

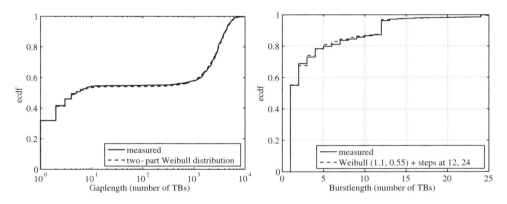

Figure 3.12 Comparison of empirical cdfs of gap- and burstlengths; measurements versus two-state Weibull renewal process; 384 kbit/s bearer, 'static' scenario, 'mobile 1'.

In the correct state, the number of subsequent error-free TBs (gaplength) is calculated via a two-part Weibull distributed random number with a probability of $p_{sg} = 0.55$ to produce short gaps. The number of subsequent erroneous TBs (burstlength) in the error state is also calculated via a Weibull distributed random number (with a probability p_{Wb}) but with additional burstlengths of 12 and 24 with estimated probabilities of $p_{12} = 0.09$ and $p_{24} = 0.01$, respectively. For bearers other than the 384 kbit/s bearer or for different transport formats these peaks are found at other burstlengths (for example, at 8 and 16 TBs for the 128 kbit/s bearer used in this work).

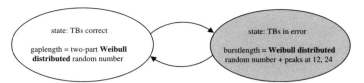

Figure 3.13 Two-state model for the UMTS DCH 'static' scenario – two-state alternating Weibull renewal process ('static-model 1').

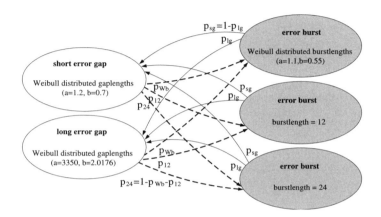

Figure 3.14 Equivalent five-state model illustration ('static-model 1').

As depicted in Figure 3.14 this two-state model can be equivalently presented with five states, which clearly shows that to model the error characteristics of the UMTS DCH for the 'static' scenario in this way, a total number of nine parameters is required. These include: four parameters for the two Weibull distributions of the correct states (error gaps) and one parameter for their separation. Additionally, two parameters are necessary to select burstlengths of either 12 or 24 or according to a Weibull distribution which also adds two parameters.

From Figure 3.14, it becomes clear that this model represents a renewal process, meaning that the probability, for example, of producing a short (long) gap does not depend on the length of the last error burst and is also independent of previous gaplengths. On the other hand – as shown in Chapter 2 – in the measured link error characteristics there is statistical dependency between subsequent gaps, subsequent bursts and also between a gap and the following burst. Therefore, in order to capture these correlation properties, a non-renewal model such as an HMM is required.

Instead of taking a general HMM, again special error characteristics of the UMTS DCH are considered to keep the model as simple as possible (as few parameters as possible and simple parameter estimation). Nonetheless, the model captures a major part of the dependencies between gaps and bursts by utilizing conditional probabilities of having long

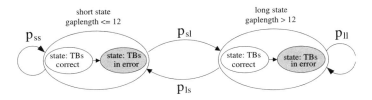

Figure 3.15 Non-renewal two-layer Markov model – Markov modulated Weibull renewal process ('static-model 2').

and short gaps and the conditional probabilities for burstlengths of 12 and 24, as shown below.

Figure 3.15 presents a non-renewal two-layer model, where dependency between gaps and bursts is introduced via the upper layer two-state Markov chain which modulates the parameters of the lower layer process. This model represents an HMM, since from the model output (erroneous/error-free TB) the current state of the model cannot be determined.

The model has one 'short state' where gaps with at most 12 TBs are generated and one 'long state' for gaps with more than 12 TBs. The corresponding transition probabilities are the estimated conditional probabilities from Chapter 2 according to Figure 2.18. After entering a new state, first the gaplength is calculated via a Weibull distributed random number for either the small or the large gaplengths according to the separation in Chapter 2 and depending on the current upper layer state. Then, while staying in the same state of the upper layer Markov chain, a burst is calculated with different probabilities of either creating a Weibull-distributed burstlength or a burst with a length of 12 or 24, again, depending on the actual upper layer state. After that, the state of the upper layer Markov chain possibly changes and the procedure begins again with calculation of the gaplength. Therefore, via the upper layer Markov chain, the model introduces dependency between subsequent gaps and also between a gap and the following burst. In Figure 3.16, a detailed illustration of this two-layer model is presented whereof it becomes clear that the following 11 parameters are required for a complete definition of the model.

Two parameters are necessary to determine the transition probabilities of the upper layer Markov chain ($p_{ss} = 1 - p_{sl} = 0.53$, $p_{ls} = 1 - p_{ll} = 0.56$). Two additional parameters are used for the Weibull distributions of each of the correct states and only two parameters more for the Weibull distribution of both erroneous states as the Weibull distributions are the same for the erroneous states of both upper layer states. The probability of generating Weibull-distributed burstlengths depends on the current state of the upper layer Markov chain in the case of the erroneous states. For these probabilities ($p_{Wb|s} = 0.99$, $p_{Wb|l} = 0.78$) and the separation between a burstlength of 12 and 24 ($p_{12|x}/p_{24|x} = 9$) three additional parameters are required ($p_{Wb|x} + p_{12|x} + p_{24|x} = 1$).

Despite the fact that this model is of type HMM, all the parameters can be determined directly from a measured link error trace without the necessity of difficult parameter estimation algorithms.

The two-state Weibull renewal process ('static model 1') and the non-renewal two-layer Markov model ('static model 2') produce correct estimated total TB error probabilities of 0.0027 and 0.0026, respectively. Also, the mean clusterlengths (10.50, 10.35), the mean

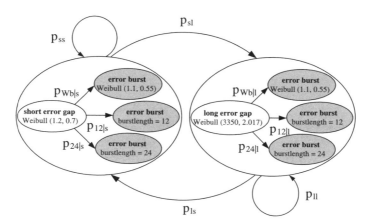

Figure 3.16   Detailed illustration of the non-renewal two-layer Markov model ('static-model 2').

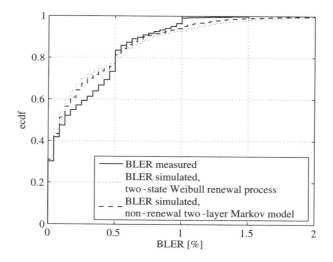

Figure 3.17   Comparison of BLERs – measurement versus models, 'static 1', 'mobile 1', 384 kbit/s bearer.

numbers of erroneous TBs per error cluster (7.90, 7.83) and the mean numbers of error bursts per cluster (2.17, 2.18) are very close to the measured results for the scenario 'static 1'. Furthermore, as presented in Figure 3.17, both models are capable of generating an error trace with an appropriate BLER. However, there is only a little improvement from the non-renewal two-layer Markov model over the two-state Weibull renewal process ('static model 1') for the case of the 'static 1' scenario.

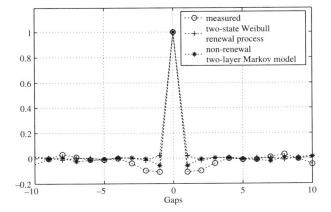

Figure 3.18   Comparison of autocorrelation functions – measurement versus models, 'static 2', 'mobile 1', 384 kbit/s bearer.

The potential of introducing correlation between subsequent gaps and bursts of the non-renewal two-layer Markov model becomes evident when analysing the autocorrelation function of, for example, the sequence of gaps (see Figure 3.18) for the scenario 'static 2' which includes a higher grade of dependency between the gaps. While the two-state Weibull renewal process does not produce any dependency between successive gaps, we observe in Figure 3.18 that the non-renewal two-layer Markov model captures the correlation properties between neighbouring gaps.

3.3   Impact of Channel Modelling on the Quality of Services for Streamed Video

Quality of Service (QoS), as well as the performance of methods and algorithms within wireless communication systems, is greatly affected by the error characteristics of the underlying channels; in particular, systems that include cross-layer algorithms to introduce network awareness (Karner et al. 2006a, 2007; Nemethova et al. 2007a,b) and therefore consider or even exploit specific properties of the link errors. Also, when utilizing a plain protocol stack there is a strong impact of the second-order error statistics of the channel on the performance of the higher layer protocols as shown in Zorzi and Rao (1999).

As the error characteristics of the channel have a great impact on the higher layer protocols, different link error model properties are also affected. In this section the impact of UMTS DCH link error modelling directly on the quality of H.264/AVC streamed video (Wiegand et al. 2003) is investigated (Karner et al. 2006b) by comparing the link- and network-layer characteristics as well as the resulting video quality simulated with link error models and with measured link layer error traces (384 kbit/s bearer, 'static' scenario, 'mobile 1').

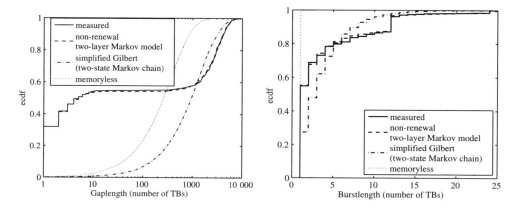

Figure 3.19 Comparison of empirical cdfs of link layer gap- and burstlengths; 'memoryless' model, 'simplified Gilbert' model, 'static model 2' versus measured trace, 384 kbit/s bearer, 'static' scenario, 'mobile 1'.

3.3.1   Compared Models

In the literature, various link error models are used for simulating video streaming over wireless networks. For example in Lo *et al.* (2005), a memoryless error model was used for evaluating video streaming over UMTS DCH, motivated by simulation results in Necker and Saur (2005). There, uncorrelated block errors are assumed for UMTS networks. However, this contradicts the measured results presented in Karner and Rupp (2005). Others (Qu *et al.* 2004) use the simplified Gilbert model (Gilbert 1960) representing a two-state Markov chain. In many publications, the authors quote GEC, which, after some assumptions for simplification, also ends in the simplified Gilbert model, thus a two-state Markov chain. Referring to these commonly used models, in our investigation a memoryless channel, the simplified Gilbert's model (two-state Markov chain), and a Markov modulated Weibull renewal process ('static model 2') have been applied.

Although all of the models were set to produce the same estimated link error probability (0.266%) as measured in the live UMTS network, they show significant differences in their error characteristics. This can be observed in Figure 3.19, where the cdfs of the gaplengths (number of error-free link layer packets between two errors) and the cdfs of the burstlengths (number of subsequently received errors) are presented.

3.3.2   Experimental Setup

For transmission of a video stream over a UMTS network the following procedure of packetization has to be performed. Each frame of the video is first subdivided into smaller parts (slices) which are then encoded. As shown in the schematic illustration in Figure 3.20, the encoded video slices or equivalently the Network Abstraction Layer Units (NALU) containing the slice data are encapsulated into Real-time Transport Protocol (RTP) packets with a header of 12 bytes. Each RTP packet is then transmitted within the UDP protocol which adds a header of 8 bytes. If no segmentation is needed, the UDP packets are further

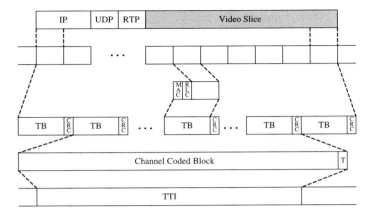

Figure 3.20 Schematic illustration of a packetization example for the transmission of a video slice over UMTS.

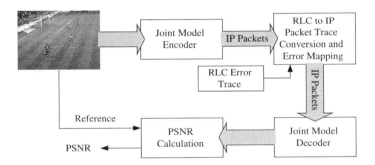

Figure 3.21 Scheme of the experimental setup for simulating H.264 encoded video over error-prone links.

packed within IP packets with a header size of 20 bytes for IPv4 and 40 bytes for IPv6. These IP packets are then transmitted over the UTRAN, where the layer-three data packets are further processed as described in Chapter 1.

Figure 3.21 displays the experimental setup. The Joint Model (JM) H.264/AVC (Joint Model Software) was adapted by adding the interface for the IP error traces and by implementing a simple error concealment scheme at the decoder: for I frames weighted averaging (Sun and Reibman 2001) was used and for P frames the corresponding location was copied from the previous frame. The encoder is modified to deliver the IP packet lengths for mapping the IP packets onto the link layer packet trace.

For the experiments a 'soccer match'(Nemethova 2007) video sequence was used with SIF (QVGA) picture resolution (320 × 240 pixels) and a frame rate of 10, containing a soccer match with different scenes. The sequence was encoded using I and P frames only (every 40th frame is an I frame) according to the baseline profile with slicing mode two and

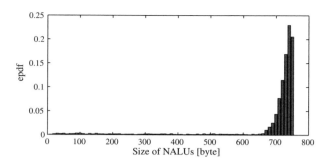

Figure 3.22 Empirical pdf of NALU sizes (size of encoded video packets).

at most 750 bytes per slice. The quantization parameter was set to 26 and the rate-distortion optimization was disabled. To obtain reliable results, the video was decoded several times resulting in ≈10 hours of video stream.

The resulting packet sizes (sizes of the NALUs including NALU header of 1 byte) in the data stream of the video sequence, encoded with the mentioned parameters are presented in Figure 3.22. For the resulting IP packet sizes, 40 bytes of headers (RTP 12 bytes, UDP 8 bytes, IPv4 20 bytes) have to be added.

To evaluate the improvements in end-to-end video quality, the peak signal-to-noise ratio of the luminance component (Y-PSNR) was used, given for the nth luminance frame $\mathbf{Y}_n$ by

$$\text{Y-PSNR}(n) = 10 \cdot \log_{10} \frac{255^2}{\text{MSE}(n)}, \tag{3.5}$$

$$\text{MSE}(n) = \frac{1}{N \cdot M} \sum_{i=1}^{N} \sum_{j=1}^{M} [\mathbf{Y}_n(i, j) - \mathbf{F}_n(i, j)]^2, \tag{3.6}$$

where $\text{MSE}(n)$ denotes the mean square error of the nth luminance frame $\mathbf{Y}_n$ compared to the luminance frame $\mathbf{F}_n$ of the reference sequence. The resolution of the frame is $N \times M$, indexes i and j address particular luminance values within the frame. The *non-compressed, non-degraded* original sequence served as reference sequence.

3.3.3 Simulation Results for H.264 Encoded Video over Error Prone Links

In Figure 3.23 histograms of the Y-PSNR per video frame for the considered models are presented. It was observed that despite having the same estimated link layer packet error probability[1] (see $\overline{P}_e$(TB) in Table 3.2) only the 'static model 2' (non-renewal two-layer Markov model) meets the measured statistics.

---

[1] As the memoryless channel and the simplified Gilbert model are defined by the mean error probability (and the mean error burstlength), they show exactly the same estimated mean error probability as the measured traces. The parameters of 'static model 2' are defined by curve fitting of the gap- and burstlengths and via the correlation between gaps and bursts. Thus, the resulting estimated link error probability differs slightly from the measured as a consequence of the modelling.

Table 3.2 Comparison of estimated link-layer and IP packet error probability.

Model	$\overline{P}_e$(TB) (%)	$\overline{P}_e$(IP) (%)
measured	0.266	0.888
'static model 2'	0.262	0.892
simplified Gilbert	0.266	1.009
memoryless	0.266	4.501

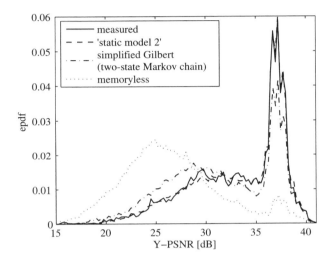

Figure 3.23 Y-PSNR per frame for the various link layer error models.

This is due to the fact that only the 'static model 2' shows almost the same link error characteristics as the measured traces. All the other models and especially the memoryless channel have completely different statistics of the burst- and gaplengths and thus result in a different number of erroneously received link layer packets (TBs) within one IP packet, as shown in Figure 3.24. This in turn results in a much higher IP error probability, for example in the case of the memoryless link layer characteristics as shown in Table 3.2. From these results we conclude that it is important to have the correct higher order statistics when modelling the link layer – not only when simulating video streaming.

Contrary to the conclusions for the link layer modelling, we now show that for error modelling in the network layer it is not important to meet the correct higher order statistics to evaluate the quality of streamed video with common parameter settings. In Figure 3.25 we observe that by applying a memoryless channel model in the network layer with appropriate IP BLER (0.888% instead of 4.501% – see Table 3.2), the measured statistics of the streamed video quality (Y-PSNR per frame) is met with a memoryless model in the network layer, despite having completely different network layer error characteristics.

These differences in the network layer error characteristics can be seen in Figure 3.26 where the pdfs of the IP-gaplengths (number of error-free IP packets between two errors)

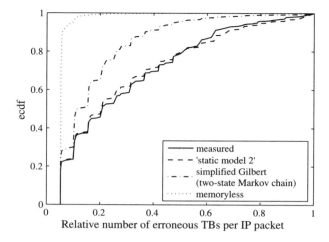

Figure 3.24 Relative number of erroneous TBs per (erroneous) IP packet.

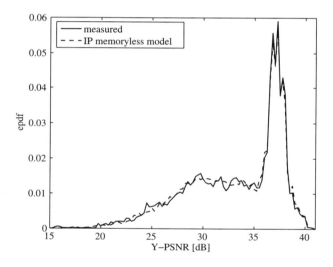

Figure 3.25 Y-PSNR per frame for IP memoryless model versus measurements.

and the pdfs of the IP-burstlengths (number of subsequently erroneous IP packets) are compared. It is observed in Figure 3.26 (left) that the pdfs of the IP-gaplengths exhibit a maximum gaplength at around 150 in the case of the measurements and the 'static model 2', whereas the memoryless network layer model (IP memoryless) shows the maximum at short IP-gaplengths. Furthermore, the memoryless network layer model only generates IP-burstlengths of length one, as shown in Figure 3.26 (right). On the other hand the measured statistics also show IP-burstlengths of up to three subsequently erroneous IP packets.

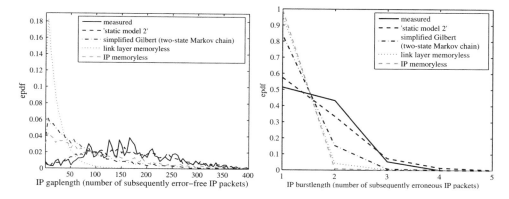

Figure 3.26   Comparison of IP gap- and burstlengths for the various error models; 'memoryless' model, 'simplified Gilbert' model, 'static model 2' versus measured trace, 384 kbit/s bearer, 'static' scenario, 'mobile 1'.

The fact that the different statistics of the IP-burstlengths and IP-gaplengths in the measured traces and the memoryless network layer model result in the same video streaming quality becomes even more interesting when observing the difference in the relative number of erroneous IP packets per (erroneous) video frame in Figure 3.27. There, it can be observed that the longer IP-burstlengths do indeed result in more erroneous IP packets per (erroneous) video frame in the measured case and the 'static model 2' compared to the memoryless network layer model (IP memoryless). However, these results contradict Liang *et al.* (2003) where a lower video quality for longer IP error bursts is expected. The reason for having the same video quality regardless of having different network layer error characteristics is the error propagation within one Group of Pictures (GOP), as we will show.

The evaluation of the video quality was performed using H.264/AVC with a common parameter setting for video streaming over wireless networks, meaning that the video stream consists of one I and 39 following P frames where the latter are encoded predictively. This predictive coding of the P frames leads to error propagation within one GOP which in turn is the reason for having the same video quality with different IP burstlengths and also with a different number of erroneous IP packets per video frame.

When encoding the video with I frames only, there is no error propagation in the decoded video sequence. Consequently, a difference in the Y-PSNR values per frame between the memoryless network layer and the measured characteristics can be observed as shown in Figure. 3.28. There, pdfs of the Y-PSNR values per frame – only the part of the erroneous frames – are presented. Note that there is a higher peak in the pdf for the error-free frames at higher Y-PSNR values.

To conclude, with a memoryless channel model in the IP layer (equivalently a memoryless channel model in the link layer with appropriate BLER), a plain IP-UDP-RTP protocol stack above RLC layer and common parameter settings for video streaming over UMTS DCH, despite having different higher order statistics in the network layer, the statistical end-to-end quality of streamed video data will be the same as if channels with highly correlated errors are applied. Therefore, when modelling the link layer, special attention has to be paid to the

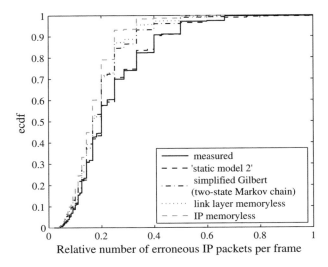

Figure 3.27 Relative number of erroneous IP packets per video frame.

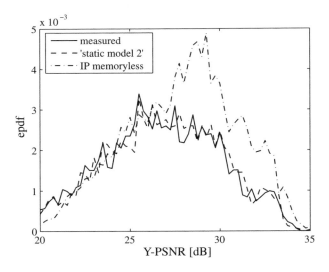

Figure 3.28 Y-PSNR per frame for only I frames (no error propagation).

error correlation properties even with plain IP-UDP-RTP protocol stack, while a model in the network layer only has to meet the correct packet error probability. Of course, it is important to mention that this analysis is based only on time-independent Y-PSNR statistics and thus different results for the perceived end user quality may be expected.

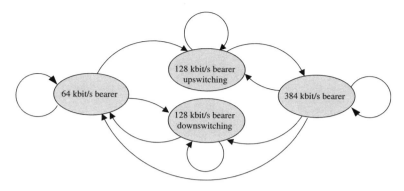

Figure 3.29 Four-state Markov model for dynamic bearer type switching.

3.4 A Dynamic Bearer Type Switching Model

3.4.1 Four-state Markov Model

Based on the measurement results of Chapter 2, the dynamic bearer switching is modelled by a four-state homogeneous Markov chain (Papoulis and Unnikrishna 2002; Turin 1999) as presented in Figure 3.29 where the valid state transitions are shown by arrows. The corresponding probability transition matrix Π has the following form:

$$
\Pi = \begin{bmatrix}
\pi_{64,64} & \pi_{64,128u} & \pi_{64,128d} & 0 \\
0 & \pi_{128u,128u} & 0 & \pi_{128u,384} \\
\pi_{128d,64} & 0 & \pi_{128d,128d} & 0 \\
\pi_{384,64} & \pi_{384,128u} & \pi_{384,128d} & \pi_{384,384}
\end{bmatrix}.
$$

Its elements denote the stationary probabilities $\pi_{i,j}$ of one-step transitions from state i to state j. The probability vector $\mathbf{p}$ having the state probabilities p_k as elements is obtained by solving the system of linear equations $\Pi^T \mathbf{p} = \mathbf{p}$, with constraint $\sum_k p_k = 1$. These probabilities (in this case $\mathbf{p} = [p_{384}, p_{128u}, p_{128d}, p_{64}]$) correspond to the measured values in Figure 2.20 (taking p_{128u} and p_{128d} as one state).

As the granularity of time in our discrete-time Markov chain is in TTIs, the jumps to the next states have to be determined every TTI via the corresponding transition probabilities. For the complete description of the model we need eight parameters. Seven parameters are in order to define the transition probabilities and one parameter is required to specify the hysteresis time of the switching algorithm (we selected 200 TTIs).

In Figures 3.30 to 3.32 the simulated runlengths (dotted black line) together with the measured distributions (solid blue line) are shown for the different bearers and the two mobiles. In the case of Figure 3.30 the measured statistics of the 'tramway' scenario are presented, for example. It is observed that the geometric distributions produced by the Markov chain are not capable of meeting the measured statistics with sufficient accuracy in all cases – for example, the runlength distribution of the 384 kbit/s bearer of 'mobile 1' (Figure 3.30 (left)). Therefore, we propose to model the dynamic bearer switching via a Weibull renewal process (Murthy *et al.* 2004) as shown below.

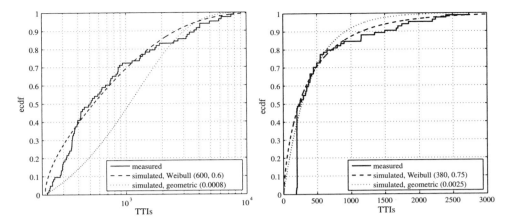

Figure 3.30 Runlength distribution of 384 kbit/s bearer (left), 128 kbit/s bearer – upswitching (right), measured versus simulated, 'tramway' scenario, 'mobile 1'.

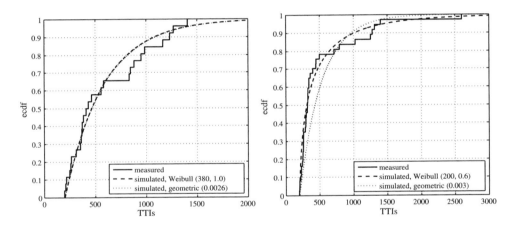

Figure 3.31 Runlength distribution of 128 kbit/s bearer – downswitching (left), 64 kbit/s bearer (right), sum of different scenarios, measured versus simulated, 'mobile 1'.

3.4.2   Enhanced Four-state Model

In Figure 3.30 (left) the empirical cdf of the runlengths of the 384 kbit/s bearer, 'tramway' scenario, is shown. The solid line shows the measurement result, whereas the dotted line shows the simulated result of the Markov model representing a geometric distribution, given by:

$$f(n|p) = (1 - p)^n p, \quad n = 1, 2, \ldots \tag{3.7}$$

with parameter p as extracted out of the measured trace ($p = 1 - \pi_{384,384}$). We can see that the geometric probability distribution is not capable of describing the measured distribution

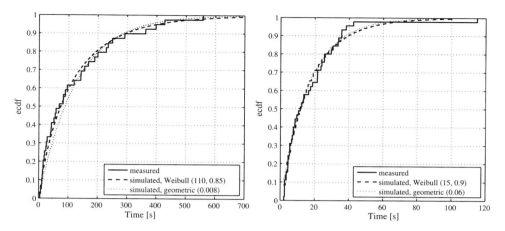

Figure 3.32 Runlength distribution of 384 kbit/s bearer (left), 128 kbit/s bearer (right), sum of different scenarios, measured versus simulated, 'mobile 3'.

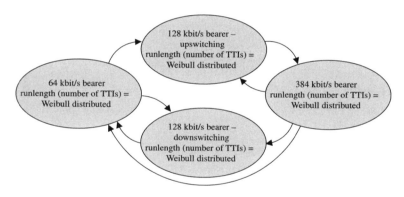

Figure 3.33 Schematic illustration of the enhanced four-state model (Weibull renewal process).

with sufficient accuracy. However, Weibull-distributed runlengths with the parameters as denoted in the figure meet the measured statistics properly (dashed line in Figure 3.30 (left)). Due to that fact a renewal process with four states and Weibull-distributed runlengths is proposed next as an enhanced model for the dynamic bearer type switching for UMTS DL DCH in the PS domain. With the high flexibility of the Weibull distribution we are capable of describing all the observed runlength distributions with adequate precision. The two-parameter Weibull cdf is given by equation 3.4.

In Figure 3.33, a schematic illustration of the model with the allowed transitions between the different states is shown. Such a Weibull renewal process (Murthy *et al.* 2004) calculates the runlengths in each state via a two-parameter Weibull-distributed random number.

The eight parameters for the four Weibull distributions (one in each state) together with the three parameters which determine the jump between the states and one additional parameter for specification of the processing time of the switching algorithm result in a total of 12 parameters for a complete specification of the model.

Note, the measured distributions of the runlengths and therefore also the model parameters depend on the network load, the service coverage and the quality of the receiver in the mobile station. For example, if there are some congested cells within the measurement route, there is less switching to the 384 kbit/s or to the 128 kbit/s bearer. In the case of varying receiver quality (varying service coverage) it was shown that the proposed model is flexible enough to describe the different runlength distributions.

References

Adoul, J-P.A., Fritchman, B.D. and Kanal, L.N. (1972) A critical statistic for channels with memory. *IEEE Trans. on Information Theory,* **18**(1), Jan.

Aldridge, R.P. and Ghanbari, M. (1995) Bursty error model for digital transmission channels. *IEE Electronic Letters,* **31**(25), Dec. 1995.

Berger, J.M. and Mandelbrot, B. (1963) A new model for error clustering in telephone circuits. *IBM Journal,* Jul.

Chouinard, J-Y., Lecours, M. and Delisle, G.Y. (1988) Estimation of Gilbert's and Fritchman's models parameters using the gradient method for digital mobile radio channels. *IEEE Trans. on Vehicular Technology,* **37**(3), Aug. 1988.

Elias, P. (1954) Error-free coding. *IEEE Trans. on Information Theory,* **4**(4), 29–37, Sep.

Elliott, E.O. (1963) Estimates of error rates for codes on burst-noise channels. *Bell Systems Technical Journal,* **42**, 1977–1997, Sep.

Elliott, E.O. (1965) A model for the switched telephone network for data communications. *Bell Systems Technical Journal,* **44**, 89–119, Jan.

Fritchman, B.D. (1967) A binary channel characterization using partitioned Markov chains. *IEEE Trans. on Information Theory,* **13**(2), 221–227, Apr.

Gilbert, E.N. (1960) Capacity of a burst-noise channel. *Bell Systems Technical Journal,* **39**, 1253–1265, Sep.

H.264/AVC Software Coordination, Joint Model Software, v.10.1, available at http://iphome.hhi.de/suehring/tml/.

Kanal, L.N. and Sastry, A.R.K. (1978) Models for channels with memory and their applications to error control. *Proceedings of the IEEE,* **66**(7), Jul.

Karner, W., Svoboda, P. and Rupp, M. (2005) A UMTS DL DCH Error Model Based on Measurements in Live Networks. In *Proceedings of the 12th International Conference on Telecommunications 2005 (ICT 2005),* Capetown, South Africa, May.

Karner, W. and Rupp, M. (2005) Measurement based Analysis and Modelling of UMTS DCH Error Characteristics for Static Scenarios. In *Proceedings of the Eighth International Symposium on DSP and Communications Systems 2005 (DSPCS'2005),* Sunshine Coast, Australia, Dec.

Karner, W., Nemethova, O., Svoboda, P. and Rupp, M. (2006a) Link Error Prediction Based Cross-Layer Scheduling for Video Streaming over UMTS. In *Proceedings of the 15th IST Mobile and Wireless Comm Summit 2006,* Myconos, Greece, Jun.

Karner, W., Nemethova, O. and Rupp, M. (2006b) The Impact of Link Error Modelling on the Quality of Streamed Video in Wireless Networks. In *Proceedings of the Third IEEE International Symposium on Wireless Communications Systems 2006 (ISWCS 2006),* Valencia, Spain, Sep.

Karner, W., Nemethova, O. and Rupp, M. (2007) Link Error Prediction in Wireless Communications Systems with Quality Based Power Control. In *Proceedings of IEEE International Conference on Communication (ICC 2007)*, Glasgow, Scotland, Jun.

Köpke, A., Willig, A. and Karl, H. (2003) Chaotic Maps as Parsimonious Bit Error Models of Wireless Channels. In *Proceedings of IEEE INFOCOM*, vol. 1, pp. 513–523, Apr.

Konrad, A., Zhao, B.Y., Joseph, A.D. and Ludwig, R. (2003) A Markov-based channel model algorithm for wireless networks. *Wireless Networks*, **9**, 189–199, Kluwer Academic Publishers.

Liang, Y.J., Apostolopoulos, J.G. and Girod, B. (2003) Analysis of Packet Loss for Compressed Video: Does Burst-length Matter? In *Proceedings of IEEE International Conference on Acoustics, Speech and Signal Processing (ICASSP'03)*, vol. 5, pp. 684–687.

Lo, A., Heijenk, G. and Niemegeers, I. (2005) Evaluation of MPEG-4 Video Streaming over UMTS/WCDMA Dedicated Channels. In *Proceedings of the First IEEE International Conference on Wireless Internet*.

Mandelbrot, B. (1965) Self-similar error clusters in communications systems and the concept of conditional stationarity. *IEEE Trans. on Comm.*, **13**(1), 71–90, Mar.

McDougall, J., Joseph, J., Yi, Y. and Miller, S. (2004a) An Improved Channel Model for Mobile and Ad Hoc Network Simulations. In *Proceedings of the International Conference on Communication, Internet and Information Technology (CIIT 2004)*, St Thomas, Virgin Islands, USA, Nov.

McDougall, J., Yi, Y. and Miller, S. (2004b) A Statistical Approach to Developing Channel Models for Network Simulations. In *Proceedings of the IEEE Wireless Communication and Networking Conference (WCNC 2004)*, vol. 3, pp. 1660–1665, Mar.

Murthy, D.N.P., Xie, M. and Jiang, R. (2004) *Weibull Models*, John Wiley & Sons, Ltd.

Necker, M.C. and Saur, S. (2005) Statistical Properties of Fading Processes in WCDMA Systems. In *Proceedings of the Second International Symposium on Wireless Communications Systems (ISWCS 2005)*, pp. 54–58, Sep.

Nemethova, O. (2007) Error Resilient Transmission of Video Streaming over Wireless Mobile Networks, PhD Thesis, Vienna University of Technology, http://publik.tuwien.ac.at/files/pub-et_12661.pdf, Austria.

Nemethova, O., Karner, W. and Rupp, M. (2007a) Error Prediction Based Redundancy Control for Robust Transmission of Video over Wireless Links. In *Proceedings of the IEEE International Conference on Communication (ICC 2007)*, Glasgow, UK, Jun.

Nemethova, O., Karner, W., Weidmann, C. and Rupp, M. (2007b) Distortion-Minimizing Network-Aware Scheduling for UMTS Video Streaming. In *Proceedings of the EUSIPCO 2007*, Poznan, Poland, Sep.

Nguyen, G.T. and Noble, B. (1996) A Trace-Based Approach for Modelling Wireless Channel Behavior. In *Proceedings of the 1996 Winter Simulation Conference*, 1996.

Papoulis, A. and Unnikrishna, P.S. (2002) *Probability, Random Variables, and Stochastic Processes*, McGraw-Hill.

Pimentel, C. and Blake, F. (1998) Modelling Burst Channels Using Partitioned Fritchman's Markov Models. *IEEE Trans. on Vehicular Technology*, **47**(3), Aug.

Poikonen, J. (2006) Half-normal Run Length Packet Channel Models Applied in DVB-H Simulations. In *Proceedings of the Third IEEE International Symposium on Wireless Communications Systems (ISWCS 2006)*, Valencia, Spain, Sep.

Poikonen, J. (2007) Parameterization of Aggregated Renewal Markov Processes for DVB-H Simulations. In *Proceedings of the 18th IEEE International Symposium on Personal, Indoor and Mobile Radio Communications (PIMRC 2007)*, Athens, Greece, Sep.

Qu, Qi, Pei, Y., Modestino, J.W. and Tian, X. (2004) Source-Adaptive FEC/UEP Coding For Video Transport Over Bursty Packet Loss 3G UMTS Networks: A Cross-Layer Approach. In *Proceedings of the 60th IEEE Vehicular Technology Conference (VTC2004-Fall)*, vol. 5, pp. 3150–3154, Sep.

Rabiner, L.R. and Juang, B.H. (1986) An introduction to hidden Markov models. *IEEE ASSP Magazine*, **3**, 4–16, Jan.

Rabiner, L.R. (1989) A tutorial on hidden Markov models and selected applications in speech recognition. *Proceedings of IEEE*, **77**(2), Feb.

Shannon, C.E. (1948) A mathematical theory of communication. *Bell Syst. Tech. Journal*, **27**, 379–423 and 623–656.

Sivaprakasam, S. and Shanmugan, K.S. (1995) An equivalent Markov model for burst errors in digital channels. *IEEE Trans. on Comm.*, **43**(2/3/4), 1347–1355.

Sun, M.T. and Reibman, A.R. (2001) *Compressed Video over Networks*, Signal Processing and Communications Series, Marcel Dekker Inc., New York.

Sussman, S.M. (1963) Analysis of the Pareto model for error statistics on telephone circuits. *IEEE Trans. on Comm. Systems*, **11**, 213–221, Jun.

Tralli, V. and Zorzi, M. (2005) Markov models for the physical layer block error process in a WCDMA cellular system. *IEEE Trans. on Vehicular Technology*, **54**(6), 2102–2113, Nov.

Turin, W. and Sondhi, M.M. (1993) Modelling error sources in digital channels. *IEEE Journal on Sel. Areas in Comm.*, **11**(3), 340–347, Apr.

Turin, W. (1999) *Digital Transmission Systems: Performance Analysis and Modelling*, McGraw-Hill, New York.

Wang, H.S. and Moayeri, N. (1995) Finite-state Markov channel – a useful model for radio communication channels. *IEEE Trans. on Vehicular Technology*, **44**(1), Feb.

Weibull, W. (1939) A statistical theory of the strength of materials. *Ingeniörs Vetenskaps Akademiens Handlingar* **151**, 1–51.

Wiegand, T., Sullivan, G.J., Bjontegaard, G. and Luthra, A. (2003) Overview of the H.264/AVC Video Coding Standard. *IEEE Trans. on Circuits and Systems for Video Technology*, **13**(7), 560–576, Jul.

Willig, A. (2002) A New Class of Packet- and Bit-Level Models for Wireless Channels. In *Proceedings of the Thirteenth IEEE International Symposium on Personal, Indoor and Mobile Radio Communications*, vol. 5, pp. 2434–2440, Sep.

Zorzi, M. and Rao, R.R. (1999) Perspectives on the impact of error statistics on protocols for wireless networks. *IEEE Personal Communications*, **6**, 32–40, Oct.

4

Analysis of Link Error Predictability in the UTRAN

Wolfgang Karner

The detailed knowledge about the link error characteristics as presented in Chapters 2 and 3 is necessary for the optimization of services (for example, their codecs or protocols) for their usage over wireless mobile communication links (Liang *et al.* 2003; Karner *et al.* 2006; Zorzi and Rao 1999). Link error statistics can also be used for performance evaluation of new (cross-layer) mechanisms (Nemethova *et al.* 2005) such as, for example, for streaming video data transmission in the UMTS network.

Statistics of link error characteristics can be used to optimize the end-to-end transmission chain in a static sense and even dynamically and adaptively to predict future link errors based on past error occurrence and to use this predictability together with cross-layer optimization procedures as shown in Chapter 8.

The analysis of the link error characteristics of the Universal Mobile Telecommunication System Dedicated CHannel (UMTS DCH) in Downlink (DL) presented in Chapter 3 shows that there is correlation between the error states of subsequently transmitted Transport Blocks (TBs), causing error bursts. There is also correlation between successive bursts and gaps resulting in error clusters. Generally, these error correlation properties of the channel can be used to predict future link errors.

To be able to predict link errors based on past error events, the transmitter has to be aware of past link errors in the forward link, meaning that there has to be a feedback link for the error status of the received data from the receiver back to the transmitter. For the UMTS DCH, this is accomplished by utilizing the error feedback of the Radio Link Control Acknowledged Mode (RLC AM). For delay sensitive services we can adjust the maximum number of allowed retransmissions for the RLC AM to zero or equivalently adjust the discard

Video and Multimedia Transmissions over Cellular Networks   Edited by Markus Rupp
© 2009 John Wiley & Sons, Ltd

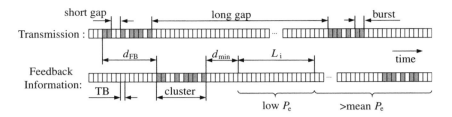

Figure 4.1 Schematic illustration of an interval with low link error probability.

timer properly. The RLC AM in the considered live UMTS networks uses 16 bits of CRC information for each TB with a size of 320 bits to detect transmission errors.

In case of no or only very small ($<$ 1 ms) feedback delay, the single TB errors within an error cluster can be predicted at the transmitter. Unfortunately, the feedback delay d_{FB} in the UMTS RLC AM ($d_{FB} \geq 30$ ms also in case of 10 ms TTIs) is in the order of the cluster size; therefore, no error prediction within an error cluster is possible.

For this reason, the focus in this work is on the prediction of error clusters rather than single error events. Moreover, a first approach detects points in time with the least probability for the occurrence of an error cluster.

4.1   Prediction of Low Error Probability Intervals

From the statistics of the gaplengths in the 'static' scenario (see Figure 2.13) it becomes clear that there are short gaps between the error bursts within an error cluster and long gaps between successive error clusters. Furthermore, there is a region in the cdf between the short and the long gaps with a very low probability of gap occurrence. This is because the cdf of the long gaps is convex between zero and the inflection point (mode) as demonstrated in Figure 3.11. Moreover, long error gaps in the case of the 'static' scenario are Weibull distributed with shape parameter $a \neq 1$, thus not geometrically distributed, resulting in memory in the link error characteristics. These properties can be used to detect intervals with low link error probability, as shown in Figure 4.1.

4.1.1   Detection of Start of Intervals

Where there is no error report via the feedback information within a certain time interval d_{min}, which has to be larger than the longest of the short gaps, we can conclude that the following TBs are outside an error cluster and the current transmission takes place within one of the long error gaps. Due to the convex distribution of the long gaps there is an interval with very low transmission error probability and we can detect its start at the minimum delay d_{min} after the last error. Of course, the total detection delay also includes the feedback delay d_{FB}. Figure 4.1 illustrates the time series of the transmitted TBs and of the error feedback information which is available at the transmitter after d_{FB}. Clusters of TB errors are marked and the start of the interval with low error probability with length L_i is shown at d_{min} after an error cluster.

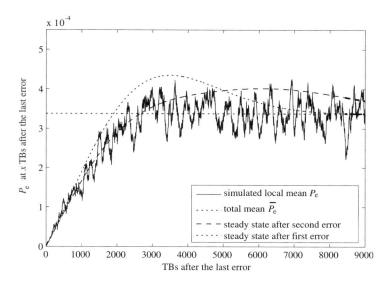

Figure 4.2 P_e at x TBs after the last error event for Weibull-distributed long gaps (Weibull scale parameter $a = 3350$, shape parameter $b = 2.176$).

4.1.2 Interval Length L_i

To determine the length L_i of the intervals with lower error probability we make use of a theoretic analysis with long gaps separating single error events only. In Figure 4.2 the local mean of the simulated estimated error probability $P_e(x)$ at a point x TBs after an error event (in practice corresponding to the occurrence of an error cluster) is presented for Weibull-distributed gaplengths as fitted to the measured distribution in Chapter 3. We conclude from this curve that the error probability is very small right after an error event, staying below the total mean error probability for approximately 2500 TBs. In practice, this allows us to estimate L_i via the intersection of this conditional mean curve with the unconditional total mean $\overline{P}_e$.

To obtain an analytic expression for L_i we first calculate the steady-state error probability $P_{e,ss}$, which is equal to the total mean $\overline{P}_e$. Let L be the random variable measuring the time (distance in number of packets = gaplength) between two error events. Assuming that the error process is stationary, Kac's lemma (Kac 1947) implies that

$$P_{e,ss} = \frac{1}{E\{L\}}, \tag{4.1}$$

where $E\{L\} = \sum_{i=1}^{\infty} i \cdot P(L = i)$ is the average recurrence time corresponding to the average gaplength with $P(L = i)$ being the probability of having a gaplength of length i.

It transpires that the stationarity assumption holds over longer ranges but does not lead to correct results in immediate succession to a known error event where the (measured) local estimated error probability is actually lower than the total mean estimated error probability $\overline{P}_e$. Therefore, one way to approximate the probability $P_e(l)$ of having an error at the lth TB after an error event is to assume that the steady state is reached before the third error after the initial error event:

$$P_e(l) \approx P(L_1 = l) + P(L_1 + L_2 = l)$$
$$+ P_{e,ss} \cdot P(L_1 + L_2 < l). \tag{4.2}$$

Here, L_1 and L_2 denote the gaplengths before the first and second errors, respectively. The result (dashed curve in Figure 4.2) shows that taking the intersection between this curve and the total mean as the end of L_i gives a good approximation whereas the assumption of reaching the steady state after the first error, calculated by

$$P_e(l) \approx P(L_1 = l) + P_{e,ss} \cdot P(L_1 < l), \tag{4.3}$$

leads to an underestimation of L_i (dash-dotted curve).

4.2   Estimation of Expected Failure Rate

For some cross-layer methods as will be shown in Chapter 8, it is not sufficient to detect intervals with lower error probability; rather, an estimation of the instantaneous probability of having transmission errors is required. As previously mentioned, an exact error prediction within the error clusters is not possible owing to the feedback delay but the location of the error clusters can be predicted. Therefore, the instantaneous link error probability is estimated by using the expected failure rate resulting from the distribution of the long gaps between successive error clusters as explained in the following.

Throughout this document the following notation is used to describe the error process. An erroneously received TB is indicated by '1' while '0' means error-free transmission. A positive integer in the exponent determines the number of consecutive erroneous or error-free TBs (for example, the sequence '000001' can be written as '$0^5 1$'). A gap with length m is defined as the number of 0's between two 1's and

$$p_M(m) = P(0^m 1 | 1) \tag{4.4}$$

for all positive integers m is the probability mass function (pmf) of the gaplengths. The conditional probability $P(B|A)$ means the probability of sequence B following sequence A. By definition,

$$\sum_{m=0}^{\infty} P(0^m 1 | 1) = 1 \tag{4.5}$$

and $P(0^0 1 | 1) = 0$, as gaps with length zero are not considered as gaplengths. The cdf of the gaplengths is then defined as

$$F_M(m) = P\{M \le m\} = \sum_{k=0}^{m} P(0^k 1 | 1). \tag{4.6}$$

The conditional link error probability $P(1|10^m)$, that is, the error probability conditioned to the number of error-free TBs since the last error, can be expressed by

$$P(1|10^m) = \frac{P(10^m 1)}{P(10^m)} = \frac{P(0^m 1 | 1) \cdot P(1)}{P(0^m | 1) \cdot P(1)} = \frac{P(0^m 1 | 1)}{P(0^m | 1)}, \tag{4.7}$$

where $P(0^m|1)$ denotes the probability of having a gaplength of at least m and can be written in terms of the complementary cumulative distribution function (ccdf) and the pmf of the gaplengths

$$P(0^m|1) = \sum_{k=m}^{\infty} P(0^k 1|1) = 1 - F_M(m) + p_M(m). \qquad (4.8)$$

Thus, the conditional link error probability $P(1|10^m)$ can be expressed as

$$P(1|10^m) = \frac{p_M(m)}{1 - F_M(m) + p_M(m)} \approx \frac{p_M(m)}{1 - F_M(m)}, \qquad (4.9)$$

also presenting the approximation via the expected failure rate (Papoulis and Unnikrishna 2002). In Chapter 3 of this book it is shown that the long link error gaps can be fitted perfectly via a Weibull distribution (Karner and Rupp 2005; Karner et al. 2007). Thus, after inserting the Weibull pmf and the Weibull ccdf in equation 4.9, we obtain the estimated conditional error probability

$$\widehat{P}(1|10^m)_{\text{Weibull}} = \frac{ba^{-b}m^{b-1}e^{-(m/a)^b}}{e^{-(m/a)^b}} = ba^{-b}m^{b-1}. \qquad (4.10)$$

In this work we estimate $\widehat{P}(1|10^m)$ by using $b = 2.018$, corresponding to the statistics of the long gaps measured in Karner and Rupp (2005). With $b \approx 2$, the Weibull distribution becomes Rayleigh and the estimated expected failure rate can be expressed by the much simpler term

$$\widehat{P}(1|10^m) = \frac{2m}{a^2}, \qquad (4.11)$$

with $a = 3350$. Thus, $\widehat{P}(1|10^m)$ increases linearly with the error-free run length m that is the number of error-free TBs received so far after the last error. It can be shown that equation 4.9 leads to an approximately linear increase in the considered region of at most 10 000 TBs even without the mentioned approximation $b \approx 2$.

The measured conditional link error probability $P(1|10^m)$ from the 'static' scenario can be observed in Figure 4.3 on the left. The linear estimator for the failure rate perfectly predicts the transmission error probability based on the number of error-free TBs since the last error (error-free runlength). In Karner et al. (2007) further analysis of the influence of the feedback delay d_{FB} and the quality of the error prediction owing to the modelling error of the gaplength statistics is presented.

While in the 'static' scenario the conditional link error probability increases linearly with the error-free runlength; in the 'dynamic' case the conditional link error probability has its maximum just after an error burst and cannot be estimated in a linear way. This can be seen in Figure 4.3 (right), representing the measurement results of the 'tramway' scenario, the estimator according to equation 4.10 and with Weibull scale parameter $a = 1202.9$ and shape parameter $b = 0.782$. However, it can be observed that the estimator meets the measured conditional link error probability properly.

Figure 4.4 shows the performance of the error estimators for different decision thresholds γ, for the 'static' scenario according to equation 4.11 on the left and for the 'dynamic' scenario according to equation 4.10 on the right.

These figures were obtained by applying the following 'hard' decision: if $\widehat{P}(1|10^m) \geq \gamma$ an error is predicted. The probability of false error P_f is the conditional probability that

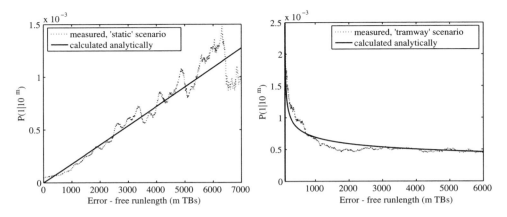

Figure 4.3 Conditional link error probability, 384 kbit/s bearer, 'static' scenario (left), 'tramway' scenario (right), 'mobile 1'.

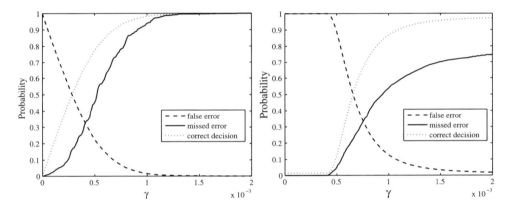

Figure 4.4 Prediction performance, 'static' scenario (left), 'tramway' scenario (right).

$\widehat{P}(1|10^m) \geq \gamma$ if no error occurred. The probability of missed error P_m is the conditional probability that $\widehat{P}(1|10^m) < \gamma$ if an error occurred. The probability of a correct decision $P_c = 1 - [(1 - P_{err}) \cdot P_f + P_{err} \cdot P_m]$ is the probability that an occurred error was predicted plus the probability that an error was not predicted and did not occur; P_{err} denotes the probability of a link error.

In most applications both low P_m and low P_f are desired. Therefore, thresholds around the crossing of P_m and P_f are relevant.

There are different optimal thresholds for the 'static' and the 'dynamic' scenarios. When using this prediction for both cases, either a suboptimal threshold can be selected to fit for the two scenarios or an additional detection for the discrimination of 'static' and 'dynamic' situations can be included in NodeB. This can be realized via monitoring the signal fading effects.

Predictability and general validity

Note, the predictability of the conditional link error probability, as well as the predictability of the intervals with low error probability, exists due to the fact that there is a quality-based closed-loop power control algorithm (the OLPC) in the UMTS DCH, which adjusts the Signal-to-Interference Ratio (SIR) target value for the Inner Loop Power Control (ILPC), as described in Chapter 2.

The quality-based power control mechanism introduces recurrence to the error process in order to meet the required quality target. For example, ideally there is a repeating sequence of 99 error-free TBs after one erroneous TB in order to reach the target BLER of 1%. Note that the real system does not behave like this, as there is fading and noise as well as error bursts and error clusters but the concept of error predictability can be applied analogously.

Therefore, due to the quality-based control mechanism, there is memory in the error process in a sense that the occurrence of the next error event depends on the number of already subsequently received error-free TBs since the last error event. This means that the error gaplengths are not geometrically distributed.

By contrast, in a mobile communication system without quality-based control algorithms, there is not necessarily the same short-term recurrence in the error process as in a system with quality based control algorithms. Therefore, for example, in the case of a very good channel, no link errors might be expected for a very long time interval in such systems and there is no predictive end of the error-free runlength as the end is caused by uncorrelated noise. The arguments mentioned lead us to assume that the predictability of link errors as described here may also be expected for currently new or future mobile communication systems, if they comprise any quality-based control mechanism.

References

Liang, Y.J., Apostolopoulos, J.G. and Girod, B. (2003) Analysis of Packet Loss for Compressed Video: Does Burst-length Matter? In *Proceedings of IEEE International Conference on Acoustics, Speech and Signal Processing (ICASSP'03)*, vol. 5, pp. 684–687.

Kac, M. (1947) On the notion of recurrence in discrete stochastic processes. *Bulletin of the American Mathematical Society*, **53**, 1002–1010.

Karner, W. and Rupp, M. (2005) Measurement based Analysis and Modelling of UMTS DCH Error Characteristics for Static Scenarios. In *Proceedings of the Eighth International Symposium on DSP and Communications Systems 2005 (DSPCS'2005)*, Sunshine Coast, Australia, Dec.

Karner, W., Nemethova, O. and Rupp, M. (2006) The Impact of Link Error Modeling on the Quality of Streamed Video in Wireless Networks. In *Proceedings of the Third IEEE International Symposium on Wireless Communications Systems 2006 (ISWCS 2006)*, Valencia, Spain, Sep.

Karner, W., Nemethova, O. and Rupp, M. (2007) Link Error Prediction in Wireless Communication Systems with Quality Based Power Control. In *Proceedings of IEEE International Conference on Communications (ICC'2007)*, Glasgow, Scotland, Jun.

Nemethova, O., Karner, W., Al-Moghrabi, A. and Rupp, M. (2005) Cross-layer Error Detection for H.264 Video over UMTS. In *Proceedings of Wireless Personal Multimedia Communications 2005 (WPMC 2005)*, Aalborg, Denmark, Sep.

Papoulis, A. and Unnikrishna, P.S. (2002) *Probability, Random Variables, and Stochastic Processes,* McGraw-Hill.

Zorzi, M. and Rao, R.R. (1999) Perspectives on the impact of error statistics on protocols for wireless networks. *IEEE Personal Communications,* **6**, 32–40, Oct.

Part III

Video Coding and Error Handling

Introduction

Video coding has evolved to quite clever techniques of lossy source coding. Rate reduction factors of 100 to 1000 are typical today, allowing moving pictures to be transmitted over wireless links with substantially lower bandwidths than the original source required. Part III of the book explains the basic principles of video coding, focusing on the newest standard H264/AVC in Chapter 5. Unfortunately, the source coding mechanisms are not designed for error prone transmissions as they occur in wireless cellular systems. Taking away the redundancies makes the pictures dependent on each other. Once an error occurs, a larger series of pictures is affected.

In many video services it is not possible to repeat lost packets owing to time or cost constraints. In such cases the user equipment has to make the best of what it has received. Error concealment methods as provided in this chapter are then of great help, in many situations allowing errors occurring more or less to disappear. Error concealment methods are not standardized and it is up to the manufacturer to decide what to implement. The contributions of this chapter show that even simple error concealment techniques can significantly improve the video quality. The better the error concealment, the smaller the corrupted area is. It is thus an important technique in conjunction with error localization techniques as described in Chapter 6.

However, error concealment is not the only means to improve videos when transmitted over wireless channels. One can also make use of the redundancy inherent in the videos. This redundancy is being used exclusively for error detection and localization, the methods of which are explained in Chapter 6. The basic trade-off is this: if more Cyclic Redundancy Check (CRC) is spent, allowing for better localizing errors and then applying error concealment techniques, more overhead is produced. An alternative is not to use additional side information such as CRC but to use the redundancy present in the video. One method is to make use of the syntax and to detect errors where syntax is incorrectly used. Another method is watermarking that works similarly to CRC but without adding side information. The drawback, however, is a small degradation in video quality; but also classical methods such as soft decoding can be applied, although requiring substantial complexity. In this chapter a final comparison of such methods is presented, showing at which channel signal-to-noise ratio what corresponding improvement in peak signal-to-noise ratio can be expected.

The work reported here by Olivia Nemethova can be found in more detail in her thesis *Error Resilient Transmission of Video Streaming over Wireless Mobile Networks*, which is available from http://publik.tuwien.ac.at/files/pub-et_12661.pdf as well as numerous publications cited within the chapter.

5

Principles of Video Coding

Olivia Nemethova

5.1 Video Compression

Nowadays for low-rate video streaming, the following video compression standards are used: H.263 standardized by the International Telecommunication Union (ITU), MPEG-4 Part II standardized by International Organization for Standardization (ISO) Moving Picture Experts Group (MPEG), and the emerging, newest and best performing H.264 (known also as MPEG-4 Part 10 and Advanced Video Coding (AVC)), standardized by the Joint Video Team (JVT) of experts from both ISO/IEC (International Electrotechnical Commission) and ITU. The principle of the compression for all mentioned codecs is very similar.

In this book, the focus is placed on H.264/AVC, especially on its baseline profile designed for low-complexity and low-rate applications; the majority of analysed experiments utilize an H.264/AVC codec. The H.264/AVC design covers a Video Coding Layer (VCL), designed efficiently to represent the video content, and the Network Abstraction Layer (NAL), which formats the VCL representation of the video and provides header information in a manner appropriate for conveyance by a variety of transport layers or storage media (Marpe *et al.* 2006). The output of the NAL is NAL Units (NALU). This structure is visualized in Figure 5.1. Data Partitioning (DP) is an optional part, allowing an encapsulation of different kinds of information elements into different packet types, as described later.

5.1.1 Video Sampling

Digital video cameras perform two kinds of sampling – sampling in the temporal domain given by the number of pictures per second (frame rate), and sampling in the spatial domain given by the number of points (pixels) in each of the pictures (picture resolution).

Frame rate, or frame frequency, is the measure of how quickly an imaging device produces unique consecutive images called frames. The frame rate is most often expressed in frames per second (f/s or frame/s) or, alternatively, in Hertz (Hz). Today's common frame rates have their origins in analogue television systems. The National Television Systems Committee

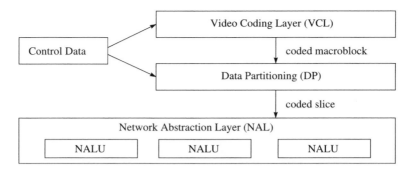

Figure 5.1 Layer structure of H.264/AVC encoder.

(NTSC) standardized the NTSC system, which is in use in Canada, Japan, South Korea, USA and some other places in South America, working with 29.97 f/s (denoted commonly as 30 f/s). In the rest of the world Phase Alternation by Line (PAL) and Séquentiel Couleur À Mémoire (SECAM) are used, operating at a frame rate of 25 f/s. In some low-rate and low-resolution applications the frame rate is reduced before the actual transmission to save data rates. Frame rate reduction can be performed by decimating the frame rate by F – maintaining each Fth frame while removing the rest. A typical example is mobile video streaming or call/conferencing with usual frame rates decimated by 2, 3, 4 or even 5. Other frame rates can be obtained by interpolation and subsequent decimation.

Furthermore, video cameras can work with two different image capture formats: 'inter-laced scan' and 'progressive scan'. Interlaced cameras record the image in alternating sets of lines: the odd-numbered and the even-numbered lines. One set of odd or even lines is referred to as a 'field', and a consecutive pairing of two fields of opposite parity is called a frame. A progressive scanning digital video camera records each frame as distinct, with both fields being identical. Thus, interlaced video captures twice as many fields/s as progressive video does when both operate at the same number of frames/s. In contrast to televisions, computer monitors generally use progressive scan, and therefore Internet/mobile video formats use it, too. H.264/AVC supports both interlaced and progressive scan.

Each frame consists of 'pixels'. Pixels of intensity pictures (black-and-white) are scalar values; pixels of colour pictures are represented by coordinates within the relevant colour space. The captured RGB picture is thus represented by three $N \times M$ colour component matrices consisting of q-bit long numbers (usually $q = 8$). In Table 5.1 some resolutions common for Internet and mobile video are summarized.

Since the human visual system is less sensitive to colour than to luminance (brightness), the bandwidth can be optimized by storing more luminance detail than colour detail. At normal viewing distances, there is no perceptible loss incurred by sampling colour details at a lower rate. In video systems, this is achieved by using the colour difference components. The signal is divided into a luminance (denoted as Y, 'luma' for short) and two colour difference (chrominance) components, denoted as U and V (or Cb and Cr, respectively), called 'chroma' for short.

The YUV signals are created from an original red, green and blue (RGB) source as follows (Richardson 2005). The weighted values of R, G and B are added together to produce

Table 5.1   Typical picture resolutions in pixels, used for Internet and mobile video applications.

Abbreviation	Size	Description
VGA	640×480	Video Graphics Array
QVGA	320×240	Quarter Video Graphics Array, also called Standard Interchange Format (SIF)
Q2VGA	160×120	
CIF	352×288	Common Intermediate Format (quarter of resolution 704×576 used in PAL)
QCIF	176×144	Quarter Common Intermediate Format

a single Y signal, representing the overall brightness, or luminance, of that spot:

$$Y = k_r \cdot R + k_g \cdot G + k_b \cdot B, \tag{5.1}$$

where k_r, k_g and k_b are weighting factors, $k_r + k_g + k_b = 1$. ITU-R recommendation BT.601 defines $k_b = 0.114$ and $k_r = 0.299$.

The U signal is then created by subtracting Y from the blue signal of the original RGB, and a scaling operation; and V by subtracting Y from the red, and then scaling by a different factor. The following formulas specify the conversion between the RGB colour space and YUV.

$$Y = k_r \cdot R + (1 - k_b - k_r) \cdot G + k_b B,$$

$$U = \frac{0.5}{1 - k_b} \cdot (B - Y), \tag{5.2}$$

$$V = \frac{0.5}{1 - k_r} \cdot (R - Y).$$

The basic idea behind the YUV format is that the human visual system is less sensitive to high-frequency colour information (compared to luminance) so the colour information can be encoded at a lower spatial resolution. The most common way of subsampling, called 4:2:0, reduces the number of samples in both the horizontal and vertical dimensions by a factor of two, that is for four luma pixels there is only one blue and one red chroma pixel. Hence, the YUV colour space subsampling is the first step to data rate reduction. The original bit rate R_{raw} of the 'raw' (uncompressed) RGB video with frame rate r_f and picture resolution $M \times N$ is given by $R_{raw\_RGB} = 3 \cdot r_f \cdot M \cdot N \cdot q$; the corresponding raw YUV video with 4:2:0 subsampling only requires the rate $R_{raw\_YUV} = 1.5 \cdot r_f \cdot M \cdot N \cdot q$. For the QCIF resolution raw video with 25 f/s and $q = 8$, the necessary bit rate is $R_{raw\_YUV\_QCIF} = 7.6$ Mbit/s. Although the bit rate is reduced to 50% of the RGB raw video, it is still not feasible for any of today's Internet or mobile applications. To overcome this, compression is employed further to reduce the data rate as described in the following sections.

5.1.2   Compression Mechanisms

The video coding layer of H.264/AVC consists of a hybrid of temporal and spatial prediction, together with transform coding. Its dataflow diagram is depicted in Figure 5.2. Each frame is

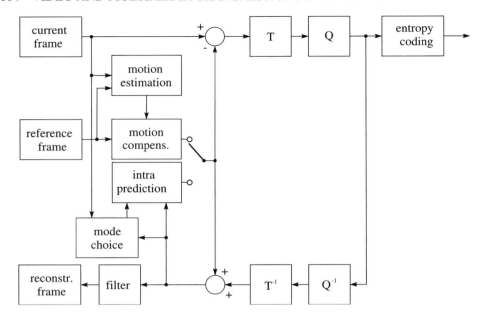

Figure 5.2 Dataflow diagram of H.264/AVC encoder.

split into non-overlapping areas – 'MacroBlocks' (MBs) – consisting of 16 × 16 samples of the luma and 8×8 samples of each of the two chroma components. The MBs are organized in 'slices', representing subsets of MBs that can be decoded independently.

Frames are called 'intra-coded' if they are encoded by means of a 'spatial prediction' without using information other than that contained in the picture itself. Typically, the first picture of a video sequence is intra-coded as well as all random access points of the video sequence (pictures that can be fast accessed without decoding previous parts of the video sequentially). Each MB in an intra-coded frame (also called intra-frame or I frame) is predicted using spatially neighbouring samples of previously coded MBs. MBs that do not have any previously encoded neighbours (for example, the first MB in the picture and MBs at the top slice boundary) are encoded without prediction. The encoder performs a mode choice – it decides which and how neighbouring samples are used for intra-prediction. The chosen intra-prediction type is then signalized within the bitstream.

For all remaining pictures of a sequence between random access points, typically 'inter-coding' is used, employing temporal prediction from other previously decoded pictures. First, the 'motion estimation' of each block is performed by searching the best matching region from the previous or following frame(s). Note that the best match is not searched for in the original (uncompressed) block, but rather in the quantized and filtered block. This prevents artefacts during the reconstruction process. The best match is taken as a prediction of the encoded block. Such a prediction is thus called 'motion compensated'. Each inter-coded MB is subject to further partitioning into fixed-size blocks (16 × 16 luma samples corresponding to no partitioning, 16 × 8, 8 × 16 or 8 × 8) used for motion description. Blocks of size 8 × 8

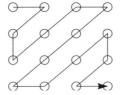

Figure 5.3 Scanning of samples in a submacroblock.

can be split again into SubMacroBlocks (SMBs) of 8×4, 4×8 or 4×4 luma samples. Chrominance parts are segmented correspondingly.

H.264/AVC supports multi-picture motion compensated prediction – more than one previously coded picture can be used as a reference. The accuracy of motion compensation is a quarter of a sample distance. The prediction values at half-sample positions are obtained by applying a one-dimensional six-tap Finite Impulse Response (FIR) filter. Prediction values at quarter-sample positions are generated by averaging samples at integer- and half-sample positions. To enable reconstruction at the receiver, the 'motion vector' (MV) between the position of the block within the frame and the position of its best match in the previously encoded frame has to be signalled as well as the mode of segmentation and corresponding reference frame(s). To avoid signalizing of the zero motion vectors and zero residuals in the cases of static picture parts, the 'SKIP MODE' allows for skipping of signalized numbers of P/B MBs. In SKIP mode neither residuals, nor motion vectors are sent. At the receiver, the MB from the previous frame is taken using predicted motion compensation.

Inter-coded frames are referred to as interframes or P and B frames; P being the frames that use only previous frames for prediction, B being the 'bi-directionally predicted' frames that also use successive frames for prediction. In H.264/AVC, other pictures can reference B frames for the motion estimation. The substantial difference between P and B MBs is that B MBs may use a weighted average of two distinct motion-compensated prediction values for building the prediction signal. Note that H.264/AVC supports frames with mixed I, P and B slices. Moreover, P and B slices may contain some I MBs.

All luma and chroma samples of an MB are either spatially or temporally predicted and the resulting 'prediction residuals' (difference between the MB samples being encoded and their prediction) are transformed. H.264/AVC uses three transforms depending on the type of residual data that is to be coded: a Hadamard transform for the 4×4 array of luma direct current (DC) coefficients in I MBs predicted in 16×16 mode, a Hadamard transform for the 2×2 array of chroma DC coefficients (in any type of MB) and a 4×4 integer Discrete Cosine Transformation (DCT) for all other blocks. The result of the transformation is a matrix of 'coefficients' corresponding to different spatial frequencies. The coefficient corresponding to the lowest frequency is DC, the others are denoted AC (alternating current) coefficients. All coefficients are further quantized. For each MB the quantization is controlled by the Quantization Parameter (QP) ranging from zero to 52. The quantization indices are scanned in zigzag order and finally entropy encoded together with other signalling information. A typical zigzag order used in H.264/AVC is illustrated in Figure 5.3.

In H.264/AVC, an MB can be coded in one of the many possible modes that are enabled depending on the picture/slice type. The 'mode decision' is performed at the encoder, that is it is not within the scope of a standard since video codec standards define the functions and features of the decoder rather than the encoder. Additional important gains in coding efficiency become possible if an MB mode decision is performed carefully. However, the additional gains can be extracted only at the expense of considerable increase in encoding complexity (for example by implementing a Rate-Distortion Optimization (RDO) at the encoder).

If the encoder parameters (QP, MV search area) are maintained during the encoding, then the number of coded bits produced for each MB will change depending on the content of the video frame and the mode decision, causing the bit rate of the encoder to vary. This variation in bit rate can cause problems especially in systems where resources are shared, as in wireless systems the resource management cannot efficiently perform the resource allocation. The Variable Bit Rate (VBR) produced by an encoder can be smoothed by buffering the encoded data prior to transmission in a First In First Out (FIFO) buffer, which is emptied at a Constant Bit Rate (CBR) matched to the channel capacity. Another FIFO buffer is placed at the input to the decoder and is filled at the channel bit rate and emptied by the decoder at a variable bit rate (since the number of bits to be extracted per frame varies over frames, but still the frames have to be rendered on the display with a constant frame rate). However, the costs are the buffer storage capacity and delay – the wider the bit rate variation, the larger the buffer size and decoding delay. Another possibility to compensate the VBR is the 'rate control' using the quantizer adaptation. However, quantizer changes need to be carefully restricted based on scene complexity, picture types and coding bit rate to maintain an acceptable end-user quality. H.264/AVC allows for an individual quantization parameter per MB, which can introduce spatial quality variations in fine texture.

An unpleasant effect of the block-based coding is blocking (also known as blockiness) of the visible block structures in the compressed picture. It is a consequence of the quantization. Regarding the perceptual quality, the user evaluate blocking as a result from lossy encoding as one of the post annoying artefacts (Winkler 2005). To provide better subjective quality, H.264/AVC defines an adaptive 'in-loop deblocking filter', where the strength of filtering is controlled by the values of several syntax elements as described by List *et al.* (2003). The blockiness is reduced without much affecting the sharpness of the content. Consequently, the subjective quality is significantly improved. Moreover, the filter reduces the bitrate by 5–10% typically while producing the same quality as the non-filtered video.

In accordance with the current H.264/AVC video coding standard, a so-called post filtering may also be applied aiming at improving the quality of the decoded image after its decoding. In order to facilitate this, additional information necessary for setting up the filter at the decoder may be provided by the encoder. The additional information may directly correspond to filter coefficients. It is an advantage to calculate the filter coefficients at the encoder, even if the filter is applied at the decoder, since the encoder disposes of the original image information. This enables, for instance, calculating the filter coefficients as Wiener filter coefficients adaptively with respect to the changing content of the video sequence. Alternatively, information for calculating the filter at the decoder may be provided such as a cross-correlation vector between the video data input to the encoding and the encoded/decoded video data. This additional post filter hint is signalized within separate packets.

0	NRI	NALU type	NALU payload

Figure 5.4 Format of a NALU and its header.

5.1.3   Structure of Video Streams

The network abstraction layer formats the compressed video data coming from the video coding layer and provides additional non-VCL information, such as sequence and picture parameters, access unit delimiter, filter data, Supplemental Enhancement Information (SEI), display parameters, picture timing and so on in a way most appropriate for a particular network. All data related to a video stream are encapsulated in NAL units. The format of a NALU is shown in Figure 5.4.

The first byte of each NALU is a header byte, the rest is the data. The first bit of the NALU header is a zero (forbidden) bit. The following two NRI (NAL Reference Identification) bits signalize the importance of the NAL unit for reconstruction purposes. The next five bits indicate the NALU type corresponding to the type of data being carried in that NALU. There are 32 types of NALUs allowed. These are classified into two categories: VCL NALUs and non-VCL NALUs. The NALU types from one to five are VCL NALUs and contain data related to the output of VCL – slices. Each encoded slice is also attached to a header containing information related to that slice.

NALUs with a NALU type indicator value higher than five are non-VCL NALUs carrying information such as SEI (which can carry, inter alia, the post-filter hint), sequence and picture parameter set, access unit delimiter and so on. Depending on a particular delivery system and scheme, some non-VCL NALUs may or may not be present in the stream containing VCL NALUs. For example, NALU type seven carries the Sequence Parameter Set (SPS), defining profile, resolution and other overall properties of the whole sequence; type eight carries the Picture Parameter Set (PPS), containing type of entropy coding, slice group and quantization properties. These sequence and picture level data can be sent asynchronously and in advance of the media stream contained in the VCL NALUs. An active SPS remains unchanged throughout a coded video sequence. An active PPS remains unchanged within a coded picture. In order to be able to change picture parameters such as picture size without the need to transmit parameter set updates synchronously to the slice packet stream, the encoder and decoder can maintain a list of more than one SPS and PPS. Each slice header contains then a codeword that indicates the SPS and PPS in use.

The NALUs can easily be encapsulated into different transport protocols and file formats, such as MPEG-2 transport stream, Real-time Transport Protocol (RTP) and MPEG-4 file format. For transmission over mobile networks, a VCL slice is encapsulated in RTP according to Wenger et al. (2005). The RTP payload specification supports different packetization modes. In the simplest mode a single NALU is transported in a single RTP packet, and the NALU header forms an RTP payload. In non-interleaved mode, several NALUs of the same picture can be encapsulated into the same RTP packet. In interleaved mode several NALUs belonging to different pictures can be encapsulated into the same RTP packet. Moreover, NALUs do not have to be sent in their decoding order. Both the non-interleaved and interleaved modes also allow for fragmentation of a single NALU into several RTP packets.

5.1.4   Profiles and Levels

Since H.264/AVC is designed for a multitude of applications, the structure of its bitstream may vary significantly. To avoid implementation of all possible stream structures by each specification-conform decoder, 'profiles' and 'levels' were defined. A profile is a subset of the capabilities including the entire bitstream syntax; a level is a specified set of constraints imposed on values of the syntax elements in the bitstream. Levels allow for standard-compliant low-complexity encoder and decoder implementations.

At present, H.264/AVC standard includes the following seven profiles, targeting specific classes of application:

- **Baseline Profile**: Primarily for lower-cost applications demanding fewer computing resources, this profile is used widely in video conferencing and mobile applications.

- **Main Profile**: Originally intended as the mainstream consumer profile for broadcast and storage applications, the importance of this profile faded when the High Profile was developed for those applications.

- **Extended Profile**: Intended as the streaming video profile, this profile has relatively high compression capability and some extra tricks for robustness to data losses and server stream switching.

- **High Profile**: The primary profile for broadcast and disc storage applications, particularly for high-definition television applications (this is the profile adopted into HD DVD and Blu-ray Disc, for example).

- **High 10 Profile**: Going beyond today's mainstream consumer product capabilities, this profile builds on top of the High Profile – adding support for up to 10 bits per sample of decoded picture precision.

- **High 4:2:2 Profile**: Primarily targeting professional applications that use interlaced video, this profile builds on top of the High 10 Profile – adding support for the 4:2:2 chroma sampling format while using up to 10 bits per sample of decoded picture precision.

- **High 4:4:4 Profile**: This profile is essentially similar to High 4:2:2 Profile. Additionally, it supports up to 4:4:4 chroma sampling, up to 12 bits per sample and efficient lossless region coding and integer residual colour transform for coding RGB video while avoiding colour-space transformation errors.

In this book the focus is on the baseline profile. Note that the baseline profile does not support B slices; only I and P slices are possible. Other baseline profile constraints will be discussed later, when necessary for the application.

5.1.5   Reference Software

All experiments with H.264/AVC codecs presented in this book were performed using the Joint Model (JM) reference software developed by JVT for testing and available under http://iphome.hhi.de/suehring/tml/. JM includes both the encoder and the decoder compliant with the H.264/AVC hypothetical reference decoder (Corbera *et al.* 2003). Settings are passed via the command line and/or the configuration files `encoder.cfg` and `decoder.cfg`.

5.2 H.264/AVC Video Streaming in Error-prone Environment

This section describes the effect of errors on the decoding of an H.264/AVC video stream. It briefly summarizes standardized error resilience means, presents open problems and summarizes the most popular frameworks for additional error resilience.

The usual way of video streaming over Internet Protocol (IP) packet networks assumes an RTP together with Real Time Control Protocol (RTCP) feedback on the application/session layer and a User Datagram Protocol (UDP) on the transport layer (Wenger 2003). In contrast to the Transmission Control Protocol (TCP), UDP does not provide any Automatic Repeat reQuest (ARQ) mechanism to perform retransmissions. It only provides a checksum to detect possible errors. The checksum is typically calculated over a rather large packet to avoid a rate increase due to packet headers.

As interactive and background services have a non-real-time nature (for example, web browsing, file transfer, email), they can make use of retransmission mechanisms, as provided by TCP – packet loss is compensated by delay. This is completely different for the real-time conversational and streaming services. The packet loss caused by limited delay does not necessarily reflect the differences in the residual bit error distribution – the distribution of bit errors within the packets containing an error. If the number of residual bit errors is low, a decoding of such stream might result in even lower distortion than using error concealment. This section describes the impact of errors in H.264/AVC on the distortion of the decoded picture and presents some standardized techniques that aim to reduce it. Finally, the most popular alternative state-of-the-art error resilience techniques are summarized.

5.2.1 Error Propagation

The compression mechanism introduced in the previous section is capable of reducing the data rate more than 100 times while keeping the user-perceived quality almost unchanged. It has, however, some drawbacks to the robustness of such an encoded video stream against transmission errors. An error in a single bit of stream may cause considerable distortion due to the error propagation.

There are three possible sources of error propagation in an H.264/AVC encoded video stream: entropy code, spatial prediction and temporal prediction. Entropy codes assign codewords to symbols to match codeword lengths with probabilities of the symbols – entropy codes are Variable Length Codes (VLCs). Hence, an error in a codeword may impact following codewords as well, if the codeword boundaries are determined incorrectly. This is also a reason for discarding the entire packet if an error is detected.

Spatially (intra) encoded MBs require the spatially neighbouring reference blocks for decoding. If these are erroneous, the MB whose reconstruction is based on them will be distorted, too. In practice, however, this problem does not occur as the entire packet is discarded if an error is detected. However, spatial error propagation becomes significant for the techniques utilizing information from damaged packets.

Temporal (inter) prediction causes propagation of errors between the frames of the same video sequence. If an error occurs in a frame, all following frames that use such a frame as a reference for the motion compensation will be reconstructed erroneously. The information in the video sequence is refreshed in each inter frame by the new information contained in the transmitted residuals; thus the sequences containing a higher amount of movement recover faster. Another source of refreshment is the I frames and the intra-coded MBs inserted (randomly) in inter-predicted slices.

5.2.2   Standardized Error Resilience Techniques

To reduce the distortion resulting from the errors and their propagation, several means were added to the H.264 standard. They are summarized nicely in Ostermann *et al.* (2004) and Wiegand *et al.* (2003). In Stockhammer and Hannuksela (2005), the benefits of the H.264/AVC error resilience features in the context of wireless transmission are emphasized. In the following, the most important of these features are summarized.

- **Slices** provide spatially distinct resynchronization points within the video data for a single frame. This is accomplished by introducing a slice header containing syntactical and semantical resynchronization information. Intra-prediction and motion vector prediction are not allowed over slice boundaries. A trivial slicing method is to use one frame corresponding to one slice. Another possibility is the subdivision of the frame into slices with a fixed maximum number of MBs, or slices with a fixed maximum number of bytes.

- **Flexible Macroblock Ordering** (FMO) allows for an arbitrary mapping of MBs to slice groups by means of MB allocation maps.

- **Arbitrary Slice Ordering** (ASO) allows the decoding order of slices within a picture not to follow the constraint that the address of the first MB within a slice is monotonically increasing within the NALU stream for a picture. This permits reduction of decoding delay in case of out-of-order delivery of NALUs.

- **Intra refreshing** helps to stop the temporal error propagation. Apart from regular inserting of entirely intra-coded frames resulting in high peaks in rate and complicating resource allocation, H.264/AVC allows for sending intra-slices and even intra-macroblocks within interframes.

- **Redundant slices** (RSs) are an error resilience feature allowing an encoder to send an extra representation of a picture region (typically at lower fidelity) that can be used if the primary representation is corrupted or lost.

- **Data Partitioning** (DP) provides the ability to separate more important and less important syntax elements into different packets of data, enabling the application of Unequal Error Protection (UEP) and other types of improvement of error/loss robustness. Header data and motion vectors are labelled 'type A', so that they can be better protected. The residuals of intra frames are labelled 'type B', while inter-predicted residuals are 'type C'.

Note that these error resilience techniques can be combined to achieve the performance required for particular scenarios. All the above-mentioned error resilience features except DP are available in the baseline profile. DP is available only in the extended profile.

5.2.3 Alternative Error Resilience Techniques

Apart from techniques standardized by H.264/AVC, there are other possibilities of enhancing error resilience of the video transmission:

- techniques considered for future standard improvements;
- methods that were considered but not selected in the final standard;
- implementation-specific mechanisms that are not in the scope of standardization;
- techniques that consider more layers/standards for joint optimization.

The most popular state-of-the-art frameworks and terms for error resilient video transmission are summarized in this section. For H.264/AVC two extensions are considered and worked on: multi-view coding and 'Scalable Video Coding' (SVC), the latter being a mean for error resilient video transmission (Schwarz *et al.* 2006). The SVC design can be classified as 'layered video coding'. A scalable representation of video consists of a base layer (providing a basic quality level) and multiple enhancement layers (serving as a refinement of the base layer, not useful without the base layer). H.264/SVC also supports three types of inter-layer prediction (motion, intra, residual). The spatial and temporal resolution scalability is provided at a bitstream level, that is, by simply discarding the corresponding NALUs. The benefit of H.264/SVC for wireless multi-user streaming is demonstrated in Liebl *et al.* (2006).

Multiple Description Coding (MDC) is an alternative to the layered coding concept. In contrast to layered coding, MDC (Goyal 2001) generates a number of equivalent-importance data streams that, all together, carry the input information. Sending of RS in H.264/AVC may also be understood as a form of MDC. To date, various MDC schemes have been proposed and evaluated in the literature for transmission over error-prone environments (Franchi *et al.* 2005). It has been argued whether layers or versions provide better means for multi-user error resilient transmission in lossy networks. The analysis in Chakareski *et al.* (2005) and Radulovic *et al.* (2004) demonstrated that layered coding schemes perform well in the cases where the packet transmission schedules can be optimized in a rate distortion (RD) sense. For the scenarios with non RD-optimized packet schedules, with somewhat relaxed constraints on aggregate rate,[1] MDC provides better results.

Of course, both concepts – layered coding and MDC – can be advantageously combined also with UEP in the lower layers. A basis for the UEP concept is the categorization of application layer data according to their importance. The importance of data is given by the distortion that their loss would cause. The categorization can be coarse, typically based on semantic information, or finer, based on distortion estimation. Distinguishing between intra- and inter-coded frames, and prioritizing the intra-coded packets, is a typical simple 'semantic-based' approach that can be found in Klaue *et al.* (2003) and Pathac *et al.* (2005). A finer, 'content-based' approach can be seen in Thomos *et al.* (2005), where an encoder applies FMO according to the calculated distortion and protects unequally different slice groups. The protection itself can also have different forms. Additional protection on the application layer can be used as in Thomos *et al.* (2005), where Reed-Solomon coding is employed. Alternatively, the mechanism may rely on the (channel) codes of the lower layers, or prioritize important packets in the scheduler as in Bucciol *et al.* (2004).

---

[1]The total rate is the sum of the rates of all streams sharing the same constraint resources.

A step forward from adapting the channel code for a given source is the joint design of those two. The well-known Shannon 'separation theorem' (Shannon 1948) states that an optimal communications system can be constructed by considering separately the source and channel coder designs. It assumes, however, that the source code is optimal, that is, it removes all source redundancy. Moreover, it considers that for rates below channel capacity, the channel coder corrects all errors. These assumptions do not really hold in practical systems with limited delay and complexity. Here, properly designed 'Joint Source-Channel Coding' (JSCC) is more beneficial. The JSCC term includes many different concepts (a concise review is provided in van Dyck and Miller (1999)) ranging from the channel-optimized (vector) quantization (Farvardin 1990), decoding exploiting the residual redundancy (Nguyen and Duhamel 2005; Sayood *et al.* 2000) up to solutions using overcomplete frame expansions (Lee *et al.* 2004). Layered coding and MDC in combination with UEP and/or RD-optimized can also be considered as JSCC.

Clearly, an appropriate design of source coding alone may also improve the robustness of transported streams against errors. Therefore, various 'reversible VLC codes' have been proposed (Takishima *et al.* 1995; Wen and Villasenor 1998) that prevent desynchronization of the VLC decoding. Alternatively, *resynchronization* information may be embedded within the video stream as in Chiani and Martini (2006) and Côté *et al.* (2000). All these mechanisms result in a rate increase. The rate increase is justified as far as the distortion for the same rate decreases. This, however, depends often on the channel characteristics and conditions. The choice of the appropriate method can thus be seen as an 'RDO' problem. Typically, solution of rate distortion problems is rather complex. On the other hand, it leads to efficient resource usage (He and Mitra 2005).

There has been a great deal of research effort on RD-optimized scheduling methods of packets over lossy networks (Chou and Miao 2006; Kalman *et al.* 2003; Kalman and Girod 2004). In heterogeneous mobile communication systems such as UMTS, not all network elements implement all protocol layers. Thus, the scheduler typically does not have information about the content transported. Such information is, however, essential for knowledge/estimation of the distortion. Hence, the strict protocol layer structure has to be evaded. In general, techniques that disregard the protocol layer structure and enable exchange of information and/or joint optimization at more protocol layers are known as 'cross-layer design'. In fact, most JSCC and UEP approaches can be considered to be cross-layer.

The Open System Interconnection (OSI) model (Zimmermann 1980) defines seven layers distinguished by their functionality. The basic idea of layering is that each layer adds value to services provided by the set of lower layers. Thus, layering divides the total problem into smaller pieces that can be designed separately, on the one hand considerably reducing the problem-solving complexity; on the other hand, limiting the information available to a particular layer, which can lead to suboptimal solutions. Lately, especially for wireless communications, the cross-layer approaches gained in importance (Chen *et al.* 2005; Choi *et al.* 2005; van der Schaar and Shankar 2005).

5.3   Error Concealment

Error concealment methods are designated to replace the parts of a video stream corresponding to its lost packets. The replacement is performed by a sort of spatial and/or temporal 'interpolation'. Spatial interpolation utilizes the correctly received (and possibly also lost but already concealed) parts of the same picture. Temporal interpolation relies on the previously received frames (preceding or following the damaged frame in the rendering order).

In general, error concealment techniques determine the error 'visibility' in the decoded video stream. In Sun and Reibmann (2001) a summary of some well-known methods is presented; apart from that, there is a vast literature proposing particular error concealment methods for various systems. The performance of error concealment depends greatly on instantaneous spatial and temporal characteristics of the video sequence as well as on the error pattern to be concealed. Furthermore, for power-limited mobile devices and real-time applications it is essential to keep the complexity as low as possible. In the following, a brief overview of error concealment methods suitable for real-time deployment is provided.

5.3.1 Spatial Error Concealment

Spatial error concealment utilizes only the information from the processed picture for the reconstruction of the lost area. The quality of reconstruction depends strongly on the spatial characteristics of the concealed picture, given merely by the distribution of edges. The approaches for concealing the missing parts with different spatial characteristics are not necessarily identical. Whereas the smooth areas can be interpolated by simple averaging or by optimization of a smoothness cost function, more sophisticated approaches are necessary to deal with areas containing edges. The smoothing should only be performed along the edges; the edges should be prolonged. The resolution of the image together with the size and shape of the lost area determine the interpolation problem to be solved.

The simplest spatial error concealment method is interpolation of the missing block by an average of the boundary pixel values. Such interpolation, however, leads to a monotone block, apparently distinguishable from the rest of the picture, especially if the missing block was not smooth. The 'weighted averaging' method (Sun and Reibmann 2001) improves this by weighting the contribution of the boundary pixels to the average according to the distance between the interpolated pixel and the opposite boundary. Consider an $M \times N$ large missing block $\mathbf{F}$, consisting of pixel values $f_{i,j}$ (of one colour component), with i being the row index and j being the column index. The top left corner corresponds to the $(0, 0)$ point. Each pixel value of the missing macroblock is then interpolated as a weighted linear combination of the nearest pixels in the west (w), east (e), north (n) and south (s) boundaries:

$$\widehat{f}_{i,j} = \frac{1}{M + N + 2}[(N - j + 1)b_i^{\mathrm{w}} + jb_i^{\mathrm{e}} + (M - i + 1)b_j^{\mathrm{n}} + ib_j^{\mathrm{s}}], \qquad (5.3)$$

with $b_i^{\mathrm{w}} = f_{i,0}$, $b_i^{\mathrm{e}} = f_{i,N+1}$, $b_j^{\mathrm{n}} = f_{0,j}$ and $b_j^{\mathrm{s}} = f_{M+1,j}$ as shown in Figure 5.5 (left). Each colour component is concealed in the same way. The possibilities of spatial prediction type based interpolation are limited by the eight predefined prediction types (Richardson 2005). A natural enhancement thereof is a pixel domain directional interpolation. In Suh and Ho (1997) the edge is first classified in one of four main directions and then smoothing in the identified direction is performed. The drawback of this method is that it supports only one main direction for the whole missing block. While this is possibly sufficient for higher resolutions, it brings limited improvement for low resolutions where often more edges meet in one missing area. Therefore, Kwok and Sun (1993) refine the edge direction determination to more directions. In Hong et al. (1999), eight directions are distinguished; smoothing is

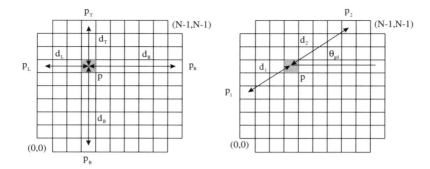

Figure 5.5   Spatial interpolation: weighted averaging (left), directional interpolation shown for the whole block (right).

performed recursively in all detected directions, from the outer pixels until filling up the gap (lost part). The edges can be detected in different ways; one of the most used is the detection using the horizontal and vertical Sobel operator (Koschan and Abidi 2005).

After detecting a main edge direction θ_{gd} with corresponding edge slope $a_d = \cot[\theta_{gd}]$, the edge is prolonged – the block or its part is interpolated in that direction, for example by means of weighted averaging as illustrated in Figure 5.5 (right):

$$f_{i,j} = \frac{1}{d_1 + d_2}[d_2 f_{i_1,j_1} + d_1 f_{i_2,j_2}], \qquad (5.4)$$

where f_{i_1,j_1} and f_{i_2,j_2} are the points at the boundaries from which the missing pixel is interpolated. They are obtained as an intersection of the block boundaries with a line of slope a_d crossing the pixel $f_{i,j}$ being concealed. Symbols d_1 and d_2 denote the distance of $f_{i,j}$ from f_{i_1,j_1} and f_{i_2,j_2}, respectively.

Note that the indexes i_1, j_1 and i_2, j_2 are rounded to integer values. This method performs well for the recovery of parts containing one dominant direction only. However, it does not provide satisfying results if there are more dominant edges entering the missing area. Moreover, if the edges crossing the missing block are not linear, pixel-based directional interpolation may cause unpleasant artefacts. In natural scene pictures with resolution as small as QCIF, there are still many blocks with more than one dominant edge. In such cases it is important to decide how the edges will be prolonged. Prolonging the edges results in partitioning of the missing block; each partition can be further recovered by smoothing. The method proposed in Kung *et al.* (2003) matches the edges entering and leaving the missing area, partitions the area accordingly and smoothes preserving the edges. However, the matching of edges cannot always be performed reliably – especially if the edges are not linear and if there are too many edges. Furthermore, such a method requires the knowledge of all four boundaries. In Nemethova *et al.* (2005), other approaches for partitioning and interpolation were investigated. The missing blocks are first subdivided into so-called regions of dominance – each such region supporting one main edge direction. Interpolation is then similar to that in the previous section, applied to each particular region separately. Another approach, based on solving an optimization problem with a quadratic smoothness cost function, was proposed in Wang *et al.* (1993) and enhanced in Zhu *et al.* (1993) to the temporal direction. In Zhu *et al.* (1998), an improved smoothness cost function is proposed, which also considers a second-order derivative to allow for sharper edge reconstruction. The optimization performed in the above-mentioned publications was performed generally for the case where some transform domain coefficients were received correctly, which may occur in layered coding.

5.3.2 Temporal Error Concealment Methods

In natural low-resolution video sequences, the correlation between two consecutive frames is higher than the correlation of the pixel values within the same frame. Hence, temporal interpolation provides often much better means for error concealment than spatial interpolation.

If the residuals are lost but the MVs have been correctly received, the simplest method is to decode the missing block by setting the missing residuals to zero. This scenario occurs if data partitioning is used. 'Decoding without residuals' performs well if the missing residuals are small. On the other hand, the residuals are small if the sequence contains mainly linear motion, which can easily be predicted.

If the whole macroblock information is lost, the simplest method, here called the 'copy-paste' method (also called zero-motion error concealment), can be used. The missing block is replaced by a spatially corresponding block from the previous frame. This only performs well for low-motion sequences. Better performance is provided by motion-compensated interpolation methods. Motion vector estimation methods estimate the motion vector of the missing block from the motion vectors of the neighbour blocks (spatially – within the same frame or temporally – from the previous frames). If the motion vectors $\underline{mv}_D$ with $D \in \mathcal{D}$ and $\mathcal{D} = \{n, s, w, e\}$ of the top (north), bottom (south), left (west) and right (east) neighbours MB are known, then the motion vector $\underline{mv}$ of a missing block may be easily approximated by an average of those:

$$\widehat{\underline{mv}} = \frac{1}{|\mathcal{D}|} \sum_{D \in \mathcal{D}} \underline{mv}_D. \tag{5.5}$$

(Here, $|\mathcal{D}|$ is the cardinality of the set $\mathcal{D}$.) The missing block is replaced by the block in the previous frame having the position indicated by $\widehat{\underline{mv}}$. The performance of this method is limited for lower resolutions as each MB can also contain parts of different objects moving in different directions. However, H.264/AVC supports motion prediction for blocks of variable size, which can be used to refine the motion estimation (Xu and Zhou 2004).

If no subblocks were used, the motion vectors on the boundary D will be the same for all the blocks. Let $\underline{mv}_D^{(k)}$ be the motion vector of the kth block in the boundary D. The estimated motion vector $\widehat{\underline{mv}}^{(i,j)}$ of the block being in the position (i, j) within the missing MB can be calculated as a weighted average of block motion vectors at the boundaries:

$$\widehat{\underline{mv}}^{(i,j)} = \frac{d_e \underline{mv}_w^{(j)} + d_w \underline{mv}_e^{(j)} + d_n \underline{mv}_s^{(i)} + d_s \underline{mv}_n^{(i)}}{d_e + d_w + d_n + d_s}. \tag{5.6}$$

This method is directly applicable only for P and B frames. Each missing block is then concealed by the part of the previous picture corresponding to $\widehat{\underline{mv}}^{(i,j)}$.

Motion vector interpolation can lead to blocking artefacts, especially for those video sequences containing inhomogeneous movement and/or nonlinear movement. Therefore, in Chen *et al.* (1997) an improved method was proposed that sets the weights in equation 5.6

according to the 'side match criterion'. The side match criterion measures the differences between the pixel values at the boundary and within the block obtained by MV interpolation. The parts with a higher match get higher weights. In the JM implementation, the side match criterion is used to choose the motion vector – a vector of the block with best side match is chosen.

The motion vectors are not only correlated in the spatial, but also in the temporal domain. Therefore, in addition to the spatial motion vector interpolation or in the cases where the spatially neighbouring MVs are not available as in I frames, temporal motion vector interpolation is an option. The efficiency of the temporal MV interpolation depends strongly on the temporal characteristics of the video stream. The performance of this interpolation method decreases considerably with a reduction of frame rate. Utilization of a mean of the previous and the following frames can improve the results, but leads to increased delay and storage.

Boundary and block matching methods do not require the knowledge of any motion vectors; they search the best matching neighbourhood of the missing block in the previous frames. An evaluation of such methods is provided, for instance, in Suh and Ho (2002) for MPEG-2 video sequences with higher resolution.

A very popular error concealment method is a so-called *boundary matching*. Let $\mathbf{B}$ be the area corresponding to a one pixel wide boundary (containing the top and/or bottom and/or left and/or right boundary pixels) of a missing block in the nth frame $\mathbf{F}_n$. Motion vectors of the missing block as well as those of its neighbours are unknown (occurring mostly for I frames, but also in specific cases for P frames with some inserted I MBs) or are not taken into account. The task is to find the coordinates $(\hat{x}, \hat{y})$ of the best match to $\mathbf{B}$ within the search area $\mathbf{A}$ in the previous frame $\mathbf{F}_{n-1}$:

$$(\hat{x}, \hat{y}) = \arg \min_{x, y \in \mathbf{A}} \sum_{i, j \in \mathbf{B}} |\mathbf{F}_{n-1}(x + i, y + j) - \mathbf{B}(i, j)|. \qquad (5.7)$$

The sum of absolute differences (SAD) may be chosen as a similarity metric for its low computational complexity. The size of $\mathbf{B}$ depends on the number of correctly received neighbours M, boundaries of which are used for matching. The area $\mathbf{A}$ is an important design parameter and should be chosen with respect to the amount of motion in the sequence. The MB sized area starting at the position $(\hat{x}, \hat{y})$ in $\mathbf{F}_{n-1}$ is taken to conceal the damaged MB in $\mathbf{F}_n$. Boundary matching works well if all four boundaries are available and if the motion is linear and homogeneous (all parts moving in the same direction).

Even better results can be obtained by a so-called 'block matching', which searches for the best match for the correctly received $\mathbf{MB}_D$, $D \in \mathcal{D}$, $\mathcal{D} = \{n, s, w, e\}$, neighbouring the missing one on top (north), bottom (south), left (west) and right (east) sides, respectively:

$$[\hat{x}, \hat{y}]_D = \arg \min_{x, y \in \mathbf{A}_D} \sum_{i, j \in \mathbf{MB}_D} |\mathbf{F}_{n-1}(x + i, y + j) - \mathbf{MB}_D(i, j)|, \qquad (5.8)$$

$\mathbf{A}_D$ representing the search area for the best match of $\mathbf{MB}_D$, with its centre spatially corresponding to the start of the missing MB. The final position of the best match is given by an average over the positions of the best matches found for the neighbouring blocks:

$$\hat{x} = \frac{1}{|\mathcal{D}|} \sum_{D \in \mathcal{D}} \hat{x}_D; \quad \hat{y} = \frac{1}{|\mathcal{D}|} \sum_{D \in \mathcal{D}} \hat{y}_D. \qquad (5.9)$$

Figure 5.6 Screenshots of a part of an I frame in the 'panorama' sequence: compressed original (Y-PSNR = 35.86 dB), error pattern (Y-PSNR = 10.45 dB), weighted averaging (Y-PSNR = 18.09 dB), directional interpolation (Y-PSNR = 16.57 dB), copy-paste (Y-PSNR = 22.76 dB), boundary matching (Y-PSNR = 26.27 dB), 8×8 block matching (Y-PSNR = 30.27 dB), 2×2 block matching (Y-PSNR = 30.74 dB).

The area of MB size starting at the position $(\hat{x}, \hat{y})$ in $\mathbf{F}_{n-1}$ is taken to conceal the damaged MB in $\mathbf{F}_n$. To reduce the necessary number of operations, only parts of the neighbouring MBs can be used for the MV search. Similarly to the refined MV interpolation, better results are achieved by searching the motion vectors for such blocks that are smaller than the whole MB. For the search of MVs belonging to a subblock, blocks of the size smaller or greater than the subblock itself may be used. The MVs of blocks belonging to the missing MB can be interpolated by the estimated MVs obtained for the subblocks of the neighbours, via equation 5.6.

Figure 5.6 shows part of an I frame from a 'panorama' sequence decoded by H.264/AVC JM and concealed 'during' decoding that is always using only the top and left neighbours, using the concealed MB for the decoding and concealment of the following MBs. The spatial domain interpolation shows the worst performance. The directional interpolation method should not be used if only two neighbours are available. The 'panorama' sequence had a frame rate decimated by four (as usual for mobile communications), resulting in a higher motion. Thus, the copy-paste method does not perform well. Perceptually it causes freezing of the concealed picture parts resulting in overall jerkiness of the video. Better perceptual results were obtained using boundary matching, but having a closer look there are still some small artefacts left. The block matching algorithm leads to the best results. Perceptually, 8×8 block matching performs already sufficiently well.

Note that the performance of all methods decreases rapidly if not all neighbours are known or correctly received. After a scene change, spatial error concealment typically performs better than temporal error concealment. The optimal choice of the best performing error concealment is content dependent. Therefore, in Belfiore *et al.* (2003) a method is proposed that applies temporal error concealment whenever possible; otherwise it employs weighted averaging. The temporal error concealment technique exploits information from a number of past frames in order to estimate the motion vectors of the lost frame. The motion vectors are used to project the last frame onto an estimate of the missing (parts of the) frame.

The holes resulting from the possibly overlapping projection are concealed by a spatial error concealment (median of the surrounding pixels). The advantage of such a method is its suitability for concealing the loss of entire frames. However, concealing in scene changes and videos with reduced frame rate may result in severe impairments.

In Cen *et al.* (2003) the benefit of combining different error concealment methods for different contexts has been shown. The appropriate method is chosen by means of a classification tree. Eight individual methods are investigated. For I frames, the following methods were considered as applicable: averaging of transformation coefficients, weighted averaging in pixel domain and copy-paste. If panning was detected, the MV was estimated from previous frames and used for motion compensated reconstruction. For P and B frames, again pixel domain weighted averaging is considered and the motion compensated reconstruction for panning. Four MV interpolation methods using different combinations of top and bottom MVs were used. The performance of the eight chosen methods, however, is limited.

The implementation of error concealment mechanism provided in the JM H.264/AVC reference software reveals two generic methods for the slice concealment: pixel domain weighted averaging for the I frames and side match distortion minimizing MV interpolation for the P and B frames. The error concealment is performed 'after' decoding all the correctly received slices in a frame. This allows not only the top and left boundaries to be used, but also the bottom and right boundaries, since it does not conceal the MBs in the scanning order, but, rather, column-wise. This method is described in more detail in Wang *et al.* (2002).

The mechanism proposed in Su *et al.* (2004) improves the JM error concealment so that it allows for temporal error concealment for inter-predicted frames and spatial error concealment for intra-predicted frames. The decision about the usage of temporal and spatial error concealment is based on a scene change detector. The scene change detector relies on an MB-type matching parameter. If a scene change is detected, spatial error concealment is employed. Otherwise, the motion activity is detected. If it is low, copy-paste error concealment is used. Otherwise, for P/B frames, spatial MV interpolation is applied and for I frames the MV of the co-located MB in the previous frame is taken.

More refinements to H.264/AVC spatial and temporal error concealment were proposed in Xu and Zhou (2004). There, spatial error concealment was modified to support one main direction rather than to perform weighted averaging only. The main direction is found within 16 possible directions. Temporal error concealment is proposed to be applied to the 8×8 large SMBs rather than to the 16×16 MBs. The decision whether to apply temporal or spatial error concealment is based on a scene change detector. In Nemethova *et al.* (2006), an adaptive error concealment mechanism is proposed as an alternative to the JM error concealment mechanism and its refinements. The mechanism is based on a decision tree that considers more criteria and more error concealment methods. A scene change detector, the number of neighbouring blocks and type of frame are used for the selection of the appropriate method. Also, many more sophisticated error concealment methods are known, while we have only reported low-complexity methods suitable for lower resolutions.

5.4   Performance Indicators

The performance of error resilience methods can be evaluated by measuring the end-user distortion in the presence of transmission errors with given characteristics and by analysing the cost determined by increasing the rate and/or complexity. In this section, metrics are presented that are used throughout this book for performance evaluation.

Distortion   To evaluate the distortion within the nth video frame $\mathbf{F}_n$ with respect to a reference (distortion-free) frame $\mathbf{R}_n$, the mean square error (MSE) can be used:

$$\text{MSE}[n] = \frac{1}{M \cdot N \cdot |\mathcal{C}|} \sum_{c \in \mathcal{C}} \sum_{i=1}^{N} \sum_{j=1}^{M} [\mathbf{F}_n^{(c)}(i, j) - \mathbf{R}_n^{(c)}(i, j)]^2, \tag{5.10}$$

where $N \times M$ is the size of the frame and $\mathcal{C}$ is the set of colour components, for RGB colour space $\mathcal{C} = \{R, G, B\}$. The number of colour components corresponds to the cardinality $|\mathcal{C}|$ of the set. Row index i and column index j address the individual elements of the colour component matrix. For image distortion, the MSE is most commonly quoted in terms of the equivalent reciprocal measured Peak Signal to Noise Ratio (PSNR) defined as:

$$\text{PSNR}[n] = 10 \cdot \log_{10} \frac{(2^q - 1)^2}{\text{MSE}[n]} \text{ [dB]}, \tag{5.11}$$

where q represents the number of bits used to express the colour component values. In YUV colourspace the two chrominance parts are usually smoother than the luminance. Therefore, especially for the applications that handle all components in the same way, it may be more discriminative to compare the distortion of luminance only ($\mathcal{C} = \{Y\}$) by means of Y-MSE[n] or Y-PSNR[n]. The PSNR averaged over the whole video sequence can be defined in two ways – as an average $\overline{\text{PSNR}}$ over the PSNR[n] of all its frames, or as

$$\text{PSNR} = 10 \cdot \log_{10} \frac{(2^q - 1)^2}{\overline{\text{MSE}}} \text{ [dB]}, \tag{5.12}$$

where $\overline{\text{MSE}}$ is average MSE defined by

$$\overline{\text{MSE}} = \frac{1}{F} \sum_{n=1}^{F} \text{MSE}[n] \tag{5.13}$$

and F is the number of frames in the video sequence. The later definition is formally correct, since it averages over the linear values. Averaging over logarithmic values results in a systematic error as a consequence of Jensen's inequality (MacKay 2006), which states that if $f(\cdot)$ is a convex function and X is a random variable, then

$$E\{f(X)\} \geq f(E\{X\}), \tag{5.14}$$

where $E\{X\}$ denotes the expectation of X. A function $f(\cdot)$ is convex over an interval (a, b) if for all $x_1, x_2 \in (a, b)$ and $0 \leq \lambda \leq 1$ applies

$$f(\lambda \cdot x_1 + (1 - \lambda) \cdot x_2) \leq \lambda \cdot f(x_1) + (1 - \lambda) \cdot f(x_2), \tag{5.15}$$

that is, all points of $f(x)$ between the arbitrary chosen $x_1, x_2 \in (a, b)$ lay below or on the line connecting them. Since $\log_{10}(1/x) = -\log_{10}(x)$ is convex, $\overline{\text{PSNR}}$ will in general provide 'better' results (meaning higher quality) than PSNR. In practice and in literature, both ways of calculating PSNR can be found. For instance, JM outputs the $\overline{\text{PSNR}}$ value. Thus, in this book, the Y-PSNR values averaged per sequence correspond also to $\overline{\text{PSNR}}$ in order to ease the comparison with literature where the JM experiments are presented. Whereas PSNR estimates the user perceptual subjective quality well for packet-loss impairments, it does not necessarily reflect the user evaluation of compression impairments. Nevertheless, it is still the most used distortion metric and therefore also one that allows the results to be compared with other methods without having to implement them.

Data Rate After compression, the video sequence is represented with a corresponding string of binary digits (bits) denoted by $\underline{b}$. The objective is to keep its length b as small as possible. The average bit rate R_b is the average number of bits used per unit of time to represent a continuous medium such as audio or video. It is quantified in bits per second (bit/s)

$$R_b = \frac{b}{T} \text{ [bit/s]}, \tag{5.16}$$

where T is the time needed for the transmission of $\underline{b}$ from sender to receiver. The rate increase ΔR will be used to quantify the change in rate caused by a new method resulting in the rate R_{new} compared to a reference rate R_{ref}

$$\Delta R = \frac{R_{\text{new}} - R_{\text{ref}}}{R_{\text{ref}}} \cdot 100 = \frac{b_{\text{new}} - b_{\text{ref}}}{b_{\text{ref}}} \cdot 100 \text{ [\%]}, \tag{5.17}$$

under the assumption that T has to be the same for the reference bitstream and the bitstream after the application of the new method. However, for some applications as in resource allocation the average bit rate is not as important as the peak bit rate. The frame rate determines the time of the picture rendering. The play-out buffer compensates partially for the variable bit rate.

References

Belfiore, S., Grangetto, M. and Olmo, G. (2003) An Error Concealment Algorithm for Streaming Video. In *Proceedings of IEEE International Conference on Image Processing*, vol. 3, pp. 649–652, Sep.

Bucciol, P., Masala, E. and De Martin, J.J. (2004) Perceptual ARQ for H.264 Video Streaming over 3G Wireless Networks. In *Proceedings of IEEE International Conference on Communications (ICC)*, vol. 3, pp. 1288–1292.

Cen, S., Cosman, P. and Azadegan, F. (2003) Decision Trees for Error Concealment in Video Decoding. *IEEE Trans. on Multimedia*, **5**(1), 1–7, Mar.

Chakareski, J., Han, S. and Girod, B. (2005) Layered coding vs. multiple descriptions for video streaming over multiple paths. *Multimedia Systems*, Springer, online journal publication, DOI: 10.1007/s00530-004-0162-3, 2005.

Chen, M.J., Chen, L.G. and Weng, R.M. (1997) Error Concealment of Lost Motion Vectors with Overlapped Motion Compensation. *IEEE Trans. on Circuits and Systems for Video Technology*, **7**(3), 560–563, Jun.

Chen, J., Lv, T. and Zheng, H. (2005) Joint Cross-Layer Design for Wireless QoS Content Delivery. *EURASIP Journal on Applied Signal Processing*, **2**, 167–182.

Chiani, M. and Martini, M.G. (2006) On Sequential Frame Synchronization in AWGN Channels. *IEEE Trans. on Communications*, **54**(2), 339–348.

Choi, L.U., Ivrlac, M.T., Steinbach, E. and Nossek, J.A. (2005) Bottom-up Approach to Cross-layer Design for Video Transmission over Wireless Channels. In *Proceedings of IEEE 61st Vehicular Technology Conference (VTC)*, vol. 5, pp. 3019–3023.

Chou, P. and Miao, Z. (2006) Rate-distortion Optimized Streaming of Packetized Media. *IEEE Trans. on Multimedia*, **8**(2), 390–404.

Corbera, J.R., Chou, P.A. and Regunathan, S.L. (2003) A Generalized Hypothetical Reference Decoder for H.264/AVC. *IEEE Trans. on Circuits and Systems for Video Technology*, **13**(7), 574–587.

Côté, G., Shirani, S. and Kossentini, F. (2000) Optimal Mode Selection and Synchronization for Robust Video Communications over Error-prone Networks. *IEEE Journal on Selected Areas in Communications*, **18**(6), 952–965.

Farvardin, N. (1990) A Study of Vector Quantization for Noisy Channels. *IEEE Trans. on Information Theory*, **36**(4), 799–809.

Franchi, N., Fumagalli, M., Lancini, R. and Tubaro, S. (2005) Multiple Description Video Coding for Scalable and Robust Transmission Over IP. *IEEE Trans. on Circuits and Systems for Video Technology*, **15**(3), 321–334.

Goyal, V.K. (2001) Multiple Description Coding: Compression Meets the Network. *IEEE Signal Processing Mag.*, **18**(5), 74–93.

He, Z. and Mitra, S.K. (2005) From Rate-distortion Analysis to Resource-distortion Analysis. *IEEE Circuits and Systems Magazine*, **5**(3), 6–18.

Hong, M.C., Kondi, L., Schwab, H. and Katsaggelos, A.K. (1999) Error Concealment Algorithms for Compressed Video. *Signal Processing: Image Communications, special issue on Error Resilient Video*, **14**(6–8), 437–492.

Kalman, M., Ramanathan, P. and Girod, B. (2003) Rate-distortion Optimized Video Streaming with Multiple Deadlines. In *Proc. of IEEE International Conference in Image Processing (ICIP)*, 2003.

Kalman, M. and Girod, B. (2004) Rate-distortion Optimized Video Streaming Using Conditional Packet Delay Distributions. *IEEE Workshop on Multimedia Signal Processing*, 2004.

Klaue, J., Gross, J., Karl, H. and Wolisz, A. (2003) Semantic-aware Link Layer Scheduling of MPEG-4 Video Streams in Wireless Systems. In *Proceedings of Applications and Services in Wireless Networks (AWSN)*, 2003.

Koschan, A. and Abidi, M. (2005) Detection and Classification of Edges in Color Images. *IEEE Signal Processing Magazine*, **22**(1), 64–73, Jan.

Kung, W.Y., Kim, C.S. and Kuo, C.J. (2003) A Spatial-domain Error Concealment Method With Edge Recovery and Selective Directional Interpolation. In *Proceedings of IEEE International Conference on Acoustic, Speech, and Signal Processing (ICASSP)*, Hong Kong.

Kwok, W. and Sun, H. (1993) Multi-directional Interpolation for Spatial Error Concealment. In *Proceedings of IEEE International Conference on Consumer Electronics (ICCE)*, pp. 220–221, Jun.

Lee, C., Kieffer, M. and Duhamel, P. (2004) Robust Reconstruction of Motion Vectors Using Frame Expansion. In *Proceedings of IEEE International Conference on Acoustic, Speech, and Signal Processing (ICASSP)*, pp. 617–620.

Liebl, G., Schierl, T., Wiegand, T. and Stockhammer, T. (2006) Advanced wireless multi-user video streaming using the scalable video coding extensions of H.264/MPEG4-AVC. In *Proceedings of IEEE International Conference on Multimedia and Expo (ICME)*, pp. 625–628.

List, P., Joch, A., Lainema, J., Bjontegaard, F. and Karczewicz, M. (2003) Adaptive Deblocking Filter. *IEEE Trans. on Circuits and Systems for Video Technology*, **13**(7), 614–619.

MacKay, D.J.C. (2006) *Information Theory, Inference, and Learning Algorithms,* 5th edn. Cambridge University Press.

Marpe, D., Wiegand, T. and Sullivan, G.J. (2006) The H.264/MPEG4 Advanced Video Coding Standard and its Application. *IEEE Communications Magazine*, **44**(8), 134–143.

Nemethova, O., Al-Moghrabi, A. and Rupp, M. (2005) Flexible Error Concealment for H.264 based on Directional Interpolation. In *Proceedings of International Conference on Wireless Networks, Communications and Mobile Computing (WirelessCom)*, Maui, Hawaii, pp. 1255–1260, Jun.

Nemethova, O., Al-Moghrabi, A. and Rupp, M. (2006) An Adaptive Error Concealment Mechanism for H.264 Encoded Low-resolution Video Streaming. In *Proceedings of 14th European Signal Processing Conference (EUSIPCO)*, Florence, Italy, Sep.

Nguyen, H. and Duhamel, P. (2005) Robust Source Decoding of Variable-length Encoded Video Data Taking into Account Source Constraints. *IEEE Trans. on Communications*, **53**(7), 1077–1084.

Ostermann, J., Bormans, J., List, P., Marpe, D., Narroschke, M., Pereira, F., Stockhammer, T. and Wedi, T. (2004) Video Coding with H.264/AVC: Tools, Performance and Complexity. *IEEE Magazine on Circuits and Systems*, **4**(1), 7–28.

Pathac, B.H., Childs, G. and Ali, M. (2005) UEP implementation for MPEG-4 video quality improvement on RLC layer of UMTS. *IEE Electronic Letters*, **41**(13), 733–735.

Radulovic, I., Frossard, P. and Verscheure, O. (2004) Adaptive Video Streaming in Lossy Networks: Versions or Layers? In *Proceedings of IEEE International Conference on Multimedia and Expo (ICME)*.

Wenger, S. *et al.* (2005) RTP Payload Format for H.264 Video, IETF RFC 3984, 2005.

Richardson, I.E.G. (2005) *H.264 and MPEG-4 Video Compression, Video Coding for the Next-Generation Multimedia,* John Wiley & Sons, Ltd.

Sayood, K., Otu, H.H. and Demir, N. (2000) Joint Source/Channel Coding for Variable Length Codes. *IEEE Trans. on Communications*, **48**(5), 2000.

Schwarz, H., Marpe, D. and Wiegand, T. (2006) Overview of the scalable H.264/MPEG4-AVC extension. In *Proceedings of IEEE International Conference on Image Processing (ICIP)*, pp. 161–164.

Shannon, C.E. (1948) A mathematical theory of communication. *Bell System Technical Journal*, **27**, 379–423.

Stockhammer, T. and Hannuksela, M.M. (2005) H.264/AVC Video for Wireless Transmission. *IEEE Magazine on Wireless Communications, Special Issue: Advances in Wireless Video*, **12**(4), 6–13.

Su, L., Zhang, Y., Gao, W., Huang, Q. and Lu, Y. (2004) Improved Error Concealment Algorithms based on H.264/AVC Non-normative Decoder. In *Proceedings of IEEE International Conference on Multimedia and Expo*, Taipei, Taiwan, Jun.

Suh, J.W. and Ho, Y.S. (1997) Error Concealment based on Directional Interpolation. *IEEE Trans. on Consumer Electronics*, **43**(3), 295–302, Aug.

Suh, J.W. and Ho, Y.S. (2002) Error Concealment Techniques for Digital TV. *IEEE Trans. on Broadcasting*, **48**(4), 299–306, Dec.

Sun, M.T. and Reibman, A.R. (eds) (2001) Chapter 3: Error Concealment. In *Compressed Video over Networks,* Signal Processing and Communications Series, pp. 217–250, Marcel Dekker Inc., New York.

Takishima, Y., Wada, M. and Murakami, H. (1995) Reversible Variable Length Codes. *IEEE Trans. on Communications*, **42**(2/3/4).

Thomos, N., Argyropoulos, S., Boulgouris, N.V. and Strintzis, M.G. (2005) Error-Resilient Transmission of H.264/AVC Streams using Flexible Macroblock Ordering. In *Proceedings of 2nd European Workshop on the Integration of Knowledge, Semantic, and Digital Media Techniques (EWIMT)*, 2005.

van der Schaar, M. and Shankar, S. (2005) Cross-layer Multimedia Transmission: Challenges, Principles, and New Paradigms. *IEEE Magazine on Wireless Communications, Special Issue: Advances in Wireless Video*, **12**(4), 6–13.

van Dyck, R.E. and Miller, D.J. (1999) Transport of wireless video using separate, concatenated, and joint source-channel coding. *IEEE Proceedings*, **87**(10), 1734–1750.

Wang, Y., Zhu, Q.F. and Shaw, L. (1993) Maximally smooth image recovery in transform coding. *IEEE Trans. on Communications*, **41**(10), 1544–1551, Oct.

Wang, Y.K., Hannuksela, M.M., Varsa, V., Hourunranta, A. and Gabbouj, M. (2002) The Error Concealment Feature in H.26L Test Model. In *Proceedings of IEEE International Conference Image Processing (ICIP)*, New York, USA, Sep.

Wen, J. and Villasenor, J.D. (1998) Reversible Variable Length Codes for Efficient and Robust Image and Video Coding. In *Proceedings of IEEE Data Compression Conference*.

Wenger, S. (2003) H.264/AVC over IP. *IEEE Trans. on Circuits and Systems for Video Technology*, **13**(7), 645–657.

Wiegand, T., Sullivan, G.J., Bjontegaard, G. and Luthra, A. (2003) Overview of the H.264/AVC Video Coding Standard. *IEEE Trans. on Circuits and Systems for Video Technology*, **13**(7), 560–576.

Winkler, S. (2005) *Digital Video Quality,* John Wiley & Sons, Ltd.

Xu, Y. and Zhou, Y. (2004) H.264 Video communication based refined error concealment schemes. *IEEE Trans. on Consumer Electronics*, **50**(4), 1135–1141.

Zhu, Q.F., Wang, Y. and Shaw, L. (1993) Coding and cell-loss recovery in dct-based packet video. *IEEE Trans. on Circuits and Systems for Video Technology*, **3**(3), 248–258, Jun.

Zhu, W., Wang, Y. and Zhu, Q.F. (1998) Second-order derivative-based smoothness measure for error concealment in dct-based codecs. *IEEE Trans. on Circuits and Systems for Video Technology*, **8**(6), 713–718, Oct.

Zimmermann, H. (1980) OSI Reference Model – The ISO Model of Architecture for Open Systems Interconnection. *IEEE Trans. on Communications*, **28**(4), 425–432.

6

Error Detection Mechanisms for Encoded Video Streams

Luca Superiori, Claudio Weidmann and Olivia Nemethova

Typically, error detection capability is provided by a checksum calculated over data partitions (packets). If an error is detected within a packet, the entire packet is discarded, although it may contain useful (non-degraded) information. Checksums can be inserted into the data flow at different protocol layers. For packet video transmission, the Universal Datagram Protocol (UDP) provides such a checksum. The content of the UDP packets thus determines the lost area in the video sequence. Reducing the size of UDP packets decreases the missing picture area, but on the other hand it also increases the necessary rate due to the protocol headers overhead.

This chapter is dedicated to alternative error detection mechanisms that exploit residual and additional redundancy in the received video stream to detect errors inside the damaged video packets. First, a short overview of the entropy coding mechanism in the H.264/AVC baseline profile is given. The syntax of the coding mechanism is then utilized to localize errors within a damaged packet (Superiori *et al.* 2007). This mechanism is later enhanced by additional localization methods based on impairment detection in the pixel domain (Superiori *et al.* 2007b). To limit the size of picture areas affected by a bit error, transmission of resynchronization information is proposed and analysed. To improve error localization possibilities within damaged packets, encoder-assisted methods are investigated.

The final section takes a step beyond simple error detection by considering error correction using sequential decoding (Wicker 1995, Ch. 13). Soft channel outputs are used to compute a metric that is combined with error detection techniques in order to find the most likely syntax-compliant bitstream. Sequential decoding of baseline and extended H.264 will be analysed in some detail (Bergeron and Lamy-Bergot 2004; Weidmann *et al.* 2004; Weidmann and Nemethova 2006).

Video and Multimedia Transmissions over Cellular Networks   Edited by Markus Rupp
© 2009 John Wiley & Sons, Ltd

6.1 Syntax Analysis

Even though one packet fails the checksum test at UDP level, it could still contain valid information that can be used at the decoder side. In order to decode only the correct part of the packet and apply concealment to the damaged one, error detection mechanisms are necessary. In the following, different approaches of error detection will be presented. Since such methods analyse the content of the corrupted packets, they work at 'macroblock level', that is, they mark only the corrupted macroblocks as damaged. The first presented method is the 'syntax analysis' that exploits the expected structure of the bitstream to detect errors in the bit domain.

6.1.1 Structure of VCL NALUs

Before discussing in detail the proposed approaches, the structure of the encoded elements stored in a Video Coding Layer (VCL) Network Abstraction Layer Unit (NALU) will be introduced. A generic VCL NALU is drawn in Figure 6.1.

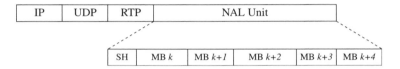

Figure 6.1 Structure of a Video Coding Layer Network Abstraction Unit (VCL NALU).

A Slice Header (SH) is stored at the beginning of the NALU. It contains basic information necessary to interpret correctly the encoded macroblocks contained in the packet. The type of prediction (spatial or temporal), the index of the first macroblock contained in the slice, the index of the Picture Parameter Set (PPS), the slice it refers to, as well as other fundamental parameters are signalized in the slice header. The misinterpretation of the slice header would cause all stored macroblocks to be incorrectly decoded.

The slice header is followed by the encoded macroblocks. The size of each encoded macroblock is not fixed, but it strongly depends on the content of the macroblock in the pixel domain. The better the prediction performs, the smaller is the amount of data that has to be sent. In the case of video streaming over UMTS networks, the NALU is further encapsulated into the Real-time Transport Protocol (RTP), User Datagram Protocol (UDP) and Internet Protocol (IP) headers. This causes an additional overhead of 40 bytes per packet.

H.264/AVC defines two types of entropy coding: Context Adaptive Binary Arithmetic Coding (CABAC) and Context Adaptive Variable Length Coding (CAVLC). In the following we will refer to the latter, since it is the only one supported by the baseline profile of the video codec (TS 26.234 2008). The standard specifies the type of entropy coding to be applied to each information element. CAVLC itself consists of several encoding strategies; in the following the main groups are presented, further details can be found in references H.264 Standard (2002) and Richardson (2003).

- **Tabulated Codewords.** H.264/AVC defines several look-up tables for different encoded elements, where a given value is associated to each codeword. The relation between the codeword and the resulting value is not immediate, but usually depends on other information available at the decoder side. Since the purpose of the entropy encoding is to associate shorter codewords to symbols with higher probability, the tables are designed accordingly, modifying the association map depending on the value of previously decoded parameters. As an example, the number of zero-runs between two non-zero residuals is encoded with respect to the total number of remaining zero residuals.

- **Exp-Golomb codewords.** Exp-Golomb is a class of parametric universal codes. In H.264/AVC the parameter k of the exp-Golomb code is fixed to zero, resulting in a structure similar to the Elias gamma code. Whereas the Elias gamma code can encode values bigger than zeros, the exp-Golomb can encode the value zero as well. The generic exp-Golomb encoded codeword has the following structure:

$$\underbrace{0_1 \ldots 0_M}_{M} 1 \underbrace{b_1 \ldots b_M}_{M} .$$

The length of the codeword is variable and equal to $2M + 1$. The first M zeros and the first one represent the 'prefix' of the codeword. The last M bits are the 'info' field of the codeword. The decoding of an exp-Golomb codeword takes place by obtaining the associated codeNum field:

$$\text{codeNum} = 2^M - 1 + \text{info}. \tag{6.1}$$

From the codeNum, the encoded value is finally reconstructed, depending on the mapping type (unsigned, signed and truncated).

Exp-Golomb encoded codewords are highly sensitive to errors. A single-bit inversion in the first $M + 1$ bits would cause the misinterpretation of the length of the codeword, resulting in an incorrectly decoded value. If this effect can be, possibly, confined to the considered parameter, the misinterpretation of the length of the codeword is a more problematic effect. Once the length of one word is incorrectly determined, the boundaries of all the following codewords up to the end of the slice will also be misinterpreted. In the following, we will refer to this effect as 'desynchronization'.

- **CAVLC level.** The CAVLC is, in a narrower sense, used to encode the transformed quantized residuals. As discussed before, the entropy coding aims to minimize the codeword length depending on the probability of the associated value. Therefore, in H.264/AVC seven different VLC-N routines have been defined. The parameter N is chosen in the range [0,6] depending on the values of the previously decoded residuals. Under the assumption that small values are associated to high-frequency components, the first residual is encoded using the VLC-0 routine. A VLC-0 codeword has the following structure:

$$\underbrace{0_1 \ldots 0_M}_{M} 1.$$

Table 6.1 Variable Length Code-N routines.

VLC-N	Threshold
0	0
1	3
2	6
3	12
4	24
5	48
6	NA

The parameter M contains both the value as well as the sign of the encoded value. For the following residuals, an appropriate VLC-N routine is chosen depending on whether the previous decoded element l_{i-1} is bigger than a given threshold, as described in Table 6.1.

The codewords have the following generic structure:

$$\underbrace{0_1 \ldots 0_M}_{M} 1 \underbrace{i_1 \ldots i_{N-1}}_{N-1} s.$$

Similarly to the exp-Golomb encoded words, the first $M + 1$ bits represent the prefix of the codeword. The info field contains $N - 1$ bits. An additional bit s is used to signalize the sign of the value.

The first $M + 1$ bits of the CAVLC words are highly sensitive to errors since, as for the exp-Golomb codewords, a single-bit inversion leads to the misinterpretation of the length of the prefix, causing desynchronization. Errors in the info field cause the incorrect decoding of the stored value. This will certainly affect the decoded picture in the pixel domain and, possibly, cause the choice of the wrong VLC-N routine, as described in Table 6.1. An error in the sign has effects only in the pixel domain.

6.1.2 Rules of Syntax Analysis

After introducing the encoding strategies defined by the baseline profile of the H.264/AVC, the proposed error handling mechanism will be discussed in this section. It has been described how the errors only affect the decoding of the current and, possibly, following codewords. The information elements preceding the error occurrence can be correctly decoded. Since the position of the error is not known, the definition of efficient error detection mechanisms is a task of major importance. In the following, an error detection mechanism based on the code syntax analysis will be presented. A similar method for the H.263 codec was proposed in Barni *et al.* (2000). However, H.263 does not make an extensive use of variable length coding as H.264/AVC does and, moreover, allows for resynchronization words between Groups Of Blocks (GOBs). Our proposal has been implemented in the standard development software JM version 10.2 (JM) and the analysed functionalities can be described as follows:

1. Evaluate the value obtained when reading a codeword.

2. Analyse how that value influences the decoding of the macroblock.

This conceptual distinction is consistent with the two different logical functions defined in the JM: READ and DECODE one macroblock, as indicated in Figure 6.2.

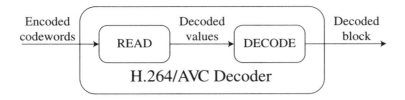

Figure 6.2 H.264/AVC decoder blocks.

In the first function, READ, each codeword is interpreted, obtaining the encoded value. In the second one, DECODE, such value is then used to reconstruct the picture in the pixel domain, applying, for example, motion compensation or correction by means of residuals.

Note that the original JM software is not able to handle errors in the bitstream, resulting in crashes of the decoder. Basically, such crashes are due to unexpected codewords or values causing exception in the reading or decoding phase. Observing the behaviour of the decoder in case of error, different typologies of exception have been defined:

- Out of Range codewords (ORs). OR occurs when the value associated to the decoded codeword lies outside an admitted range. This may happen, for example, when reading the number of skipped macroblocks. In case of an error this may lead the index of the next decoded macroblock to exceed the number of macroblocks contained in the picture.

- Illegal Codewords (ICs). ICs occur when the codewords do not find any correspondence in the appropriate look-up table. This may happen because the look-up tables can be incomplete and contain only a subset of the possible codewords with a given length.

- Contextual Errors (CEs). This kind of error arises when the value obtained in the READ phase leads the decoder to perform forbidden actions. For example, one-bit inversion may change the codeword associated to the intra-prediction mode of the first macroblock of a slice. At bitstream level this error cannot be detected. However, when applying the required prediction mode in the DECODE phase, no prediction mode different from DC (the average value of the block is predicted without referencing any neighbour) is allowed. Therefore, when obtaining a prediction mode different from DC, an exception will be raised and the error is signalized to the concealment mechanism.

These kinds of error can be tracked during the decoding, and the crashes of the decoder can be avoided, setting appropriate exception handling strategies. However, these errors may not be caused directly by bit inversions affecting the currently read codeword, but, rather, be caused by desynchronization due to errors occurring earlier. The distance between the error occurrence and the error detection will be defined as 'detection distance'. This distance can be expressed in bits, but it is more meaningful to express it in terms of macroblocks.

The information encoded between the error occurrence and the error detection will be incorrectly decoded, resulting in impairment in the reconstructed picture. The detection distance, therefore, has to be kept as small as possible. In order to fulfil this requirement, all the information elements of the encoded stream have been analysed and, for each, the typology of error that can arise has been identified. The names of the parameters are those used in the standard development software; they match, however, the parameter defined in the standard.

I Frames

Parameter Name	Enc.	Err.
mb_type	*EG*	*OR*
intra4x4_pred_mode	*TE*	*CE*

Since the spatial prediction uses reference to the surrounding macroblocks, if they are not available, not yet decoded or belonging to another slice, a contextual error is produced.

intra_chroma_pred_mode	*EG*	*OR*
coded_block_pattern	*EG*	*OR*
mb_qp_delta	*EG*	*OR*
Luma(Chroma) # c & tr.1s	*TE*	*IC*

The look-up table used to decode this value is not complete. The decoded codeword could not find reference to any legal value.

Luma(Chroma) trailing ones sign	*EG*	

The signs of the trailing ones are fixed length encoded and do not influence any of the following parameters. By means of syntax check it is not possible to detect such errors.

Luma(Chroma) lev	*VL*	*OR/CE*

Decoded macroblock pixels can only take values lying in the range [0,255]. During the READ phase, values outside the bounds are associated to errors. During the DECODE phase, the residuals are added to the predicted values and the contextual check is performed. An extended range $[-\lambda, 255 + \lambda]$ is considered due to possible quantization offsets.

Luma(Chroma) totalrun	*TE*	*IC*
Luma(Chroma) run	*TE*	*IC/OR*

Depending on the number of remaining zeros, a Variable Length Code (VLC) look-up table is chosen. For more than six remaining zeros a single table covering the zero run range [0,14] is used. Therefore, the decoder is exposed to out-of-range errors.

P Frame

Many of the parameters encoded in P frames are equivalent to those used to describe an I frame. In the following, only the parameters specific for Intra encoding are discussed.

mb_skip_run	*EG*	*OR/CE*

    The number of skipped macroblocks cannot be greater than the number of not-yet-decoded MBs belonging to the current frame.

`sub_mb_type`	*EG*	*OR*
`ref_idx_l0`	*EG*	*OR/CE*

    The index of the reference frame cannot be greater than the actual reference buffer size.

`mvd_l0`	*EG*	*CE*

Slice Header

`first_mb_in_slice`	*EG*	*OR*
`pic_parameter_set_id`	*EG*	*OR/CE*

    The VCL-NALU cannot reference a PPS index greater than the number of available PPSs.

`slice_type`	*EG*	*OR*
`frame_num`	*EG*	*OR*

    Depending on the GOP structure an out-of-range error can be detected.

`pic_order_cnt_lsb`	*EG*	*OR*
`slice_qp_delta`	*EG*	*OR*

6.1.3 Error-handling Mechanism

Once the rules of the syntax analysis have been discussed, the three compared handling mechanisms will be described. The macroblocks marked as corrupted will be concealed by means of a simple copy-paste mechanism. Assuming that the macroblock in row i and column j of frame f, $MB_f(i, j)$, is labelled as corrupted, it will be replaced by the macroblock $MB_{f-1}(i, j)$ occupying the same position in the previous picture.

1. **Straight Decoding (SD).** This approach consists of the decoding of the corrupted bitstream by means of a modified H.264/AVC decoder. Without any error detection mechanism, in case one of the errors described in section 6.1.2 is detected, the decoder replaces the invalid value with the next closest that is valid. Since no error detection is performed, the macroblocks preceding the error occurrence are correctly decoded, whereas the following ones are incorrectly decoded.

2. **Slice Level Concealment (SLC).** This approach represents the standard error-handling mechanism currently considered in the literature. In case one packet fails the UDP checksum test, it is discarded. All the macroblocks contained in the NALU are, thus, marked as erroneous and concealed. Therefore, the macroblocks preceding the error are concealed, even if this is unnecessary.

3. **MacroBlock Level Concealment (MBLC).** This handling mechanism exploits the syntax analysis described before. Even though the packet fails the UDP checksum test, it is decoded. As soon as one of the errors described in section 6.1.2 is detected, the current MB as well as all the following ones until the end of the slice are marked as erroneous. As discussed before, the MB in which the error is detected is not necessarily the one where the error occurred. This causes the MBs between the error occurrence

and the error detection to be, possibly, incorrectly decoded. However, the information preceding the error detection will be correctly exploited and the MB after the error detection marked as corrupted and concealed. The proposed method represents a smart compromise between the two considered approaches mentioned before.

In Figure 6.3, the decoding of a picture using the three proposed approaches is discussed. The considered picture is an intra-predicted frame, consisting of five slices. An error has been introduced in the second slice that contains the MBs with index from 22 to 38. The error will therefore propagate spatially until the end of the second slice, namely up to the MB 38.

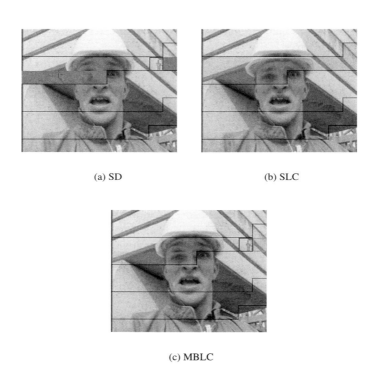

(a) SD   (b) SLC

(c) MBLC

Figure 6.3 Decoding of a corrupted frame using the different error-handling strategies.

In Figure 6.3(a) the result of the decoding of the corrupted packet using the straight decoding approach is shown. Notice that, beginning from MB 32, the impairments propagate until the end of the slice. The MBs before the error occurrence are correctly decoded.

Figure 6.3(b) shows the result of the decoding when using the slice level concealment approach. Since the packet failed the UDP checksum test, the whole slice is concealed. The MBs from 22 to 31, even if correctly decoded, have been concealed.

The result of the proposed method is shown in Figure 6.3(c). The MBs from 22 to 31 are correctly decoded. The error is detected in MB 33, therefore the MBs from 33 to 38 are concealed. In this example, a detection distance of one MB has been measured.

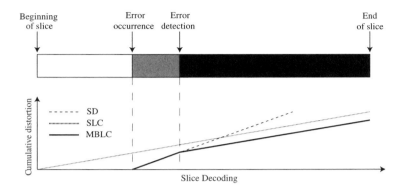

Figure 6.4 Cumulative distortion using the three considered approaches.

Figure 6.4 shows a qualitative comparison between the three proposed methods. The cumulative distortion between the slice decoded without errors and the one obtained handling the corrupted slice with the three considered mechanisms is displayed. Notice that, using the SD, the distortion starts to increase as soon as the error occurs. Using the SLC, being the concealment applied to the whole slice, a lighter grade of distortion is introduced all over the slice. Using the proposed approach, the MBLC, the distortion caused by desynchronized decoding is limited to the MBs between the error occurrence and error detection. The distortion introduced by the concealment is spread from the error detection until the end of the slice. The results of the proposed method depend strongly on the detection distance and on the position where the error occurs. For errors detected near the beginning of the slice, the MBLC acts as the SLC. Increasing detection distances affect the performance of the proposed method.

6.1.4 Simulation Setup

In order to evaluate the performance of the three considered methods, the decoding of different corrupted sequences has been simulated. The sequences were encoded using the standard development encoder without any modification. Depending on the handling method considered, different features in the JM decoder were enabled. In the case of slice level concealment, once an error has been detected within the packet, all the MBs stored in the NALU are marked as erroneous. When considering the straight decoding, all the incorrect codewords or values are turned to valid ones. For the MBLC, during the decoding, different flags are raised in case invalid codewords or values have been detected. Once one flag has been raised, the decoder stops reading and decoding the MB and conceals it. The flags are turned back to zero at the beginning of the following slice.

The sequence used for the simulation is the 'foreman'. It consists of 400 frames played at 30 frames/s in Quarter Common Intermediate Format (QCIF) resolution (176×144 pixels). The GOP size was set to 10 and different Quantization Parameters (QPs) have been investigated.

6.1.5   Subjective Quality Comparison

The first analysis performed regards the resulting quality in terms of Luminance Peak Signal to Noise Ratio (Y-PSNR). For a set of sequences obtained by encoding the foreman video with different quantization parameters, the performances of the three methods have been compared for error patterns characterized by different Bit Error Ratios (BER). Lower BERs have lower probability of errors, therefore the number of simulations increases with diminishing BERs. The graphs in Figure 6.5 show the resulting objective quality. The four lines drawn represent, respectively, the quality when considering the error-free decoding, therefore not depending on the BER and considered as the reference quality, the MBLC, the SLC and the SD.

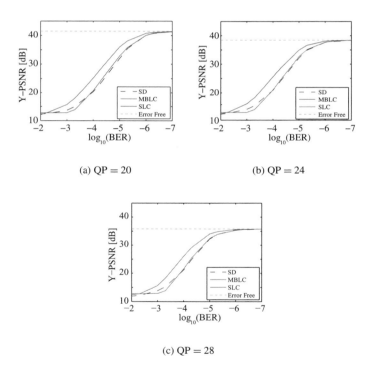

Figure 6.5 Performance of the different error handling strategies.

Notice that the proposed method performs better than the standard handling mechanism, particularly in the range of BERs from 10^{-4} to 10^{-6}. For lower BERs, the three mechanisms are close to the error-free case, since the number of errors introduced is not relevant. For higher BERs, such as 10^{-2}, the errors are introduced so frequently that the overall quality of the video is unsatisfactory independently from the chosen handling mechanism.

6.1.6 Detection Performance

As discussed, the detection probability as well as the detection distance represent key parameters for the performance of the proposed method. Since the desynchronized decoding remains limited to the single NALU, new simulations were performed introducing one error each slice. It should, therefore, be underlined that the focus of these simulations is on the detection performances and not on the quality that would also depend on the temporal propagation of the resulting errors.

The simulations were performed on a sequence encoded with a quantization parameter of 28 and NALU size limited to 700 bytes. These encoding parameters lead the I frames to be segmented into more than four slices, whereas the P frames still usually consist of a single NALU.

The probability of detecting an error in the I frames is around 60%, whereas for the P frames it is around 47%. This difference can be explained considering the information elements encoded in the two different encoded slices. I frames are self-contained, that is, the encoded information is sufficient to reconstruct the picture without referencing any other previous decoded picture. This results in a less effective prediction, the spatial one, and in a set of coefficients with higher information content. Such coefficients are much more sensitive to bit inversions or to desynchronized decoding. The packets containing encoded P frames, on the contrary, contain the information for reconstructing the picture applying motion compensation to the previously decoded pictures. Most of the information is contained in the motion vectors, describing the position of the best prediction of the considered block in the reference picture. Errors affecting the motion vectors are usually not detectable and would cause the selection of the wrong prediction block.

Figure 6.6 shows the detection distance, expressed in number of macroblocks, for the two types of predicted frame. Notice that more than 90% of the detected errors in I frames are detected within two MBs. For the P frames, this detection increases and, on average, the detection occurs within seven MBs. The detection distance for the P frames is higher because of a specific element encoded in the P frames, namely the mb_skip_run. It signalizes how many MBs have to be skipped, that is, how many MBs can be reconstructed without further encoded information. Errors affecting this parameter will modify the number of skipped MBs. Since, in most of the cases, the encoded value is zero, an error will increase the number of skipped MBs and, therefore, the detection distance.

For the undetected errors, a similar investigation was performed. In this case, the distance between the error occurrence and the end of the slice was measured. The histogram of the distribution is plotted in Figure 6.7. For I frames, 50% of the undetected errors are located in the last two MBs of the slice. Also, for the P frames a peak is recognized for the short distance between the error occurrence and the end of the slice. This shows how a considerable number of errors cannot be detected because the number of MBs to be decoded is smaller than the average detection distance.

In order to evaluate the influence of the undetected errors, their impact in terms of objective distortion was measured. In Figure 6.8, the average distortion introduced by the

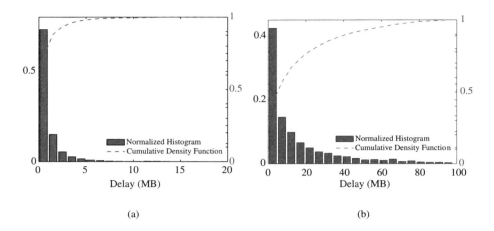

Figure 6.6 Detected errors: detection distance, (a) I frames; (b) P frames.

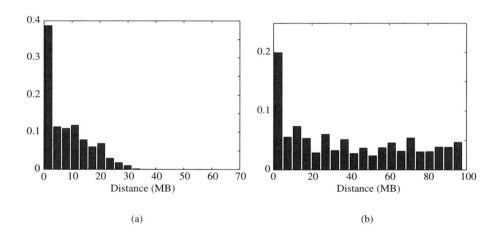

Figure 6.7   Undetected errors: distance between error appearance and end of slice, (a) I frames; (b) P frames.

missed detection is drawn as a function of the distance between the error occurrence and the end of the slice. The distortion is measured as the Mean Square Error (MSE) between the MBs affected by decoding desynchronization and the same MBs reconstructed in an error-free environment. Small MSE values signalize that, even if the error has not been detected, the effects did not impair the decoded picture. Observe that, on average, the MSE is higher for small distances between the error occurrence and the end of the slice. These high distortion values are caused by the errors that were not detected for being too close to the end of

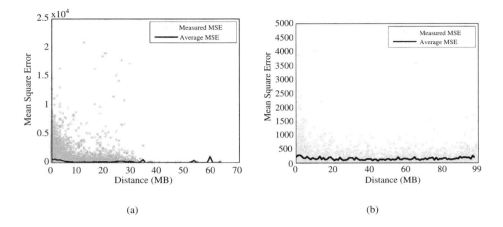

Figure 6.8 Undetected errors: resulting MSE, (a) I frames; (b) P frames.

the slice. Since the MSE is calculated over the whole area possibly affected by decoding desynchronization, the expected behaviour would have been characterized by an increasing MSE for an increasing size of the area. However, the performed simulations demonstrate that the resulting MSE does not depend on the number of affected MBs. This leads to the conclusion that, on average, the undetected errors do not significantly affect the decoded picture.

6.2   Pixel-domain Impairment Detection

As shown in section 6.1.6, the syntax analysis still suffers from a detection distance, where the encoded information is incorrectly decoded. This results in visual impairments in the pixel domain. The detection of such visual artefacts can further improve the performance of the syntax analysis, reducing the detection distance. However, the detection of impairments in the pixel domain calls for refined preprocessing techniques.

An analysis over video sequences obtained by decoding corrupted bitstreams has been performed. Notice that the effect of desynchronization is significantly different depending on the type of frame, Inter or Intra.

6.2.1   Impairments in the Inter Frames

The Inter encoded frames exploit the temporal correlation between consecutive frames. The encoded picture is reconstructed applying motion compensation to its MBs, or its Sub-MBs (SMBs). The differences between the current image and the available reference pictures are compensated by means of transformed quantized residuals. However, there is only a small dependency between the encoded information belonging to consecutive MBs of the same picture slice. This means that, in case of error, there is no significant spatial propagation of artefacts due to the wrong reconstruction of an MB. However, the desynchronization still remains a major drawback. As introduced in section 6.1.6, the desynchronization of the decoding in the inter frames occurs more rarely, if compared with the intra frames. Errors in the motion vectors, moreover, can rarely be detected by syntax check. The field `mb_skip_run` signalizes the length of the run of skipped macroblocks, and it is encoded by means of exp-Golomb coding, as described in section 6.1.1. Since, usually, one macroblock is not skipped, the associated codeword is `1`, indicating a zero run length. In case of desynchronization, the `mb_skip_run` can turn to `0`; this would be the first bit of the prefix of a value greater than zero. This causes, in the case of desynchronization, an overestimation of the number of MBs to be skipped. As a result, the impairments result in being isolated and spatially interleaved between skipped MBs.

When detecting errors in frame n, we assume that frame $n - 1$ is correct. To detect errors in P frames, we analyse the pixel-wise absolute difference map D_n between frame n and frame $n - 1$:

$$D_n(i, j) = |\mathbf{F}_n(i, j) - \mathbf{F}_{n-1}(i, j)|. \tag{6.2}$$

Since we aim to detect artefacts with the resolution of an MB, the difference map D_n is then reshaped to consider the average difference in the 16×16 pixels. However, the difference map is not only dependent on possible visual artefacts, but also on the movement between the two consecutive pictures. In order to handle this, we observed how the isolated artefacts represent out-of-context square collections of pixels. We therefore propose to use a simple edge detector to highlight the edginess in the picture. Notice that better results are obtained when observing the edge characteristics of 8×8 pixels SMBs.

The final decision whether or not one MB is detected as erroneous is then taken considering the information about both the difference map D_n as well as the edginess map E_n. They are both compared with an adaptive threshold considering the movement characteristic of the whole picture.

6.2.2   Impairments in the Intra Frames

In the Intra predicted frames the correlation between neighbouring MBs belonging to the same picture is exploited. Each block is predicted considering the luminance and chrominance component of one of the confining already encoded blocks. At the decoder side, the quality of a single block strongly depends on the correctness of the neighbouring ones. The spatial propagation of the error in I frames may also occur in case no desynchronized decoding takes place. As an example, assume the codeword of a VLC residual to affect a bit inversion in the info field, therefore causing no decoding desynchronization. The considered block will, however, be reconstructed differently from the one available at the encoder. The following blocks using that MB as a reference will suffer from spatial error propagation, since the obtained prediction will not be consistent with the one considered at the encoder side.

Therefore, as shown in Figure 6.9(a), one error in the I frames usually corrupts the affected MBs as well as the following ones until the end of the slice. The different behaviours observed in the two kinds of frame call for the design of two different detection mechanisms.

To ensure robustness in the detection mechanism, the decision is taken using a voting system. Similarly to the detection performed in the inter frames, the inputs to the voting system are the block difference map D_n and the edge map E_n. Since the detection

(a)   (b)

Figure 6.9 Visual impairments caused by undetected errors, (a) I frames; (b) P frames.

performance of the syntax analysis in the intra frames was significantly better, the error position as detected by the syntax analysis is considered also.

A voting procedure is initialized each time the difference and edginess value of a block overcomes a given threshold. This block is considered as the root of the artefacts sequence. The following blocks are further investigated and, in case their difference and edginess characteristics are compatible with those of an artefact, the vote is increased; otherwise it is decreased. One sequence of possible artefacts is terminated in the following cases:

1. The vote of the sequence overcomes a given threshold. In that case, the root of the sequence is assumed to be the MB in which the error occurred. The following MBs until the end of the slice are marked as corrupted.

2. The vote of the sequence goes below a given threshold. In that case, the sequence is handled as a false positive. The detection is restarted and a new possible root is searched in the following MBs.

3. The syntax analysis signalizes that an error was detected in the current MB. The root of the sequence is considered as the MB in which the error occurred. The following MBs until the end of the slice are marked as corrupted.

4. The end of the slice has been reached, and none of the previous condition was fulfilled. In this case the current vote is compared with a given threshold. In case the current vote overcomes it, the root is considered as the MB affected by the error. Otherwise, the whole sequence is treated as a false positive.

All the considered thresholds, as discussed for the intra predicted frames, are adaptive and depend on the movement characteristic of the sequence.

6.2.3 Performance Results

In order to measure the detection performance, the same simulation setup as described in section 6.1.6 was used. The results are discussed separately for intra and for inter encoded frames.

1. **Inter encoded frames.** The syntax analysis suffered from a significant detection distance. This caused the detected errors to be spotted after more than seven MBs on average. Performing the detection of visual impairments, such distance was slightly reduced to 6.8 MB. The average motion compensation measured in the neighbouring MBs was applied to the skipped MBs; therefore, they do not result in out-of-context MBs.

2. **Intra encoded frames.** Even though the performance of the syntax analysis was satisfactory, by means of visual artefact detection they have been further improved. The average detection distance, in particular, was reduced from 1.39 MB to 0.92 MB. Also, the detection probability was increased from 54.35% to 59.99%.

Although the detection probability does not exceed 60%, note that, usually, the errors that do not cause desynchronization do not produce any visible artefacts on the decoded picture. If we consider, for example, bit inversions in the trailing ones, they cannot be detected by means of syntax analysis, since the sign of the trailing ones is fixed length encoded and, moreover, does not influence any following parameter. Also, in the pixel domain, an error in the trailing ones can be barely spotted, since it would influence only high-frequency components. It remains questionable whether the detection of such errors would influence positively the resulting quality. The following MBs, in fact, can be correctly decoded and possible drifts in the spatial prediction would result in negligible distortion. Marking these MBs as corrupted would cause the concealment not to exploit the available valid information.

6.3   Fragile Watermarking

An alternative to the previously described bitstream domain redundancy is 'Watermarking' (WM). WM inherits its name from watermarking of paper or money as a security measure. It is a technique which allows for adding a hidden verification message in (in this context) video images. Such a message is a group of bits that is known and can be verified at the decoder. WM is typically used for authentication and security applications; it can be 'robust' or 'fragile' against attacks, such as changes of video content.

WM can also be used advantageously for error detection. A WM is inserted in each picture region (that is, M (S)MBs) in which the presence of an error should be detected. After transmission over an error-prone channel, the watermark is checked at the receiver; if it is distorted, an error must have occurred. However, insertion of a WM deteriorates the quality already at the transmitter. This can be understood as an implicit rate increase: in order to obtain the quality of a watermarked image equal to the quality of the same non-watermarked image, the rate has to be increased. Nevertheless, by inserting watermarks in a video transmitted over an error prone channel, it is expected that, after decoding, the quality of the watermarked video will be higher. Clearly, such a method is only beneficial for channels with an error rate higher than a certain threshold.

Watermarking is usually applied to the image either before or after transformation and quantization. Inserting of WMs before transformation and quantization would require a semi-fragile watermarking, robust against transformation and quantization but fragile against errors. Such watermarking at MB level is difficult to design (Cox *et al.* 1997). Insufficient robustness would affect especially videos encoded with higher QPs and, generally, would

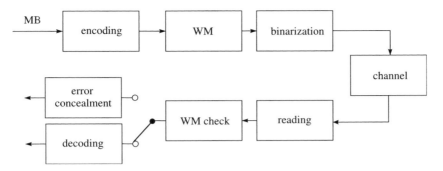

Figure 6.10 Functional scheme of fragile watermarking for error detection.

lead to false detections. Therefore, fragile watermarking inserted after the transformation is considered as more suitable for the purpose of error detection. Figure 6.10 depicts a generic block scheme of such an error detection mechanism. After transformation and quantization, each MB (or an M-MB region) is watermarked, binarized (entropy encoded and encapsulated in packets) and transmitted over a channel. After reading of the stream, the watermark is extracted and checked. If an error occurs, error concealment is applied. If not, the MB is decoded in the usual way. The choice of watermarking method determines the probability of detection as well as the rate increase, as exemplified in the following.

Force Even Watermarking (FEW)

The authors Chen *et al.* (2005) were the first to propose watermarking as a means for error detection. They chose 'force even watermarking' and inserted it in each $N \times N$ pixel large H.263 block containing Discrete Cosine Transform (DCT) coefficients a_j, $j \in [1, N^2]$ after quantization (the experiments in Chen *et al.* (2005) were performed for $N = 8$). The WM is not necessarily inserted in the entire bandwidth. The DC coefficient is never changed, the AC coefficients are modified from the position p on, in the order of the zigzag scan (see, for example, Figure 5.3 of section 5.1.2). Each a_i with $i = p \ldots N^2$ is transformed into a watermarked coefficient $a_i^{(w)}$ according to the following relation

$$a_i^{(w)} = \begin{cases} a_i, & |a_i| \bmod 2 = 0 \\ a_i - \mathrm{sign}(a_i), & |a_i| \bmod 2 = 1, \end{cases} \tag{6.3}$$

where mod 2 denotes the modulo 2 operation and

$$\mathrm{sign}(a_i) = \begin{cases} 1, & a_i \geq 0 \\ -1, & a_i < 0. \end{cases} \tag{6.4}$$

In summary, all odd coefficients are set to their lower even absolute values. Note that the coefficients with value ± 1 become zero, allowing for higher entropy coding gain. Thus, the rate of the watermarked sequence will typically be lower than the rate of the original.

In H.264/AVC the DCT transform is applied to blocks of size 4×4. In order to test FEW for H.264/AVC, it was implemented into the JM and applied to these blocks. The probability of detection using FEW strongly depends on the start position p and on the content of the video sequence since, due to efficient prediction mechanism, most of the coefficients (especially those at higher frequencies) in the blocks are typically zero. The probability mass function (pmf) of the coefficient values is shown in Figure 6.11 for QP $= 30$. Corresponding to the results in Bellifemine *et al.* (1992), the pmf follows a zero-mean Laplace distribution $f(x) = \alpha \exp(-2\alpha|x|)$ with parameter $\alpha > 0$ depending on the value of QP. For QP $= 30$, the value of $\alpha = 0.5$ was obtained by least-squares fitting. As zeros are encoded in a run length manner, no watermarking can be applied in such a case.

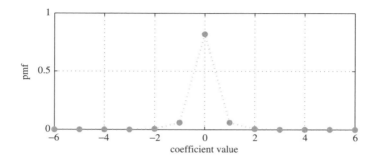

Figure 6.11 Distribution of the quantized transformation coefficients values for QP $= 30$ and the QVGA 'carphone' video sequence.

In Chen *et al.* (2005) it was already shown that FEW offers a significant improvement over the simple syntax check mechanism. However, its detection possibilities are limited, too. If an error occurs resulting in an even coefficient, it remains undetected. Moreover, FEW allows the distortion to be controlled only by means of changing the start position p. The influence of parameter p on the probability of detection for different QPs is illustrated in Figure 6.12 for the 'carphone' video sequence and BER $= 10^{-6}$. Here, the errors detected within the same MB are considered as detected (even if WM was applied to the SMBs). An unpleasant effect of FEW is the decrease of detection probability with an increasing QP. In general, changes to coefficient values in a picture encoded with higher QP have a greater impact on the quality. Syntax analysis had a negligibly small probability of error detected within the same MB; the corresponding detection probability achieved by impairment detection lies between 30 and 40% for QP $= 28$. Hence, FEW with $p < 7$ clearly outperforms the impairment detection in the sense of detection probability. Note that like the syntax analysis, there are no false detections possible with fragile watermarking. The price paid for the enhanced detection probability is the distortion at the encoder that can be equivalently represented as a rate increase necessary to achieve the original (without WM) quality. For instance, the degradation at the encoder is about 1 dB of Y-PSNR for QP of 28, $p = 7$ and the 'carphone' video sequence.

Whereas the usage of higher values of p is feasible for lower QPs, the typical video streaming scenarios (QP in the interval of 26–30) require $p < 7$. Finally, Table 6.5 illustrates the rate reduction caused by FEW. There is also a possibility of defining an additional stop position, not handled here, since the later coefficients are with high probability zero

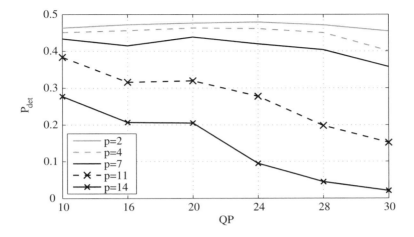

Figure 6.12 Probability of error detection for various values of p and QP.

especially for higher QPs. The lower is p, the higher is the distortion at the encoder. From the point of view of the error detection probability, however, inserting the watermark in the significant regions (low frequencies) is beneficial. In the following section a novel scalable fragile watermarking approach is proposed to improve the error detection probability and to allow for better control of the distortion.

Table 6.5 Average bit rate reduction in % after applying FEW to the 'carphone' video.

QP	$p = 2$	$p = 4$	$p = 7$	$p = 11$	$p = 14$
10	30.13	27.26	22.08	14.40	7.49
16	42.45	37.32	20.82	16.99	8.01
20	65.92	31.78	20.82	11.40	4.85
24	30.75	24.54	17.17	9.02	3.49
28	31.97	24.10	16.41	7.69	2.73
30	31.48	23.15	15.05	6.94	2.01

Relation Based Watermarking

The main problem with FEW is its low scalability. Setting all odd coefficients to even has an effect similar to the effect of higher QP.

This problem is solved in Nemethova *et al.* (2006) by introducing 'relation based watermarking' (RBW). This fragile watermarking method in general modifies n chosen DCT coefficients $\underline{c} = (c_1, c_2, \ldots, c_n)$ to $\underline{\tilde{c}} = (\tilde{c}_1, \tilde{c}_2, \ldots, \tilde{c}_n)$ so that $\underline{\tilde{c}}$ fulfils the predefined equation of the extracting function $f(.)$:

$$f(c_0, \tilde{c}_1, \tilde{c}_2, \ldots, \tilde{c}_n) = K,\qquad(6.5)$$

where $K \in \mathbb{R}$ is a constant known to both encoder and decoder. Note that c_0 is the DC component that is used in the extracting function. It remains unchanged to avoid the high distortion its change may cause. It is assumed that equation 6.5 has more than one solution. From the possible solutions for $\tilde{c}$ the one that minimizes the distortion is chosen

$$\tilde{c} = \arg \min_{\tilde{c}} \|IDCT(\tilde{c}) - IDCT(c)\|, \tag{6.6}$$

where IDCT denotes the operation of re-scaling and inverse DCT. Selection of the extracting function $f(\cdot)$ determines the resulting distortion and detection probability. In order to minimize the distortion, a function is required that provides many solutions near to the values of the original coefficients. The number n of the coefficients that can be modified to match the desired extracting function influences the minimum distortion also. The more coefficients can be modified, the higher is the chance to achieve it with lower distortion. On the other hand, a higher n also means higher computational complexity. The n coefficients might be chosen as an arbitrary subset of the (S)MB coefficients. Nevertheless, it makes sense to choose the low-frequency coefficients, since the probability that they are zero is lower. By modifying the zero-valued coefficients, the entropy coding process changes and possibly results in a higher required rate. The influence of n on the error detection probability is not easy to recognize. Although higher n means more coefficients involved in the calculation – thus higher error detection probability – it means, also, more ways in which equation 6.5 can be fulfilled in spite of an error, thus reduced fragility. Fragility directly influences the probability of error detection. Functions that change their output value by a single error are necessary, being robust against multiple errors. To meet all these criteria a modulo M operation (mod M) over the sum of the selected coefficients was chosen:

$$f(c_0, \tilde{c}_1, \tilde{c}_2, \ldots, \tilde{c}_n) = \left(c_0 + \sum_{i=1}^{n} \tilde{c}_i \right) \mod M = K. \tag{6.7}$$

Increasing M will improve the fragility of the scheme against multiple errors but on the other hand it will increase the distortion. The choice of K has no influence on the performance.

Apart from fragility, error detection probability, rate and distortion, computational complexity has to be taken into account when deciding what n and M to choose for a real application. The main source of complexity in the RBW is the RD-optimized WM embedding. The number of WM embedding possibilities $\mathcal{O}_{n,M}$ is given by

$$\mathcal{O}_{n,M} = \binom{n + M - 2}{M - 1} = \frac{(n + M - 2)!}{(M - 1)!(n - 1)!}. \tag{6.8}$$

In order to choose an RD-optimized solution, $\mathcal{O}_{n,M}$ decodings of each block are necessary. For example, for $n = 4$ and $M = 4$, 20 encodings are required. This can be reduced by excluding some possibilities that are likely to bring high distortion (for example, modifying one coefficient only by a high value). Note that for streaming applications, the complexity of WM embedding is not necessarily critical as far as the decoder remains simple. The decoder using RWB only requires calculating equation 6.7 and checking if the resulting value equals K. The value of K is fixed and does not need to be signalled.

In order to evaluate the influence of the parameters n and M as well as QP and BER on the probability of error detection, several experiments were conducted. The video sequence 'carphone' in QVGA resolution was chosen again to enable a fair comparison with FEW. Figure 6.13 shows the probability of error detection for BER $= 1 \times 10^{-6}$ and various values of n, M and QP. The results show a very pleasant feature of RBW – the probability of correct detection increases with increasing QP. This is mainly caused by a lower number of possibilities such that an error results in fulfilling the embedding equation. The absolute value of the detection probability is also considerably higher than for FEW.

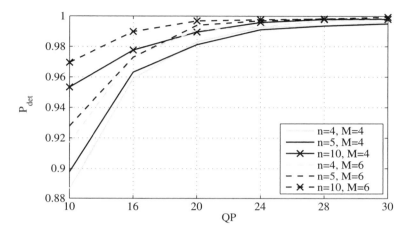

Figure 6.13 Probability of error detection for various values of n, M and QP.

The distortion at the encoder side, caused by the RBW with $M = 4$ and various values of n, corresponds approximately to the distortion caused by FEW with $p = 7$.

Note that for all these experiments, the RBW was only embedded in one SMB of the MB. Of course, the embedding with a higher n and over more SMBs could provide better results in terms of distortion at the encoder. However, it would also be connected with a higher embedding complexity. The consequences of RBW embedding on the resulting rate can be seen in Table 6.6. As expected, for some cases, especially for higher QPs, the rate increase is considerable. It is caused by forcing different and sometimes higher values to the DCT coefficients. For those cases, MB-based checksums may provide a competitive gain. For a fair comparison in which we consider the rate increase caused by using RBW resp. rate decrease using FEW, we show in Figure 6.14 the Y-PSNR at the decoder for BER $= 10^{-6}$ in relation to the size of the compressed and watermarked stream and the Y-PSNR at the decoder for BER $= 10^{-6}$. To visualize the picture areas containing undetected errors that would lead to a decoder crash we replaced them by a monotone green colour. If an error is detected, error concealment is applied from that position until the end of the slice. Nevertheless, RBW brings between 2 and 15 dB gain against FEW. The gain is slightly higher for higher error probabilities and is expected to decrease for decreasing BER. This gain is given by the higher error detection probability of RBW. For example, when encoding with QP $= 20$ we obtained an error detection probability of 98.2% for BER $= 1 \times 10^{-6}$, 98.3% for BER $= 1 \times 10^{-5}$ and 98.6% for BER $= 1 \times 10^{-4}$ with RBW ($M = 4$, $n = 4$). With FEW ($p = 4$) the probabilities

Table 6.6 Average bit rate increase in per cent after applying RBW to the 'carphone' video.

| | $M = 4$ | | | $M = 6$ | | |
QP	$n = 4$	$n = 5$	$n = 10$	$n = 4$	$n = 5$	$n = 10$
10	1.22	2.01	3.09	1.87	1.73	4.38
16	9.30	10.95	5.79	14.44	12.12	17.74
20	25.03	27.85	36.05	44.04	37.70	43.02
24	32.78	34.34	46.57	72.94	63.04	72.07
28	39.66	41.54	55.38	111.62	95.56	107.35
30	47.69	50.23	64.58	151.85	131.25	147.69

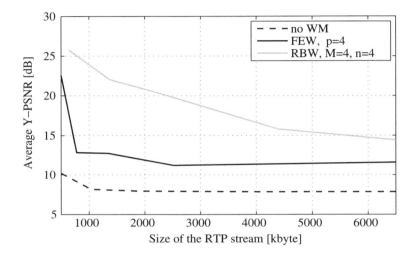

Figure 6.14   Comparison of FEW, RBW and packet-loss concealment: Y-PSNR at the decoder over the size of the encoded and watermarked RTP stream and BER $= 1 \times 10^{-6}$.

of error detection were about 90%, similar for all three BER values. Single-bit parity checks only result in about 49% of error detections.

Please note that this high gain is likely to decrease slightly if watermarking is employed together with the syntax check, enabling further detection of errors and decoding of the stream despite the errors which remained undetected. Nevertheless, it can be concluded that RBW provides more flexibility and better probability detection than FEW. Thanks to its parameters n and K, RBW may be adapted to the deployment scenario.

6.4   VLC Resynchronization

6.4.1   Signalling of Synchronization Points

The problem of the VLC desynchronization as presented in section 6.1 can be solved by different means. In H.264/AVC, the VLC starts at the beginning of each packet. In some literature (Wenger 2003) it is therefore suggested to keep the packet length as low as approximately 100 bytes to achieve robust transmission over error prone channels. The transmission over an RTP/UDP/IPv4 protocol stack is, however, connected with an overhead of 40 bytes per packet (neglecting the slice header, NALU header and padding; without any other assumption on the underlying system). For instance, the 100-byte long packets result in 29% of the rate only for packet headers. An overhead lower than 10% of the data is achieved for packets larger than 400 bytes; an overhead lower than 5% of the data is achieved for packets larger than 800 bytes. Clearly, smaller packet sizes are not an efficient solution for the desynchronization problem.

The usage of reversible VLC codes (Takishima *et al.* 1994) helps to prevent the desynchronization of the decoder. However, the shortcoming of the majority of VLCs with good resynchronization properties is their lower lossless compression performance. A fair trade-off between compression gain and good resynchronization properties represents reversible exp-Golomb (Wen and Villasenor 1998) adopted in MPEG-4 and H.263+ (Annex D), but not into H.264/AVC where context adaptive VLC schemes provided higher compression gain than a single (R)VLC table.

Another popular way to increase robustness of source coding schemes is by insertion of Synchronization Markers (SMs) (Kiely *et al.* 2000). Synchronization marks allow for bit synchronization of the decoder, preventing error propagation. An RD-optimized mode selection and SM insertion is proposed in Côté *et al.* (2000). The SMs there correspond to slice header insertion. Apart from the overhead caused by 'in-stream' insertion, sending the SMs within the same stream increases the risk of errors resulting in their misinterpretation. In order to avoid these problems, an alternative resynchronization mechanism using 'Length Indicators' (LIs) is proposed in this section. The LIs are sent 'out-of-stream', for example in dedicated NALUs, and require fewer bits for their encoding.

The aim of resynchronization is to limit the size of the distorted picture area which in turn results in better performance of error concealment. In Nemethova *et al.* (2005) a method of sending the position indicator of a VLC codeword start every M MBs has been proposed. This method enables resynchronization of the VLC stream at the position of the first VLC codeword that starts within the M MBs. The LIs allow the VLC decoder to restart at known positions and thus to confine the propagation of decoding errors, which are thereby also easier to detect and conceal.

A trade-off has to be found between the overhead and the frequency of the LIs determining the size of the erroneous area in (macro)blocks. The overhead O_{MB} caused by sending a LI of m bits after each M MBs is

$$O_{MB} = \frac{m}{M \cdot E\{L\}}, \tag{6.9}$$

where L is a random variable representing the size of an MB in bits. The distribution of L differs considerably for I and P frames. Furthermore, it varies also for different QPs as well as for different video sequence contents. This is illustrated in Figure 6.15, obtained by analysis of the 'soccer match' video sequence encoded by JM in baseline configuration. The mean P MB size $\bar{l}_P$ calculated for the 'soccer match' sequence is 86.24 bits for QP = 30 and 269.37 bits for QP = 20. The epmfs for QP values between 20 and 30 (used for mobile communication) would lie in-between. The mean I MB size $\bar{l}_I$ is 240.3 bits for QP = 30 and 591.75 bits for QP = 20.

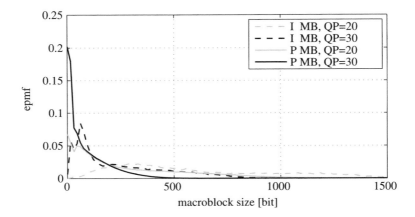

Figure 6.15 Distribution of the MB size for intra and inter coded MBs.

To convey the LI information to the decoder without errors, it could be appended to the A packets if DP is used; however, that is not standard-compliant and might break compatibility with other receivers. A better way is to use a reserved NALU type, which will be ignored by receivers not knowing how to handle it. The additional packetization overhead is smaller if the side information can be precomputed to generate larger packets. The decoder needs only buffering capability for one additional packet. To estimate numerically the overhead, the necessary length m has to be determined.

There are two possibilities for representing resynchronization information. It can be represented as an absolute 'position' of M-MB segments within the packet or as a 'length' of the M-MB segments (and slice header at the beginning) in bits. The first representation is more robust against errors, as the positions are independently represented; the second representation may lead to lower signalling overhead, depending on the applied binarization method. In this section, the signalling of length will be discussed further.

A trivial way to binarize the LIs is Fixed Length Coding (FLC). The range of values for both position- and length-based LIs is limited by the maximum size of the packet, that is, in general the Maximum Transmission Unit size (MTU, typically 1500 bytes) minus the length of all packet/slice headers. This corresponds to 14 bits necessary for LI encoding, which is still smaller than an average slice header (approximately 25 bits long). A bit error in such an FLC codeword will lead to an erroneous decoding of that codeword only. The compression gain of a VLC depends on the distribution of the source symbols, the source symbols being the values of LI.

The empirical probability mass function (epmf) $p(l_i)$, shown in Figure 6.15, corresponds to the LI distribution if LI represents the length of an MB and $M = 1$. Distributions for $M > 1$ are flatter, and there will be fewer occurrences of small LIs and more of the larger LIs.

6.4.2   Codes for Length Indicators

The non-uniform distribution of LI values may provide the opportunity for lossless compression. Let i be an integer value of LIs and w_i the corresponding codeword. For an optimal

Huffman binary prefix-free code (Sayood 2000) and for a source $\mathcal{S}$, the mean codeword length $\bar{l} = \mathrm{E}\{\ell(w_i)\}$ is bounded by

$$H(\mathcal{S}) \leq \bar{l} \leq H(\mathcal{S}) + 1, \tag{6.10}$$

with

$$H(\mathcal{S}) = - \sum_i p(x_i) \log_2 p(x_i) \tag{6.11}$$

being the 'entropy' of the source $\mathcal{S}$ and $p(x_i)$ the probability of the occurrence of the ith codeword. However, designing an optimal Huffman code is not feasible since the LI distribution differs considerably for different types of frame, QPs and values of M as well as for different sequence contents. If the probability of the source is not known/varying, 'codes for integers' (MacKay 2006) are more suitable. In data compression, a code for integers is a prefix code that maps the positive integers i onto binary codewords, with the additional assumption that the distribution is monotone (that is, $p(i) \geq p(i+1)$ for all positive i). Such code is called universal if for any distribution in a given class, it encodes into an average length that is within some factor of the ideal average length.

There have been several coding schemes for integers proposed so far. The simplest 'unary coding' encodes an integer i into a codeword w_i as a sequence of i ones followed by a zero (or equivalently a sequence of i zeros followed by a one). Unary codes correspond to Huffman codes for the semi-infinite alphabet $i \in [0, \infty)$ with probability model $p(i) = 2^{-i}$ and it is, thus, optimal for this probability model. Each codeword w_i has length $i + 1$.

Another large class of codes combines a unary prefix with other types of code. An example is the 'Golomb code' family (Golomb 1966) with an integer parameter $g > 1$. Each integer $i \in [0, \infty)$ is represented by two numbers $q = \lfloor i/g \rfloor$ and $r = i - qg$, where q is represented in unary form and the reminder r takes values in $[0, g-1]$. If g is a power of two, r is represented by a $\log_2(g)$ bit binary representation. Correspondingly, each codeword w_i has the length $\lfloor i/g \rfloor + \lceil \log_2(g) \rceil + 1$. If g is not a power of two, $\lceil \log_2(g) \rceil$ bits can be used. Alternatively, the required number of bits may be reduced if a $\lfloor \log_2(g) \rfloor$-bit binary representation of r is used for the first $2^{\lceil \log_2(g) \rceil} - g$ values, and the $\lceil \log_2(g) \rceil$-bit binary representation of $r + 2^{\lceil \log_2(g) \rceil} - g$ for the rest of the values. The Golomb code is optimal for the probability model

$$p(i) = (1-q)^{i-1} q, \qquad g = \left\lceil -\frac{1}{\log_2(1-q)} \right\rceil. \tag{6.12}$$

Elias codes γ, δ and ω were first proposed in Elias (1975). Elias-γ is equivalent to the exponential Golomb code shown in section 6.1.1. Note that exp-Golomb codes are different from Golomb codes. In the literature, these two codes are often confused. Richardson (2003), for example, refers incorrectly to Golomb (1966) as a source for exponential Golomb codes. Elias-γ can be further regarded as a Golomb code with parameter g adapting to i as $g = \lfloor \log_2(i) \rfloor$. The codeword length of an Elias-γ encoded integer i is given by $\ell(w_i) = 1 + 2 \cdot \lfloor \log_2(i) \rfloor$.

An Elias-δ code is also composed of two parts: the Elias-γ encoded length of the binary representation of the integer i and the binary representation of i without its most significant bit. The codeword length of an Elias-δ encoded integer i is given by $\ell(w_i) = 1 + \lfloor \log_2(i) \rfloor + 2 \cdot \lfloor \log_2(1 + \lfloor \log_2(i) \rfloor) \rfloor$.

Table 6.7 Example codewords w_i for chosen integers $i = \{1, 2, 3, 4, 15, 32\}$ encoded by unary, Golomb and Elias codes.

Code	$i = 1$	$i = 2$	$i = 3$	$i = 4$	$i = 15$	$i = 32$
Unary	10	110	1110	11110	$\underbrace{1\cdots1}_{15}0$	$\underbrace{1\cdots1}_{32}0$
Golomb(2)	00	01	100	101	111111100	$\underbrace{1\cdots1}_{15}01$
Golomb(4)	000	001	010	011	111010	1111111011
Golomb(8)	0000	0001	0010	0011	10110	1110111
Elias-γ	1	010	011	00100	0001111	00000100000
Elias-δ	1	0100	0101	01100	00100111	0011000000
Elias-ω	0	100	110	101000	1111110	101011000000

An Elias-ω code is composed of a variable number of groups of bits. The right-most bit is zero. The first group to its left is the binary encoding of the integer i. From there on, each next group to the left is the binary encoding of the length in bits of the preceding group, minus one. This iterative procedure stops when the length to be encoded is one. The encoding process halts on the left with group of length 2. The length of an Elias-ω encoded integer i is given by the recursive formula $\ell(w_i) = 1 + \sum_{j=1}^{k}(l_j(i) + 1)$, with $l_1(i) = \lfloor(\log_2(i))\rfloor$, $l_{k+1}(i) = l_1(l_k(i))$, $k \in \mathbb{N}^+$. The summation stops with integer k such that $l_k(i) = 1$.

In Table 6.7 example codewords of all introduced codes are presented. It can be seen that the codeword lengths $\ell(w_i)$ of the encoded integer i vary for each code. The distribution of LI values of I frames for various values of M is not monotone, and is rather flat. Therefore, for I frames, FLC encoded LIs will probably be the best choice. For P frames, VLC encoding could help in reducing the rate.

Table 6.8 contains the expected codeword lengths for all discussed VLC codes applied to the LIs for $M = 1$. The ideal Golomb code is the Golomb code with g resulting in the smallest $E\{\ell(w_i)\}$. This was different for I and P frames with different QP values: $g_{I20} = 256$, $g_{P20} = 127$, $g_{I30} = 128$ and $g_{P30} = 62$. Golomb codes approach the bound since the distribution of LIs is rather flat. Therefore, the very short codewords are not as beneficial. The codeword set for Golomb codes with high g is similar to an FLC – the length of codewords stays constant over larger ranges of i. Thus, VLC Golomb coding still benefits over FLC and is clearly worth using, as it saves 3–6 bits per LI in our test cases.

Figure 6.16 shows the average Y-PSNR if VLC resynchronization is applied together with syntax analysis. Note that the absolute values of Y-PSNR for $M = 99$ are lower than the values obtained for syntax analysis before. This is because, in this experiment, the number of MBs per slice/NALU was limited unlike the size in bytes in previous experiments. The sequence 'foreman' was sent over a BSC channel with BER $= 10^{-5}$; for each configuration (maximum number of MBs per slice, QP) 20 realizations of the channel were used.

VLC resynchronization may perform even better if combined with a more reliable error detection mechanism, such as watermarking or checksum, applied to intervals of the same length as the resynchronized intervals.

Table 6.8 Calculated average codeword length for chosen integer codes, calculated based on the distribution of LIs estimated from the 'soccer match' video sequence and $M = 1$.

Code	$\bar{l}_I$, QP $= 20$	$\bar{l}_P$, QP $= 20$	$\bar{l}_I$, QP $= 30$	$\bar{l}_P$, QP $= 30$
Unary	490.22	266.29	228.90	84.59
Golomb (ideal)	10.45	9.61	9.34	7.83
Elias-γ	16.52	13.82	13.86	9.85
Elias-δ	14.33	12.29	12.43	9.20
Elias-ω	15.30	13.27	13.50	9.98
Entropy	10.25	9.23	9.17	7.40

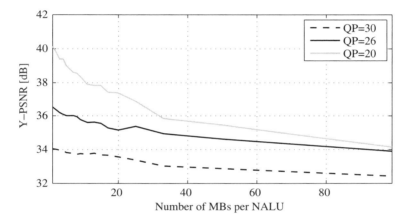

Figure 6.16 Y-PSNR after error detection by syntax analysis, depending on the size of the NALU in MB.

6.5   From Error Detection to Soft Decoding

By detecting violations of constraints on the form of the encoded stream, syntax analysis demonstrates that there is still a possibility of detecting errors in the H.264/AVC CAVLC. This section shows how these constraints can also be used to correct errors by means of sequential decoding, as presented in Weidmann *et al.* (2004) and Weidmann and Nemethova (2006). The following decoding mechanism assumes that H.264 Extended Profile with 'data partitioning' is used, which puts header data and motion vectors of a slice in 'type A' NALUs, which will be better protected and thus assumed to be received error-free. The CAVLC-encoded 'residuals' (prediction differences) of intra frames are in type B NALUs, while inter-predicted residuals are in type C NALUs. Sequential decoding will only be used for B/C packets (NALUs), although the general principle could also be applied to baseline CAVLC without data partitioning, as well as main profile Context Adaptive Binary Arithmetic Coding (CABAC) (Jamaa *et al.* 2006).

6.5.1   Sequential CAVLC Decoder

The encoding of residual coefficients depends causally on previous data in the same B/C packet and on the relevant information from the A packet. This causal code structure is well matched to sequential decoding, which was originally proposed for convolutional channel codes and later refined in Fano (1963). A list of partial decoding paths is kept in memory and each is labelled with a metric that allows paths of different length to be compared. Since the list size shall be limited, the decoder needs to decide which paths to explore further, based on the path metric. Several strategies exist; one of the simplest involves storing the paths in a stack which is sorted according to the metric. The top path (with the highest metric) is replaced by its extensions (a corresponding number of low-metric paths will be dropped from the stack) and the stack is sorted again. These steps are repeated until the top path has the required length and can be output as the decoded path. The choice of metric determines the performance of sequential decoding. In Massey (1972) it was shown that the heuristic metric introduced by Fano (1963) minimizes the error probability, provided the so-called 'random tail assumption' holds.

Consider a message w that is encoded with the binary variable-length codeword $\underline{x}_w = x_{w,1} x_{w,2} \ldots x_{w,\ell(w)}$ of length $\ell(w)$ and transmitted over a binary-input memoryless channel with transition probabilities $P(y|x)$. The received vector $\underline{y}$ is assumed to be longer than the codeword $\underline{x}_w$. Then the random tail assumption states that the bits following the codeword (and belonging to the next codeword) are chosen i.i.d. with some pmf $Q(x)$. For a good binary source code this is approximately satisfied with equal probabilities of $1/2$ for zero and one. Then the a-posteriori probability that message w has been sent is

$$\Pr(w|\underline{y}) = P(w) \prod_{i=1}^{\ell(w)} \frac{P(y_i|x_{w,i})}{P_0(y_i)}, \qquad (6.13)$$

where $P(w)$ is the a-priori probability that w has been sent and $P_0(y_i) = \sum_x P(y_i|x)Q(x)$ is the marginal channel output distribution induced by $Q(x)$. The metric is now simply the logarithm (usually of base two) of $\Pr(w|\underline{y})$:

$$L(w, \underline{y}) = \log P(w) + \sum_{i=1}^{\ell(w)} \log \frac{P(y_i|x_{w,i})}{P_0(y_i)}. \qquad (6.14)$$

Using $Q = (1/2, 1/2)$ (uniform distribution), the argument of the right-hand 'channel term' can be computed directly from the soft inputs, for example the log-likelihood ratios (LLRs) $\log(P(y_i|1)/P(y_i|0))$. Extending the metric to sequences $w_1^k = w_1 w_2 \ldots w_k$ is straightforward: the a-priori term $\log \Pr(w_1^k)$ can be decomposed into the sum $\sum_{i=1}^{k} \log \Pr(w_i|w_1^{i-1})$, which takes care of dependencies on past message symbols, for example due to syntax and/or semantics of the H.264/AVC CAVLC. The channel term $\log \Pr(w_1^k|\underline{y})$ is clearly additive; its summands will have to be conditioned on w_1^{i-1}, since the choice of VLC codebook for w_i may depend on past symbols.

The a-priori probabilities $P(w)$ must be known in order to compute this MAP metric. Assuming that the compression is efficient, the probability of emitting a codeword will be exponentially related to its length:

$$P(w) = \frac{2^{-\ell(w)}}{\sum_w 2^{-\ell(w)}}. \qquad (6.15)$$

It can be seen that, by this assumption, equation 6.14 reduces to the Maximum Likelihood (ML) metric.

A key difference to sequential decoding of convolutional codes is the fact that not all syntactically valid paths correspond to a valid decoding of a NALU, since the header information imposes additional constraints. Only paths that have the correct length in bits *and* encode the correct number of SMBs (in the slice) are valid decoder outputs. This yields some error correction capability, since semantically invalid paths can be eliminated from the decoder stack. Several kinds of redundancy in the encoded stream of residuals (B/C partitions) can be exploited by the sequential decoder to correct errors:

- mismatch between the actual source and its model in the encoder (for example, using a memoryless model for a Markov source);

- mismatch between ideal and actual codeword lengths (for example, integer codeword lengths, unused leaves/codewords in the VLC tree);

- semantic side information about the encoded content.

The first two kinds of redundancy are of little importance in H.264/AVC: on the one hand, the CAVLC syntax has been closely matched to the correlation structure present in the residual data. Consequently, having more precise a-priori source probabilities, whether transmitted as side information or estimated online, yields only minor performance improvements, if at all. Efficient use of statistical model redundancy is also hampered by typically relatively small NALU sizes. On the other hand, although there are some unused codewords in some of the VLC tables, they generally differ in just one position from the longest codeword in a given table; thus, they are unlikely to occur as a result of a bit error, which would otherwise be detected. In summary, little redundancy is left in the residuals data stream that could be easily exploited by the decoder to correct the errors; the statistical properties of the encoder output are close to those of a binary symmetric source. These observations together with the analysis in section 6.1 suggest that the main source of redundancy is the semantic side information contained in the A packets; that is, outside the actual residual data stream in the B/C packets. That information puts semantic constraints on the contents of the residual packets. Of these constraints, the simplest to exploit in a sequential decoder is the knowledge of the number of (S)MBs that are encoded in a given packet of length n.

6.5.2   Additional Synchronization Points

The side information sent out-of-band (cf. section 6.4) can also be used to signal additional synchronization points to a sequential decoder. The boundaries of MBs in the code stream are natural candidates for resynchronization, since the decoder already checks for the correct number of MBs at the end of a packet. The decoder is informed of the current bit position every mth MB in a slice; that is, it knows the length n (in bits) and the position of a segment of the code stream encoding m MBs (or less, at the end of a slice). Note that this is not the same as having smaller packets of at most m MBs, due to the prediction mechanism that causally affects the encoding of MBs within a slice. The sequential decoder uses the extra side information simply to discard paths that do not line up correctly at the synchronization points.

Providing a synchronization point every m MBs has two shortcomings: first, the rate of side information is still limited by its granularity on the MB level (possibly SMB). This could be problematic for video encoded at high rates with a small quantization parameter, resulting in large intervals between (S)MB boundaries in the code stream. However, that is unlikely to be a problem in low- to medium-rate wireless applications. The other shortcoming is more serious: increasing the frequency of synchronization points does not necessarily increase the capability of correcting (or detecting) VLC errors, regardless of decoder complexity. For example, it cannot prevent confounding two codewords of the same length, for the same reason that errors in the FLC fields of the code stream go unnoticed. Fortunately, both problems can be mitigated by proper postprocessing – impairment detection and error concealment.

6.5.3   Postprocessing

The CAVLC for the residuals contains also FLC fields to code sign and mantissa of coefficient values. The decoding metric assigns uniform probabilities to FLC fields; hence FLC errors remain undetected and will cause some distortion as in the detection based on syntax analysis. A more severe decoding error occurs when the sequential decoder outputs a wrong VLC path. This is more likely to occur within I frame slices, since these will utilize the maximal packet size. Moreover, decoding a wrong VLC path causes a larger distortion for I frames than for P and B frames. The distortion in the P and B frames is negligibly small in most cases due to the efficiency of the temporal prediction and use of data partitioning. In natural scene video sequences the residuals of P and B frames are usually rather small, if no scene change occurs. Using data partitioning, motion vectors are correctly received and thus the distortion caused by wrong residuals is very small. Spatial prediction used for I frames has lower efficiency than the temporal prediction and, therefore, the residuals usually have higher values. This leads to considerable visual artefacts if the residuals are wrong. Furthermore, the blocks that use the first block with wrong residuals as a reference for decoding will also be wrongly decoded, which even worsens the situation. Spatial error propagation will result in block-shaped artefacts over several MBs, differing in colour and/or intensity from the rest of the picture.

Such decoding errors typically result in impairments corresponding to those caused by decoding of the entire slice out of a damaged packet. Hence, they can also be detected by means of impairment detection. Due to data partitioning, for P and B frames there are no errors in motion vectors. As the errors in P and B frame residuals typically have a very small visual importance, it does not make sense to perform artefact detection for them.

For FLC errors, the soft LLR inputs are used to compute an estimate of the expected distortion, which is input to a voting-based impairment detection procedure. After detecting the errors, boundary matching temporal error concealment is used for I frames. Factors that make the temporal error concealment more difficult, such as scene cuts, transitions and fast zooming in/out, are not present in the tested 'foreman' video sequence.

6.5.4   Performance

All simulations are based on the 'foreman' sequence in QCIF resolution, encoded with JM using data partitioning. Type A and IDR packets are protected by a stronger channel code

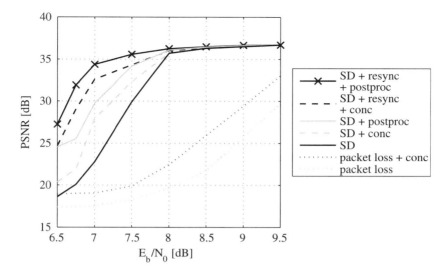

Figure 6.17 Comparison of all presented decoder configurations, wherein SD is sequential decoding, 'conc' is error concealment, 'resync' is VLC resynchronization and 'postproc' is postprocessing.

than type B/C, so that one can assume that all A and IDR packets are received without error. However, the weaker (or absent) protection of B/C packets causes them to be received with random errors. It is assumed that the physical layer provides the soft information (LLR) for the bits in these packets.

The frame rate of 'foreman' was reduced by two resulting in 200 frames of video at 15 f/s. The video was encoded with $QP = 30$, GOP of size 10 frames without B frames, maximal packet size 750 bytes, and then transmitted over a Binary Input Additive White Gaussian Noise (BIAWGN) channel. The total file size was 169 509 bytes (corresponding to about 100 kbit/s), of which 69 197 were in error-free A packets. The remaining 100 312 bytes were in B/C packets and were decoded with the sequential decoder (on average, ten different realizations of noise were taken into account per E_b/N_0 point).

Synchronization points were added with a frequency of $m_I = 4$ in I frames and $m_P = 20$ in P frames, resulting in the same average side information rate of 2% in both frame types, or 15 837 bits in total. To model error-free transmission of this information, 3×40 bytes of headers were added and assumed a channel code of rate 0.9, which in the practical decoder operating region above 6 dB E_b/N_0 is at more than 3 dB from capacity. All together, this corresponds to an increase of 0.1 dB in E_b/N_0 compared to a reference system without additional synchronization information. To make comparisons possible, this shift is included in the plots shown.

The computational complexity of sequential decoding becomes exponential when operating above the cut-off rate. Therefore, the complexity was limited by letting the decoder drop at most 100 invalid paths before declaring a slice 'erasure'. More details on the performance of sequential decoding can be found in Weidmann and Nemethova (2006).

Figure 6.17 depicts the YUVYUV-averaged sequence PSNR in dependency of the channel E_b/N_0. Results are shown for different decoders and different postprocessing scenarios. Lost or erased packets (slices) are either left as is or concealed with copy-paste from the previous frame. The best results are obtained when combining the sequential decoder with the artefact detector that generates concealment requests. The additional synchronization information results in about 0.1–0.2 dB gain in E_b/N_0 (after the 0.1 dB penalty). The quality improvement resulting from using the resynchronization is for the same rate about 5 dB for $E_b/N_0 = 7$ dB. The postprocessing adds another improvement of 2 dB.

Figure 6.18 illustrates the resulting visual gains for $E_b/N_0 = 6.75$ dB. The resynchronization decreases the occurrence of the large visual artefacts considerably. The impairment detection further reduces the visually apparent impairments in both cases.

Figure 6.18   Screenshots from 'foreman': original and concealed packet loss (left); sequential decoding without and with postprocessing (middle); sequential decoding with resynchronization, without and with postprocessing (right).

For the practical deployment at power-limited mobile terminals, sequential decoding still remains too complex. The assumption of the error-free A packets is a rather strong assumption. On the other hand, the packet loss ratio in a static UMTS scenario is approximately 0.088%, which is quite low and can even be improved by a stronger channel code (less puncturing). This rate increase can be recouped by sending the B/C packets uncoded. Another practical problem is the need for soft channel information, which is typically not available in the architecture of current wireless mobile systems. However, in cross-layer designed systems this is easy to solve, since both the physical layer and the application layer are implemented in the same network element, the terminal. Sequential decoding would be beneficial especially for broadcasting and multicasting applications, where retransmissions are inefficient or unfeasible.

References

Barni, M., Bartolini, F. and Bianco. P. (2000) On the Performance of Syntax-based Error Detection in H.263 Video Coding: a Quantitative Analysis. In *Proceedings of SPIE Conference on Image and Video Communications (Electronic Imaging 2000)*, San Jose, CA, Jan.

Bellifemine, F., Capellino, A., Chimienti, A., Picco, R. and Ponti, R. (1992) Statistics analysis of the 2D-DCT coefficients of the differential signal of images. *Signal Processing: Image Communications*, **4**(6), 477–488.

Bergeron, C. and Lamy-Bergot, C. (2004) Soft-input decoding of variable-length codes applied to the H.264 standard. In *Proceedings of IEEE Workshop on Multimedia Signal Processing (MMSP)*, Siena, Italy, Sep.

Chen, M., He, Y. and Lagendijk, R.L. (2005) A fragile watermark error detection scheme for wireless video communications. *IEEE Transactions on Multimedia*, **7**(2), 201–211, Apr.

Côté, G., Shirani, S. and Kossentini, F. (2000) Optimal mode selection and synchronization for robust video communications over error-prone networks. *IEEE Journal on Selected Areas in Communications*, **18**(6), 952–965, Jun.

Cox, I.J., Kilian, J., Leighton, F.T. and Shamoon, T. (1997) Secure spread spectrum watermarking for multimedia. *IEEE Transactions on Image Processing*, **6**(12), 1673–1687, Dec.

Elias, P. (1975) Universal codeword sets and representations of the integers. *IEEE Transactions on Information Theory*, **21**(2), 194–203, Mar.

Fano, R.M. (1963) A heuristic discussion of probabilistic decoding. *IEEE Transactions on Information Theory*, **9**, 64-73, Apr.

Golomb, S.W. (1966) Run-length encoding. *IEEE Transactions on Information Theory*, **12**, 399–401, July.

ITU-T Rec. H.264/ISO/IEC 11496-10, (2002) Advanced Video Coding, Final Committee Draft, Document JVTE022, Sept.

Jamaa S. Ben, Kieffer, M. and Duhamel, P. (2006) Improving Sequential Decoding of CABAC Encoded data via Objective Adjustment of the Complexity-efficiency Trade-off. In *Proceedings of SPIE*, San Jose, USA, Jan.

H.264/AVC Software Coordination, Joint Model Software, ver. 10.2, available at http://iphome.hhi.de/suehring/tml/.

Kiely, A., Dolinar, S., Klimesh, M. and Matache, A. (2000) Error Containment in Compressed Data Using Sync Markers. In *Proceedings of International Conference on Information Theory (ICIT)*, Jun.

MacKay, D.J.C. (2006) *Information Theory, Inference, and Learning Algorithms,* 5th edn. Cambridge University Press.

Massey, J.L. (1972) Variable-length codes and the Fano metric. *IEEE Transactions on Information Theory*, **18**, 196-198, Jan.

Nemethova, O., Canadas, J. and Rupp, M. (2005) Improved Detection for H.264 Encoded Video Sequences over Mobile Networks. In *Proceedings of International Symposium on Communication Theory and Applications (ISCTA)*, Ambleside, UK, July.

Nemethova, O., Forte, G.C. and Rupp, M. (2006) Robust Error Detection for H.264/AVC using Relation Based Fragile Watermarking. In *Proceedings of 13th International Conference on Systems, Signals and Image Processing*, Budapest, Hungary, Sep.

Richardson, I.E. (2003) *H.264 and MPEG-4 Video Compression*, John Wiley & Sons.

Sayood, K. (2000) *Data Compression,* 2nd edn. Morgan Kaufmann Publishers.

Superiori, L., Nemethova, O. and Rupp, M. (2007) Detection of Visual Impairments in the Pixel Domain. In *Proceedings of Picture Coding Symposium (PCS 2007)*, Lisboa, Portugal, Nov.

Superiori, L., Nemethova, O. and Rupp, M. (2007b) Performance of a H.264/AVC error detection algorithm based on syntax analysis. *Journal of Mobile Multimedia*, **3**, 337–345, Dec.

Takishima, Y., Wada, M. and Murakami, H. (1994) Reversible Variable Length Codes. *IEEE Trans. on Communications*, **42**(2/3/4), 158–162.

3GPP TS 26.234 2008 (2008) Transparent end-to-end Packet-switched Streaming Service (PSS); Protocols and Codecs, v.7.3.0, Jun. 2008.

Weidmann, C., Kadlec, P., Nemethova, O. and Al-Moghrabi, A. (2004) Combined Sequential Decoding and Error Concealment of H.264 Video. In *Proceedings of IEEE Workshop on Multimedia Signal Processing (MMSP)*, Siena, Italy, Sep.

Weidmann, C. and Nemethova, O. (2006) Improved Sequential Decoding of H.264 Video with VLC Resynchronization. In *Proceedings of 15th IST Mobile and Wireless Communications Summit*, Myconos, Greece, Jun.

Wen, J. and Villasenor, J.D. (1998) Reversible Variable Length Codes for Efficient and Robust Image and Video Coding. In *Proceedings of IEEE Data Compression Conference*, Snowbird, Utah, USA, Apr.

Wenger, S. (2003) H.264/AVC over IP. *IEEE Transactions on Circuits and Systems for Video Technology*, **13**(7), 645–656, Jul.

Wicker, S.B. (1995) *Error Control Systems for Digital Communication and Storage*, Prentice-Hall, Inc., 1995.

Part IV

Error Resilient Video Transmission over UMTS

Introduction

Part IV starts in Chapter 7 with an overview on 3rd Generation Partnership Project (3GPP) video services including video codecs, transport protocols and bearer configurations. It is shown that adequate Quality of Service (QoS) can be achieved by applying appropriate bearer service configurations, content delivery protocol options, video error resilience means, or through combinations of the three. It is demonstrated that although the 3GPP QoS concept does not encourage cross-layer design, it is indeed possible to improve the performance of 3GPP video services substantially by applying the appropriate combination of methods. On three typical service scenarios (Multimedia Telephony Services, Multimedia Download Delivery and Multimedia Streaming Services over MBMS) the potential of cross-layer designs is exemplified.

Chapter 8 continues the topic of cross-layer approaches and presents several improvements. Following the results from Chapter 6, lower-layer packet-based error detection as well as VLC resynchronization mechanisms are applied by a so-called *Link Layer Error Detection*. These mechanisms are then further improved by using link error prediction from Chapter 4 in a so-called *Link Error Prediction Based Redundancy Control* in which side information is only attached if channel conditions are poor. The content awareness is further exploited by a link-layer scheduling mechanism prioritizing the more important packets according to the channel conditions called *Semantics Aware Scheduling*. The decisions based on video semantics are further enhanced by employing a novel distortion model named *Distortion Aware Scheduling*.

7

3GPP Video Services – Video Codecs, Content Delivery Protocols and Optimization Potentials

Thomas Stockhammer and Jiangtao Wen

This chapter provides an overview of 3rd Generation Partnership Project (3GPP) video services including video codecs, transport protocols and bearer configurations. A general discussion as well as examples of multimedia telephony and multimedia broadcast are provided. We show that adequate Quality of Service (QoS) might be achieved by applying appropriate bearer service configurations, content delivery protocol options, video error resilience means, or through combinations of the three. We demonstrate that, despite the fact that the 3GPP QoS concept does not encourage complex cross-layer design, it is nevertheless possible to improve the performance of 3GPP video services tremendously by using the appropriate combination of tools. Selected experimental results are provided to support our claims and to illustrate optimization potentials for such services.

7.1 3GPP Video Services

7.1.1 Introduction

Multimedia and especially video services have been a hot area for both academic research and business development for over a decade. A large number of research papers have been published to address specific aspects of video transmission over mobile networks. However,

Video and Multimedia Transmissions over Cellular Networks   Edited by Markus Rupp
© 2009 John Wiley & Sons, Ltd

looking at 3G mobile video services, although many industrial standards have emerged that support operation in mobile environments, compared with the large body of research efforts, surprisingly little has made it into specifications or standardization reports; even fewer have been deployed in the field. On the other hand, the landscape of commercial multimedia services and products remains highly fragmented and small compared to what has been envisioned, and not all of them conform to 3GPP specifications. On the video codec side, although the majority of 3GPP multimedia services utilize the Moving Picture Experts Group (MPEG)-4 Part 2 and Part 10 (aka H.264/AVC (H264 2003)) video coding standards, legacy or proprietary formats such as FLash Video (FLV) or older codecs such H.261, H.263 or even MPEG-2 can still be found. On the network side, protocols range from Real Time Streaming Protocol (Schulzrinne *et al.* 1998) (RTSP) over User Datagram Protocol (Postel 1980b) (UDP), RTSP over Transmission Control Protocol (Postel 1980a) (TCP), progressive Hypertext Transfer Protocol (Fielding *et al.* 1997) (HTTP) download, to even peer-to-peer networking approaches.

In this chapter, we analyse how practical constraints of the 3G architecture, such as strictly defined interfaces, layering, and so on, often make it difficult to apply most of the proposed cross-layer optimization strategies developed in academic research. We point out options within the specified solutions to optimize the system, for example by taking into account practical cross-layer constraints.

7.1.2   System Overview

Mobile handheld devices are nowadays equipped with colour displays and cameras. They also have sufficient processing power to allow recording, encoding, decoding and presentation of video sequences. In addition, emerging mobile systems provide sufficient bandwidth to support increasingly sophisticated video communication services, even as bandwidth remains a scarce resource in wireless transmission environments due to physical bandwidth and power limitations; thus, efficient video compression is required.

The 3GPP architecture supports a significant amount of packet-switched services. The sessions are typically established end-to-end between two terminals, or a terminal and a central server. The virtual connection is split logically into core and radio access networks. Figure 7.1 shows the high-level architecture and the corresponding network nodes and protocol layers. Obviously of major interest is the radio-related part, as it is most critical in terms of bandwidth, resource consumption and reliability.

Although any Internet Protocol (IP) (Postel 1981) based application can be considered as a packet-switched service, 3GPP restricts itself to a specific set of multimedia services that are supported by the network architecture. This greatly simplifies interoperability and allows efficient and high-quality integration of real-time multimedia services. Among others, this is achieved by the provision of network QoS, which means that a suitable radio (access) bearer with clearly defined characteristics and functionality is established on the air interface. The bearer definition includes all aspects for enabling the provision of a contracted QoS, including control signalling, user plane transport and QoS management functionality. On top of such radio access bearers, various end-to-end video multimedia services are supported, including conversational services, streaming services, download services as well as broadcast and multicast services. However, different video applications generally have different service

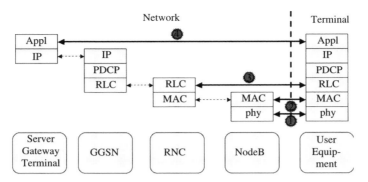

Figure 7.1 Conversation video transmission in RTP-based mobile environment.

constraints, leading to the selection of different codec settings, delivery protocols and radio
bearer settings to support the required end-to-end QoS.

A service generally requires certain QoS requirements that must be met to satisfy the user.
To achieve this, the radio bearer in particular must provide adequate means at the physical,
Medium Access Control (MAC) or Radio Link Control (RLC) layers, as shown in Figure 7.1.
Furthermore, QoS can also be provided end-to-end within the application layer. Initially, we
omit any cross-layer design aspects, and concentrate on the appropriate selection of different
tools in the radio layer and the application layer for different applications. Such design
follows a strict layer-separation-based approach, and is commonly preferred in complex
3GPP network architectures.

Although shared radio channels such as Enhanced GPRS (EGPRS) or High Speed
Downlink Packet Access (HSDPA) may exploit radio resources more efficiently, dedicated
channel modes are used for conversational as well as other low-delay applications. This is
because it is essential to provide bearers with a certain QoS in terms of guaranteed bandwidth,
delay, as well as error rate. Hence, we will in the following concentrate on Universal Mobile
Telecommunications Systems (UMTS) dedicated channels that can guarantee a requested
bit-rate and in case of UMTS also a fixed error rate by using fast power control and diversity
techniques such as multiple antenna systems in combination with forward error correction.
As already discussed earlier in this book, two different bearer modes are supported:
Unacknowledged Mode (UM) and Acknowledged Mode (AM). Typical properties of UMTS
dedicated channels for the transmission of packet-based applications have already been
introduced. In contrast to the transmission over wired IP networks, four major differences
apply in the characteristics when transmitting over UMTS dedicated channels:

1. The IP packet length impacts the loss rates as a longer packet is more likely to be hit
   by bit-errors or be part of an erroneous link layer packet.

2. In contrast to wired IP connections when transmitting over dedicated mobile links the
   delay of the IP-packet is also influenced by its length due to the transmission delay.
   Therefore, for low-delay applications a Constant Bit Rate (CBR)-like video rate control
   is critical.

3. The end-to-end principle is no more completely fulfilled. In entry gateways to the mobile system, as in support nodes, the IP-packets might be modified by applying header compression. Therefore, costly header overheads can be reduced for mobile transmission systems.

4. Finally, a mobile link layer can provide different QoS in terms of reliability and delay by the use of power control, Forward Error Correction (FEC) or retransmission protocols.

The impacts of these additional aspects for video transmission systems are discussed in the following, but first we will briefly review standardized video codecs in 3GPP video services.

7.1.3   Video Codecs in 3GPP

Standardized Video Codecs

Since the finalization of the first digital video compression standard H.120 in 1984 (H120 n.d.), mainly two organizations, the International Telecommunications Union – Telecommunications Sector (ITU-T)'s Video Coding Experts Group (VCEG) and International Organization for Standardization/International Electrotechnical Commission (ISO/IEC) MPEG, have been working on the development of new video coding standards. The main focus of the VCEG is on video coding standards for communication applications, namely H.261 (H261 n.d.a) and H.263 (H263 n.d.b), whereas MPEG originally concentrated on different higher quality applications for storage (MPEG-1 (MPG 1993)) and digital video broadcast of television signals within MPEG-2 (MPG 1994). Whereas the VCEG group worked on improvements of the H.263 (H263 n.d.b) in terms of compression efficiency and error resilience, ISO/IEC launched a new project within the MPEG-4 framework to specify a very general multimedia coding standard. Thereby, more emphasis was put on functionalities such as the coding of video objects with arbitrary shape, sprite coding and scalability rather than on coding efficiency.

In 1997, the ITU-T's VCEG started working on a new video coding standard with the internal denomination H.26L.[1] For H.26L, a completely new design was targeted rather than enhancing existing standards. The lessons learned from the standardization of H.263, namely that extensive optional modes limit the interoperability, and from MPEG-4, that the most important functionality of a video coding standard is compression efficiency, were the main driving forces in the standardization process of H.26L. In addition, the integration of video coding standards into networks has been considered from the very beginning. Therefore, most proposals were evaluated based on three primary criteria, (i) improved coding efficiency, (ii) improved network adaptation, and (iii) simple syntax specification. In August 1999 a remarkably simple and familiar draft model[2] was adopted and was evolved into a Test Model Long-term (TML) reference design. The draft model basically only contained features known from previous standards and was enhanced by adding additional features, especially for advanced motion compensated prediction (Motion Compensated Prediction (MCP)). In late 2001, MPEG and VCEG decided to cooperate within the Joint Video Team (JVT) and to

---

[1]The 'L' addressed the idea of a long-term project.

[2]The initial model document only contained fewer than 30 pages of description for the entire codec.

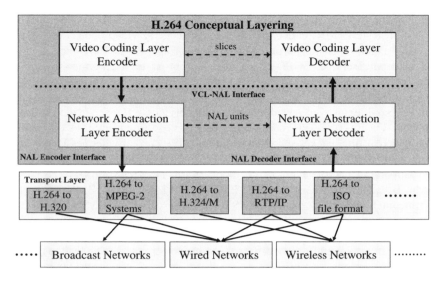

Figure 7.2 Layering concept of H.264/AVC and integration in network environments.

create a single technical design for a forthcoming ITU-T Recommendation and for a new part of the MPEG-4 standard based on the committee draft version of H.26L at the time. Finally, Recommendation H.264/AVC (H264 2003; Wiegand *et al.* 2003) was approved by the ITU-T in May 2003 and shortly after by ISO/IEC. The normative part of a video coding standard only consists of the appropriate definition of the order and the semantics of syntax elements and the decoding of error-free bit-streams.

In terms of network friendliness, network specific features have been addressed by a conceptual separation of compression-related tools in the Video Coding Layer (VCL) and network integration features in the Network Abstraction Layer (NAL). This includes the public Internet with best effort delivery as well as wireless networks. Therefore, the H.264/AVC design distinguishes between two different conceptual layers, the VCL and the NAL as shown in Figure 7.2. Both the VCL and the NAL are part of the standard. The NAL concept has been introduced (Stockhammer *et al.* 2002) with the vision that H.264/AVC will be integrated in many applications and transport protocols and that many of these new transport protocols will be packet based. The encapsulation as well as the transport in different transport protocols are not specified in the H.264/AVC standard but by the responsible standardization bodies for transport protocols such as MPEG-2 systems, RTP/IP, MP4 file format or H.32X. The NAL decoder interface is normatively defined in H.264/AVC coding standard, whereas the interface between the VCL and the NAL is conceptual and helps in describing and separating the tasks of the two layers.

Video Codecs in 3GPP Services

Video is an important component in many 3GPP services. It is vital to the continuous success of mobile broadband networks, as it allows full utilization of the provided bandwidth and complements existing services. In contrast to speech and audio codecs, such as the Adaptive Multi Rate (AMR) codec, 3GPP does not specify video codecs for its services, but references existing general-purpose video codec specifications. Referencing here means that, for a specific service, 3GPP either mandates or at least recommends the support of a video decoder conforming to a certain profile and level, possibly with some further restrictions or clarifications.

Figure 7.3 provides an overview of selected video services, referenced video specifications and protocols. On top of 3GPP radio access bearers, end-to-end multimedia services are supported. For real-time packet-switched services, 3GPP currently defines Packet Switched Conversational (3GPP TS 26.235 2007; 3GPP TS 26.236 2007) (PSC) services, Packet Switched Streaming (3GPP TS 26.234 2006; 3GPP TR 26.937 2004) (PSS) services, Multimedia Broadcast/Multicast Service (3GPP TS 26.346 2005; MBMS 2003) (MBMS) and Multimedia Telephony Service for IMS (3GPP TS 26.144 2007) (MTSI). Within these service specifications, different applications are supported, of which a majority include digital video, for example, in video telephony, video streaming or mobile TV applications. However, different applications generally have different service constraints, which lead to the selection of different codec settings, delivery protocols and radio bearer settings to support the required end-to-end QoS.

Specifically, within Release 7 of 3GPP specifications, only the ITU-T H.263 Profile 0 and Profile 3, ISO/IEC MPEG-4 simple profile and the baseline profile of the H.264/MPEG-4 AVC codec are referenced. Each service may and generally does recommend multiple video codecs. The usage of one or other codec in a specific session is established through capability exchange and session establishment procedures. Note that compared to speech and audio codecs, the specifications of video codecs in 3GPP are quite compact. Encoder settings, as in bit-rate control, error resilience or motion estimation processes, are generally completely left open. Some discussions on encoder configurations are provided in the remainder of this chapter. As 3GPP relies on existing multimedia protocols, not only the video codec itself is of essence, the support of the encapsulation and packetization into service specific media formats and protocols also need to be defined. In this case, 3GPP also almost exclusively relies on outside work, mainly on specification in Internet Engineering Task Force (IETF) Audio/Video Transport (AVT), Multiparty Multimedia Session Control (MMUSIC), and Reliable Multicast Transmission (RMT) group, ITU-T SG16, and ISO MPEG for File Formats (FFs). Similarly, the specific profiles and options of general-purpose specifications are referenced.

Furthermore, recognizing the lack of supplementary information of video performance and configurations in different environments, 3GPP has compiled several technical reports to address this issue. Reference 3GPP TR 26.937 2004 addresses the use of RTP for PSS. In particular, this document considered the impacts of the underlying network configurations and how the streaming mechanism itself could be optimized. Among others, trade-offs between radio usage efficiency and streaming QoS, stream and transmission adaptation, packetization, application layer QoS methods and client buffering to ease the QoS requirements on the network and enable more flexibility have been considered. However, this document is slightly outdated for emerging radio access systems such HSDPA. Reference 3GPP TR 26.902 2007 comprises a technical report on video codec performance for packet-switched video-capable multimedia services and also collects methods and tools for measuring the video performance in typical 3GPP service environments.

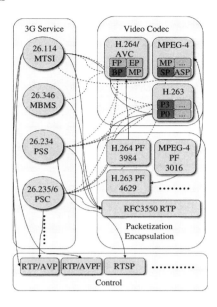

Figure 7.3 Video codecs in 3G service specifications.

7.1.4 Bearer and Transport QoS

The mobile radio channel places fundamental limits on the performance of a wireless communication system. Unlike wired channels for which the terminals are mostly stationary and channel conditions slowly varying, radio channels exhibit extremely volatile behaviour. In mobile systems, the radio transmission path between the transmitter and the receiver can vary from a simple line-of-sight to one that is severely obstructed by buildings, mountains, foliage and so on. The speed of the mobile terminal also has a great impact on the received radio signal. As a result, in contrast with wired channels, where packet losses are usually caused by congestion and bandwidth starvation, for wireless channels, both bandwidth starvation and radio interferences cause packet losses. Therefore, technologies and services developed for multimedia services over wired line channels (as in monitoring and in some cases predicting channel bandwidth variations and adapt streaming bandwidth accordingly) may not be deployed directly in the case of wireless. After all, simply reducing the media content bandwidth by itself will not combat losses caused by radio interferences.

Figure 7.4 illustrates an example of the movements of four users in a cell for different speeds and different starting points during an interval of 10 min. The movements of the users are depicted in the left-hand side of the figure. Notice that the vehicular user undergoes a much larger distance due to his speed of 30 km/h, while the pedestrian users moving at 3 km/h cover less distance in the same time period. The right-hand side of Figure 7.4 contains the respective Signal-to-Noise and Interference Ratio (SNIR) values over time for each of the users. The SNIR variance is caused by large-scale effects, such as distance-dependent attenuation and shadowing at large objects, as well as by short-term effects, such as multipath

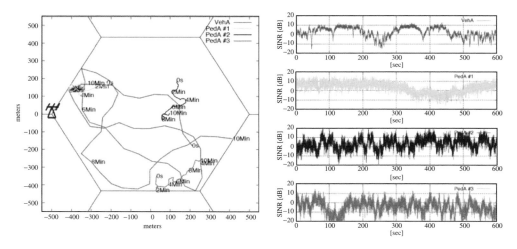

Figure 7.4 Example of user movement over 10 minutes and corresponding SNIR without power control.

propagation and Doppler spread. Note also that for the vehicular user, the SNIR shows faster variations than for the pedestrian ones.

Despite these challenging reception conditions, the radio network must still be able to provide radio bearers with the appropriate QoS as requested by the service. To this end, the radio resource management typically relies on a number of QoS methods at different layers. However, it is also important to understand that the radio efficiency is strongly affected by both the requested QoS and the varying channel quality: for example, a QoS request combining high bit rates, low error rates and low latency generally requires a huge amount of transmission resources. Therefore, it is important that the available QoS methods are adapted to the service constraints to efficiently deliver multimedia and video applications with the least amount of resources.

Existing and emerging 3G systems, such as UMTS, High Speed Packet Access (HSPA), Long Term Evolution (LTE) or Multimedia Broadcast/Multicast Service (3GPP TS 26.346 2005; MBMS 2003) (MBMS) offer a variety of different QoS methods. At the physical layer, this includes, for example, fast power control, rate-adaptive forward error correction and modulation based on fast feedback from the receiver, or variable interleaving depth. These physical layer QoS methods are complemented by MAC layer strategies, such as hybrid Automatic Repeat reQuest (ARQ) and multi-user scheduling. Finally, any remaining block errors may be cleaned by RLC retransmissions. In combination, these methods enable multimedia services with a certain QoS guarantee, if the radio parameters are chosen appropriately. Together with the optimal selection of any application layer QoS tools and parameters, this already constitutes a quite complex system design problem. However, due to the limited amount of side information that can be passed across the interfaces between the layers, potential cross-layer optimization is often neglected. Later in this chapter we will discuss several 3rd-generation (3G) multimedia services in more detail and highlight such cross-layer optimization potentials, taking into account QoS methods at both the radio and application layers.

7.1.5 QoS using Video Error Resilience

Modern video coding standards such as H.264/AVC provide a large set of video coding tools for applications in error-prone environments. These tools include slice-based coding, data partitioning, Flexible Macroblock Ordering (FMO) and so on (Ostermann *et al.* 2004; Stockhammer *et al.* 2003). Other tools such as long-term memory motion estimation, Switching Intra (SI) and Switching Predictive (SP) frames and so on, when used appropriately, can also provide improved error resilience to combat packet losses and resulting video decoding error propagation. Interestingly, however, few of these tools have found practical applications in deployed systems for a number of reasons: first and foremost is the lack of decoder support in deployed mobile devices. Error resilient video coding tools such as FMO, data partitioning and slice-based coding require additional buffering and complicate memory access patterns in the encoding and decoding devices, which can lead to significant increase in chip size, power consumption and therefore cost. As a trade-off for improved error resilience, these video coding tools also have lowered video coding efficiency than comparable, less error resilient tools, which also makes them less attractive for usage in bandwidth-starving mobile applications. Secondly, although these error-resilient video coding tools can significantly improve tolerance of the coded video bitstream to transmission losses, the benefits of using these tools are not easily quantifiable or even predictable. The gains when using these tools are not guaranteed and are functions of many factors including the information field (header, motion and texture data) of the compressed bitstream that has been lost, the Macroblock (MB)(s) that have been impacted (for example, foreground versus background areas, static versus moving regions), the characteristics of the lossy channel (for example, uniform versus bursty packet losses, the duration of typical bursts of losses and so on). These factors make error-resilient video coding tools nice 'add-ons' to other means of provisioning QoS in the network and transport layers, as opposed to serving as standalone tools for guaranteeing QoS.

7.2 Selected QoS Tools – Principles and Experimental Results

In this section we will provide an overview of available QoS tools, analyse their applicability to different applications and provide selected use cases.

7.2.1 3G Dedicated Channel Link Layer

The integration of mobile video services into 3G networks requires the mapping of the video data units on packet-based bearers. Figure 7.5 shows a typical processing of a multimedia data unit, for example a NAL unit in the case of H.264, encapsulated in RTP/UDP/IP (Wenger *et al.* 2005) through the protocol stack of a wireless system. We will in the following concentrate on UMTS terminology; the corresponding layers for other systems are shown in Figure 7.5. After Robust Header Compression (RoHC) (Hannu *et al.* 2001) the IP/UDP/RTP packet is encapsulated into one Packet Data Convergence Protocol (PDCP) packet, then

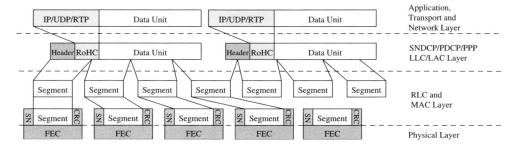

Figure 7.5 Processing of a multimedia data unit through the protocol stack of a wireless system such as GPRS, EGPRS, UMTS or CDMA–2000.

becomes an RLC-Service Data Unit (SDU). This SDU typically has a larger size than a RLC-Packet Data Unit (PDU) and is segmented into smaller units or link layer packets of length k_{LLP} which serve as the basic units to be transmitted within the wireless system. The length of these segments depends on the bearer of the wireless system and the Coding Scheme (CS) or Modulation and Coding Scheme (MCS) in use. The RLC layer in wireless systems can operate in one of basically two different modes, Unacknowledged Mode (UM) and Acknowledged Mode (AM). For all RLC modes, sequence numbering and Cyclic Redundancy Check (CRC) error detection is performed. However, whereas the UM is unidirectional and data delivery is not guaranteed, in the AM an ARQ mechanism is used for backward error correction. In the latter mode, usually a persistent ARQ is used, that is, retransmissions of erroneous packets are performed until the link layer packet is correctly received. The physical layer generally adds FEC to RLC-PDUs depending on the coding scheme used such that a constant length channel-coded block is obtained. This channel-coded block is further processed in the physical layer before it is sent to the far end receiver. There, error correction and detection are performed, and retransmissions may be requested. In general, the detection of a lost segment of an IP packet results in the loss of the entire packet.

The transmission process of these link layer packets can be abstracted by a very simple model, namely by its size k_{LLP}, its Transmission Time Interval (TTI) T_{TTI}, its loss process $\mathcal{L}_{LLP}$ and its constant transmission delay δ_{LLP}. In other words, the wireless system transmits the application data in chunks of packets of size k_{LLP} every T_{TTI} seconds. By the use of sufficiently long block check sequences the probability of non-detected errors is extremely low and link layer packets can be assigned one of two different states at the receiver: they are correctly received or they are lost. The loss process $\mathcal{L}_{LLP}$ indicates both the loss probability of link layer packets and also the time correlation of the loss process. Independent link layer packet losses can in general not be assumed, but are appropriate for some systems as discussed in the following. We summarize this model described by the four parameters of the link layer of a wireless system model as $\mathcal{W}$, that is, $\mathcal{W} \triangleq \{k_{LLP}, T_{TTI}, \mathcal{L}_{LLP}, \delta_{LLP}\}$.

The modelling of 3G dedicated channels is of specific interest as they will be used for video and multimedia services in wireless environments. In the standardization for the H.264/AVC video coding standard, an environment had been proposed which allows appropriate modelling of 3G systems such as UMTS and CDMA–2000 (Roth *et al.* 2001).

Table 7.1 Characterization of 3G link layer patterns (Roth *et al.* 2001).

Pattern	Bit rate R (kbit/s)	Speed (km/h)	T_{TTI} (ms)	k_{LLP} (bit)	BER	LER	Mode
1	64	3	10	640	9.3×10^{-3}	1.2×10^{-1}	AM
2	64	3	10	640	2.9×10^{-3}	3.6×10^{-2}	AM
3	64	3	10	640	5.1×10^{-4}	1.1×10^{-2}	UM
4	64	50	10	640	1.7×10^{-4}	3.7×10^{-3}	UM
5	128	3	5	640	5.0×10^{-4}	1.2×10^{-2}	UM
6	128	50	5	640	2.0×10^{-4}	3.5×10^{-3}	UM

3G systems are designed to guarantee a certain QoS. Therefore, it can be assumed that channel characteristics for the duration of the transmission are stationary, and the provided network QoS in terms of bit-rate and packet loss rate is assumed to be nearly constant throughout the session. To model the radio channel conditions and loss characteristics, error patterns are used that were captured in different real or emulated mobile radio channels. The error patterns reflect different speeds and different QoS requests on the link layer. They have been captured above the physical layer and below the link layer such that in practice they act as the physical layer simulation. As only entire link layer packets of size $k_{LLP} = 640$ bit are checked, the bit error pattern can be mapped to a loss process of user data of length $k_{LLP} = 640$ bit. The TTI depends on the supported data rate; for the most common service with bit-rate $R = 64$ kbit/s the TTI results in $T_{TTI} = 10$ ms. The characteristics of Bit Error Ratio (BER) and LLC error rate (LER) for the different bit error patterns, together with the target transmission mode, are shown in Table 7.1.

Patterns 1 and 2 are mostly suited for use in conjunction with the AM for video streaming applications; LERs up to 12% are typical. Patterns 3 to 6 are meant to simulate a more reliable, lower error rate bearer that is required in conversational applications with LERs around and below 1%. Note that with higher speed (50 km/h) the channel tends to have lower error rate than in the case of the walking user (3 km/h) as the fading process is faster and the channel is more ergodic. For the AM, the retransmission delay δ_{RT} depends on the system configuration, but is in general below 100 ms.

7.2.2 Experimental Results for Conversational Video

Figure 7.6 shows the use of a hybrid video codec such as H.264/AVC in a Real-time Transport Protocol (RTP) (Schulzrinne *et al.* 2003) based conversational environment in UMTS. For the following simulations H.264/AVC error resilience tools, for example introduced in Stockhammer (2006) and Stockhammer *et al.* (2003) and references therein, are now applied also to mobile communication environments. The detailed setup and parameters are presented in the following. The simulations are based on Roth *et al.* (2001).

The experimental results based on the test conditions in Roth *et al.* (2001) demonstrate the impact of the selection of different error resilience and error concealment features on the quality of the decoded video for MPEG-4 as well as H.264/AVC. The bearer cases (3) and (4) according to Table 7.1 are used. As a performance measure the average luminance Peak Signal to Noise Ratio ($\overline{PSNR}$) is chosen as suggested in Roth *et al.* (2001).

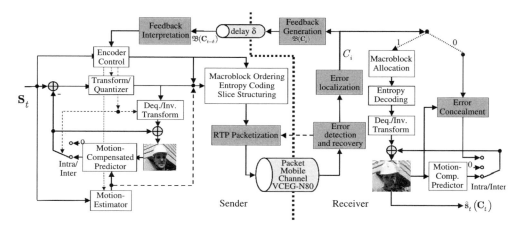

Figure 7.6 Low-delay video transmission in RTP-based mobile environment.

The average is taken over at least 200 transmission and decoding runs. Protocol overheads are appropriately taken into account according to Roth *et al.* (2001). Results are shown for the Quarter Common Intermediate Format (QCIF) test sequence 'foreman' of length 10 s coded at a constant frame rate of 7.5 f/s with a single intra frame in the beginning and only P-frames afterwards. Encoding is performed using slice-structured encoding with different maximum slice sizes S_{max}. In Figure 7.7, we present the results in $\overline{PSNR}$ over the applied end-to-end delay at the decoder, Δ. Thereby, for the relevant delay components in a video transmission system only the encoder buffer delay $\Delta T^{(\mathcal{B}_e)}$ and the transmission delay $\delta^{(C)}$ on the physical link is considered, that is, $\Delta \triangleq \Delta T^{(\mathcal{B}_e)} + \delta^{(C)}$. Additional processing delay as well as transmission delays on the backbone networks may add in practical systems, but are not further considered.

Figure 7.7(a) shows the performance for bearer case (3) according to Table 7.1, whereby all curves (1)–(6) use the UM. The system designs according to curves (1)–(4) do not make use of any fast feedback channel, but it is assumed that the encoder has knowledge of a link layer loss rate of about 1%. Therefore, these modes may also be applied in video conferencing with multiple participants. Specifically, in curves (1), (2) and (3) CBR encoding with bit rates 50, 60 and 52 kbit/s, respectively, is applied to match the bit rate of the channel taking into account the overhead. Curve (1) relies on slices of maximum size $S_{max} = 50$ bytes only; no additional intra updates to remove error propagation are introduced. Curve (2) in contrast neglects slices, but uses optimized intra updates with $p = 4\%$, and curve (3) uses a combination of the two features with $S_{max} = 100$ bytes and $p = 1\%$. The transmission adds a delay of about 170 ms for the entire frame. For initial delays lower than 170 ms, NAL units are lost due to late arrival.

When initial playout delays are higher than this threshold, only losses that are introduced by link errors will occur. If small initial playout and end-to-end delay is less critical, a similar performance can be achieved by Variable Bit Rate (VBR) encoding combined with FMO with five slice groups in checkerboard pattern as well as optimized intra mode selection with $p = 3\%$ as shown in curve (4). However, the VBR encoding causes problems for low-delay

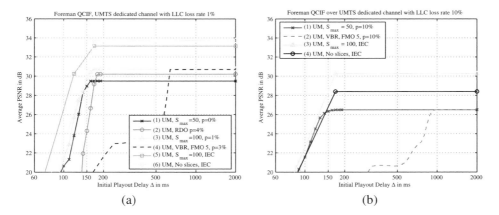

Figure 7.7 Average PSNR for different low-delay video transport systems over initial playout delay Δ for UMTS dedicated channel with link layer error rates of (a) 1% and (b) 10% LLC loss rate (see Table 7.1, Bearers (3) and (1), respectively).

applications in mobile bottleneck links and, therefore, a CBR-like rate control is necessary. Systems according to curves (5) and (6) assume the availability of a feedback channel from the receiver to the transmitter, which is capable of reporting the loss or acknowledging the receipt of NAL units: Interactive Error Control (IEC) with a feedback delay of roughly 200 ms ($d = 2$) is applied. For the slice mode with $S_{max} = 100$ bytes as shown in curve (5) significant gains can be observed for delays suitable for video telephony applications, but due to the avoided error propagation it is even preferable to abandon slices and only rely on IEC as shown in curve (6). The average PSNR is about 3 dB better than the best mode not exploiting any feedback.

In summary, the application of advanced error resilience schemes in standard-based environments can improve the video quality significantly. If feedback information is available and completely lost frames are tolerable to some extent, it is better to avoid slice structured coding so as to achieve better compression efficiencycompression efficiency. In the case of available feedback, the simple IEC system design without requiring the expected decoder distortion and slice structuring outperforms many highly sophisticated error resilience schemes as long as the delay of the feedback is reasonably low.

7.2.3 Experimental Results for Moderate-delay Applications

For applications for which short delay is less critical, other modes may be utilized to enhance the quality over UMTS networks. In Figure 7.8 we have added some system configurations which generally introduce higher delays. Specifically, Figure 7.8(a) shows in addition to the modes in Figure 7.7(a) the performance for bearer case (3) according to Table 7.1 using the UM for curve (7) and the AM for curves (8) and (9). For the system according to curve (7) the feedback is exploited for application layer retransmission of RTP packets and VBR encoding is used. It is obvious that this mode is not suitable

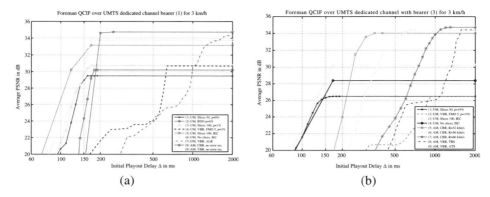

Figure 7.8 Average PSNR for different advanced video transport systems over initial playout delay Δ for UMTS dedicated channel with link layer error rates of (a) 1% and (b) 10% LLC loss rate (see Table 7.1, Bearers (3) and (1), respectively).

for low-delay applications, but if end-to-end delay is of less relevance, it provides better performance than any other scheme relying on methods in the video layer. In addition, it is important to note that no error resilience is necessary in this case. Finally, curves (8) and (9) show the performance of CBR encoded video and VBR encoded video, respectively, with matching bit rates for the acknowledged mode. The performance of the CBR mode is excellent even for lower delays, but at least 200 ms of initial playout delay must be accepted, which makes the applicability for conversational modes critical, but not infeasible, if the system supports fast retransmissions. It is also observed that for VBR encoding low-delay applications cannot be well supported, but if initial playout delays of a few seconds can be accepted, VBR encoding with acknowledged mode on the link layer provides the best overall performance.

For the UMTS bearer case (3) with 10% link layer loss rate, advanced transport system enhances the overall system performance significantly. Figure 7.8(b) shows that significantly better performance can be achieved by the use of the acknowledged mode, but only for initial playout delays well over 300 ms according to curve (5) with CBR and bit rate 52 kbit/s. Interestingly, if the initial playout delay is increased, one can also support higher bit rates resulting in higher quality. This behaviour has been exploited in the HRD specification of H.264 where it was recognized that an encoded stream is contained not just by one but by many leaky buckets (Ribas-Corbera *et al.* 2003). Finally, the systems according to curves (8) and (9) show the performance for VBR encoded video in the case of Timestamp Based Streaming (TBS) (real-time streaming whereby packets are sent according to their RTP timestamp) and Ahead-of-Time Streaming (ATS) (progressive download for which packets may be sent faster) over the AM mode. It is interesting that with ATS low playout delays can be achieved, but ATS is obviously restricted to non-live sequences and requires more buffering in the client. In addition, in practical systems some start-up delay might occur due to TCP-like congestion control. It is also worth noting that the performance of video over the 10% link layer loss bearer does not differ significantly from the 1% one if the initial playout delay constraints are not stringent.

Table 7.2 Proposed video and transport features for different applications with performance in terms of delay and average PSNR for different RLC-PDU loss rates and QCIF video sequence 'foreman' coded at 7.5 f/s.

Video application	Video features	Transport features	1% RLC-PDU loss rate		10% RLC-PDU loss rate	
64 kbit/s UMTS transmission scenario			Delay	PSNR (dB)	Delay	PSNR (dB)
Download-and-play On-demand streaming	VBR, no error res., playout buffering	ATS, AM on RLC	>15 s	35.2	>1 s >10 s	34.4 35.1
Live streaming	CBR/VBR, no error res., playout buffering	TBS, AM on RLC	>250 ms	34.7	>400 ms >1.5 s	34.0 34.7
Broadcast	VBR, regular IDR, no other error resilience	FEC, long TTI, fragmentation	>5 s	34.5	>5 s	32.0
Conferencing	CBR, intra updates, slices	UM	>150 ms	30.7	150 ms	26.5
Telephony	CBR, IEV, no slices	UM	>150 ms	33.7	>150 ms	30.2

7.2.4 System Design Guidelines

The obtained results allow different options to be compared for different applications. A summary of proposed video and transport features for the video test sequence 'foreman' over a 64 kbit/s UMTS link with RLC-PDU loss rates of 1% and 10% is provided in Table 7.2. For more details we refer to (Stockhammer 2006). In addition to the discussed schemes, we have also added guidelines and results for the case of broadcast applications. Details on the system design for broadcast applications, especially the use of erasure-based FEC, are discussed in Stockhammer (2006) and Stockhammer *et al.* (2008).

For download-and-play as well as on-demand streaming applications with initial playout delays beyond one and two seconds, the video should be encoded primarily for compression efficiency, that is, relatively relaxed VBR rate control and no explicit error resilience features. The reliability should be provided in the link layer by using AM on the radio link layer. The resulting delay jitter can be compensated with playout buffering. ATS can be applied if the receiver buffer has sufficient size. For higher error rates, the quality even scales with the initial playout delay as the jitter can be better compensated for larger delays. For live applications, TBS must be applied. For lower requested delays in the range of 250 to 500 ms, CBR-like rate control is also preferable. Video error resilience is still not essential as the AM provides sufficient QoS.

For broadcast applications without any feedbacks, it is proposed to apply extensions of the FEC. This can be accomplished by longer TTIs in the physical layer and/or application layer FEC. In addition, we suggest using regular IDR frames for random access and error resilience, but a relatively relaxed rate control. For low RLC-PDU loss rates, the FEC based scheme is almost as good as the one relying on feedback mechanisms. However, for higher loss rates the acknowledged mode with RLC layer retransmission outperforms application layer FEC by about 2–3 dB. In any case, the application layer FEC adds delay.

For low-delay applications , the video encoder should apply CBR-like rate control and the transport and link layer basically must operate in UM without any retransmission or deeply interleaved FEC schemes. The video application itself must take care to provide sufficient robustness. In case that feedback is not available or only limited to reporting statistics, for example, in conferencing applications, more frequent intra-MB updates based on robust mode decision as well as slice-structured coding are proposed.

However, compared with the acknowledged mode, significant degradations in the video quality must be accepted, especially if the RLC-PDU loss rates are high. Therefore, the physical layer must provide sufficient QoS to support these applications. For video telephony, the fast feedback channel can be exploited for Interactive Error Control (IEC). No additional means of error resilience are necessary. In this case and for low loss rates, the achieved video quality is significantly better, about 3 dB, when compared to video error resilience without feedback. The degradation compared to reliable download-and-play applications is only about 1.5 dB. The system design for video communication in emerging multi-user systems such as HSDPA and MBMS requires new design principles. More details on this subject can be found in Jenkač *et al.* (2004), Liebl *et al.* (2004), Stockhammer (2006), Luby *et al.* (2007) and Stockhammer *et al.* (2008). Some further specific service examples are discussed in the following.

7.3   Selected Service Examples

Based on the previous general discussions demonstrating significant room for optimization, we will in the following discuss three service examples in some more detail. We focus on techniques improving performance by exploiting information across 3GPP layers, while maintaining standard compliance.

7.3.1   Multimedia Telephony Services

Service Overview

3GPP has just recently completed the definition of Multimedia Telephony Service for IMS (3GPP TS 26.144 2007) (MTSI), including the definition of media handling and interaction. An MTSI client supports conversational speech, video and text transported over RTP with the scope to deliver a user experience equivalent to or better than that of circuit-switched conversational services using the same amount of network resources. The basic MTSI client specifies media codecs for the three different media components, all of which are transported over RTP as shown.

The specification defines QoS profiles for bidirectional voice, video and text. For a 128 kbit/s video a QoS profile with packet loss rate of at most 0.7% is proposed, which is considered a reasonable trade-off between radio efficiency and service quality under stringent delay constraints. The service also supports extended RTCP-based feedback messages to facilitate reporting of receiving conditions to the sender. For audio, the service defines a number of packet loss handling mechanisms, such as redundant audio packets. For video, however, no packet loss handling is defined. Obviously, the service requires low end-to-end

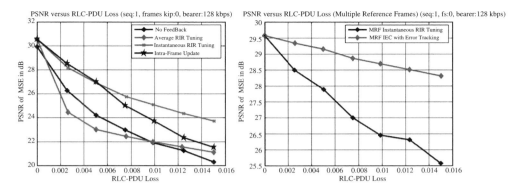

Figure 7.9 PSNR of MSE over loss rate for different settings.

delay to support telephony applications, which limits the applicability of QoS methods at the radio level.

Cross-Layer Optimization Potentials

Although transmission errors and losses are tolerable to some extent in conversational video, they will always degrade service quality. On the other hand, unduly requesting lower error rates would be likely to result in unreasonably large resource consumption and thereby make such a service commercially unattractive. The main problems created by errors and losses in video transmission are not necessarily the immediate effects of the packet loss, but the resulting error propagation (Stockhammer *et al.* 2003). Simplistic application of typical error resilience tools, such as slices or intra updates, as a result of the lowered entropy coding efficiency of these tools, will either degrade video quality after encoding in the error-free case, or lead to non-adaptive and often insufficient error protection when errors occur, given the highly varying nature of wireless transmission.

As an example, we have investigated several error resilience methods in the context of multimedia telephony services making use of frequent feedback messages (Stockhammer 2006) under strict constraints according to the framework of reference 3GPP TS 26.144 2007. Feedback delay, bit-rate restriction and signalling constraints are taken into account. Figure 7.9 shows video performance in terms of PSNR of average MSE over the loss rate of PDUs at the RLC layer. The left-hand side shows the performance in the case of no feedback, but Random Intra macroblock Refresh (RIR) of 5% MBs. The drop in quality is significant and, for the proposed low loss rates of 0.7%, the degradation will be noticeable. Adaptive intra refresh methods are also shown. Average RIR tuning attempts to adapt its RIR rate based on long-term feedback, a method that is quite unstable and does not provide sufficient results. For the remaining two cases on the left-hand side, upon receiving a report of error, the encoder introduces a significant amount of intra information, either as an entire intra update, or as very high RIR ratio that is then slowly decreased over time. As a constant bit rate needs to be maintained, the RIR refresh may perform better as the rate control is less challenging.

More advanced methods are applied on the right-hand side: here the encoder uses more than one reference frame, such that IEC can be applied. In the case of a reported error the encoder does not use lost frames for further referencing. This method shows very good performance over the whole range of loss rates and is significantly better than even the best RIR tuning method. However, this method requires fast and timely feedback and larger memory, but can tolerate much higher loss rates. The potential of cross-layer design is obviously significant, as now radio resources may be saved based on the available memory in encoder and decoder, the timely feedback and so on. The support of such features is achieved by appropriate usage of defined protocols, but obviously comprehensive evaluation and characterization of such methods is required before they are included in standards and commercial deployments.

7.3.2   Multimedia Download Delivery

Service Overview

3GPP has recognized that point-to-point delivery of multimedia content might be quite costly in terms of radio resource consumption, if the same content is distributed to several or many users. Therefore, MBMS (3GPP TS 26.346 2005; MBMS 2003) has been introduced to provide an efficient means for reliable distribution of multimedia content over 3G networks. Content that is intrinsically not time-sensitive can be provided via a file push delivery service. The appealing characteristic of these subscription based services is that the content can be enjoyed by users at their convenience, anywhere and anytime. In this case, the transmission bit rate, latency and transfer delay are rather unimportant, as the content is consumed off-line anyway.

To deliver a file in a broadcast session, File Delivery over Unidirectional Transport (FLUTE) (Paila *et al.* 2004) provides mechanisms to signal and map the properties of a file to the Asynchronous Layered Coding (ALC) protocol, such that receivers can assign these parameters to the received files. A file is partitioned in one or several so-called source blocks. Each source block represents a single channel coding word for the application layer FEC. Each source block is split into k source symbols. For each source block, additional repair symbols can be generated by applying Raptor encoding (Luby *et al.* 2007). Symbols are mapped to FLUTE payloads and distributed over the IP multicast MBMS bearer. MBMS bearers are defined by the transport format size and number of transport blocks that are to be protected by a physical layer channel code at every TTI. The TTI is adaptive and can be selected from the set {10 ms, 20 ms, 40 ms, 80 ms}. Higher values mean longer physical layer interleaving and/or longer codeword size, but at the expense of higher latencies and higher decoder memory requirements. The receivers collect correctly received FLUTE packets, and with the information available in the packet header and the session setup, the Raptor decoder attempts to recover the source block from all received data. If all source blocks belonging to the file are received correctly, the entire file is recovered.

The MBMS system applies FEC on both the application and the physical layers in a complementary way. Other interesting radio bearer parameters are the required transmit power, as well as the achievable transmission bit rates, especially for non-time-sensitive applications. For the assessment of different system configurations, basically two aspects are of major interest: the user perception of the multimedia delivery, as well as the total resources consumed at the physical layer. The latter are quite suitably measured by the time it takes to distribute the file and the power assigned during this interval. The product of these two values, namely the assigned transmit power times the 'on-air time' represents the total energy to distribute the file. It is interesting to determine the minimum energy required to support 95% of the user population and to identify the parameter settings for achieving this minimum.

Cross-Layer Optimization Potentials

Distribution of multimedia files is generally not delay-sensitive, and allows the use of long FEC code words that adapt to the content to be delivered. In this case, the physical layer provides a basic QoS, but the recovery of the data happens on the application layer by the use of the application layer FEC. The question is, how much error correction capability should be spent at which layer without harming the overall efficiency? In Luby *et al.* (2007) the MBMS download delivery service for the UMTS was investigated under realistic physical transmission conditions to determine optimized system parameters at radio and application layers. Figure 7.10 provides some selected results for a 240-kbaud UMTS radio bearer, a file size of 512 kbyte (corresponding to a 30 s video clip) and an ensemble of 1000 users moving randomly with 30 km/h. Different system configuration points were investigated, namely, the transmit power has been varied from 0.5 W to 16 W and the physical layer code rate was varied from 0.33 to 1. The application of a combination of transmit power and physical layer code rate results in a certain RLC-PDU loss rate after physical layer channel decoding. For each of the 1000 users, the observed RLC-PDU loss rate was evaluated, along with the amount of application layer FEC overhead that was necessary for perfect file reconstruction, and the largest delay reached until the file could be reconstructed. In addition, the worst among 95% of the supported users was determined, and its respective RLC-PDU loss rate is shown on the x-axis of Figure 7.11. On the y-axis, the necessary energy to support this user is plotted. Note that the transmit time is determined basically by the applied code rate on the physical layer and the overhead on the application layer. Figure 7.10 shows these plots for different transmit powers. For an operation point with typically requested bearer QoS on the upper-left corner of the figure with physical layer FEC code rate 0.33 and high transmit power of 8 W, the RLC-PDU loss rate is rather low at around 5%. Therefore, to support 95% of the users, only 15% Raptor overhead is necessary. The file is delivered in 65 s and the required energy is therefore about 520 J. In contrast, if a different operation point with FEC code rate 0.43 and low transmit power of 0.5 W is chosen, then, to support 95% of the users, 175% Raptor overhead is necessary. The file delivery in this case takes about 120 s. Nevertheless, despite the fact that the delivery time is increased by about a factor of two, the overall energy consumption is reduced by a factor of ten compared to the other operation point.

This example shows the benefits of combining application layer FEC with physical layer FEC. The awareness and exploitation of QoS schemes on different layers can improve the system performance significantly, especially of the service constraints and the effects on the radio resource consumptions.

7.3.3   Multimedia Streaming Services over MBMS

Service Overview

The real-time MBMS streaming service targets more classical mobile TV services. In this case, timely delivery is much more relevant than for download delivery services, to allow, for example, interactivity and fast channel switching. For these services the same multicast

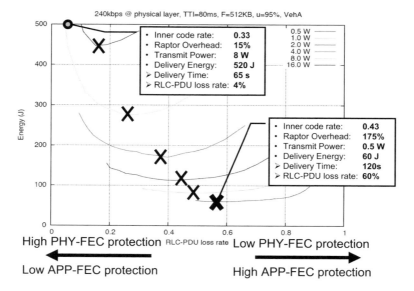

Figure 7.10 Selected results for MBMS download.

radio bearers as for the MBMS download delivery are used. For delivering the content the MBMS streaming framework including Raptor codes (Shokrollahi 2006) provide application layer QoS. The streaming framework operates on RTP packets, or more precisely on UDP flows transmitted over different ports. The UDP payloads form a source block of size k symbols, from which $n - k$ repair symbols are generated which are transported in repair packets. The number of source packets included in one source block and the service bit rate defines the protection period P of the FEC code, and the number of repair symbols determine the overhead as $O = (n - k)/k - 1$. Both parameters can be adjusted. In addition to the AL-FEC parameters and the radio parameters for the IP multicast bearer, the video coding parameters can also be selected based on the bit rate and the error resilience and tune-in properties determined by the random access point frequency. Further criteria are the achievable video quality at the decoder, the target channel switching times and the required system resources. The selection of the parameters obviously should be such that the user satisfaction is maximized for as many users as possible in the service area with the least amount of required resources for a given environment and user mobility.

Cross-layer Optimization Potentials

The optimization potentials for mobile TV services are manifold, especially as the service quality is more complex than download delivery. For example, for optimizing channel switching times, Internet Streaming Media Alliance (ISMA) (Singer *et al.* 2006) has defined a summary of interesting technologies, which include practical cross-layer aspects. In terms of radio resource usage, similar trade-offs are of interest as for download delivery. Therefore, comparable investigations as for the file delivery case have been performed for the same set of radio parameters and the video sequence party from reference 3GPP TR 26.902 2007. The

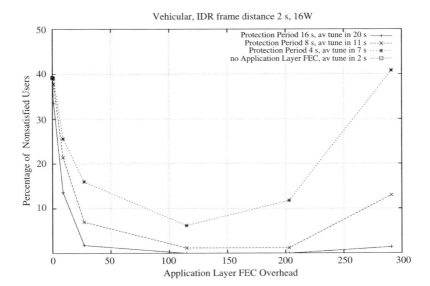

Figure 7.11 Percentage of non-satisfied users as a function of application layer overhead for different system configurations.

30 s sequence was looped 15 times. The sequence was encoded with IDR frame distances of 2 s and the target quality for a satisfied user was set to an average PSNR of at least 32 dB. The resulting bit rate of the video stream was approximately 100 kbit/s. Furthermore, the applied protection periods for the application layer FEC were $P = 4$, 8 and 16 s, and the AL-FEC overhead was selected optimally to fill the IP bearer for a selected physical layer code rate of $r = \{0.24, 0.24, 0.26, 0.3, 0.5, 0.7, 0.9\}$ resulting in $O = \{0, 1, 10, 27, 115, 203, 291\}\%$ overheads. It transpired that only full power usage of 16 W provided satisfactory results. Each experiment was carried out for 500 users that were all assumed to move at a speed of 30 km/h in the serving area. For video quality evaluation, a percentage of Degraded Video Duration (pDVD) (for the respective definition, see reference 3GPP TR 26.902 2007) of 5% was considered a satisfying quality. The benefits of application layer FEC are investigated along with the influence of the protection period. Figure 7.11 shows the percentage of non-satisfied users over application layer overhead for constant system resources, IDR frame distance 2 s and different protection periods P compared to no application layer FEC. Along with the different configurations for the protection periods, the average tune-in delays are also reported.

Without application layer FEC and with only physical layer FEC, the performance of the system is not satisfactory, that is, only 60% of the users can be supported despite the relatively low physical layer code rate chosen. With the use of application layer FEC, more users are supported. The number of non-satisfied users decreases for a fixed protection period of 4 s, which amounts to shifting some of the FEC part from the physical layer to the application layer. A reasonably good operation point is when physical layer code and application layer code use about the same rate of 0.5, which corresponds to the 100% overhead case. If the physical layer code rate gets too high (corresponding to high application layer overhead), then the performance decreases again. It is also obvious from the result that, with longer protection periods, more and more users can be supported. With 16 s protection period and code rate 1/2 for each layer, almost all users observe satisfying quality. However, the introduction of application layer FEC increases the tune-in delay. Hence, this trade-off needs to be taken into account in the system design. In addition, with some methods proposed in Singer *et al.* (2006), the channel switching delay can be significantly reduced.

7.4   Conclusions

This chapter provides an overview of 3GPP video services and some specific examples for cross-layer optimizations in existing and emerging systems. Although advanced video error resilience tools have not been widely deployed in existing 3G video services, it is demonstrated that with suitable cross-layer designs utilizing adequate QoS means on different layers, significant performance gains can be achieved when video error resilience tools are applied. We have omitted some further interesting use cases, including, for example, cross-layer aspects in Packet Switched Streaming (3GPP TS 26.234 2006; 3GPP TR 26.937 2004) (PSS). Promising results have been presented (3GPP TS 26.234 2006; Schierl *et al.* 2005; Stockhammer 2006) in this case as a result of the more relaxed time constraints, despite real-time delivery. Most of the provided references include additional references for further study of these interesting and compelling services. However, it is also worth mentioning that many mobile video services are accomplished nowadays using standard Internet video platforms such as YouTube or others. To investigate and analyse managed and unmanaged video services it is of great importance to use adequate test, simulation and evaluation tools for multimedia services over emerging and future mobile systems – simple tools are not sufficient to reach relevant conclusions. For example, the Nomor Research 3GPP application test and simulation platforms (see http://www.nomor.de) offer both real-time and non-real-time investigation of various cross-layer aspects and provide the option of implementation and assessment for practical system and cross-layer optimizations for 3GPP services and also other IP based applications.

References

3GPP TS 26.114 2007 (2007) *Multimedia Subsystem (IMS); Multimedia telephony; Media handling and interaction (Rel. 7)*, Mar.

3GPP TS 26.234 2006 (2006) *Transparent end-to-end Packet-switched Streaming Service (PSS); Protocols and codecs (Rel. 6)*, Apr.

3GPP TS 26.235 2007 (2007) *Packet switched conversational multimedia applications; Default codecs (Rel. 7)*, Mar.

3GPP TS 26.236 2007 (2007) *Packet switched conversational multimedia applications; Transport protocols (Rel. 7)*, Mar.

3GPP TS 26.346 2005 (2005) *Multimedia multicast and broadcast service (MBMS); Protocols and codecs (Rel. 6)*, Sept.

3GPP TR 26.902 2007 (2007) *Video Codec Performance (Rel. 7)*, June.

3GPP TR 26.937 2004 (2004) *Transparent end-to-end packet switched streaming service (PSS): RTP usage model*, Mar.

Fielding, R., Gettys, J., Mogul, J., Frystyk, H. and Berners-Lee, T. (1997) *Hypertext transfer protocol –HTTP1.1, Request for Comments (standard) 2068*, Internet Engineering Task Force (IETF).

H120 n.d. (1984, 1988) *Codec for Videoconferencing Using Primary Digital Group Transmission.* Ver. 1, 1984; ver. 2, 1988.

H264 2003 (2003) *Advanced Video Coding for Generic Audiovisual Services.*

H261 n.d.*a* (1984, 1988) *Video Codec for Audiovisual Services at p×64 kbit/s.* Ver. 1, Nov. 1984; ver. 2, Mar. 1988.

H263 n.d.*b* (1998, 2000) *Video Coding for Low Bit Rate Communication.* Ver. 1, Nov. 1995; ver. 2, Jan. 1998; ver. 2, Nov. 2000.

Hannu, H. *et al.* (2001) *Robust header compression (ROHC): Framework and four profiles: RTP, UDP, ESP, and uncompressed, Request for Comments (standard) 3095*, Internet Engineering Task Force (IETF).

Jenkač, H., Stockhammer, T. and Liebl, G. (2004) H.264/AVC video transmission over MBMS in GERAN. In *Proceedings of IEEE 6th Workshop on Multimedia Signal Processing*, pp. 191–194.

Liebl, G., Jenkač, H., Stockhammer, T. and Buchner, C. (2004) Radio Link Buffer Management and Scheduling for Video Streaming over Wireless Shared Channels. In *Proceedings of International Packet Video Workshop*, Irvine, CA, USA.

Luby, M., Gasiba, T., Stockhammer, T. and Watson, M. (2007) Reliable multimedia download delivery in cellular broadcast networks. *IEEE Trans. on Broadcasting*, **53**(1), 235–246, Mar.

Luby, M., Watson, M., Shokrollahi, A. and Stockhammer, T. (2007) Raptor forward error correction scheme for object delivery. Request for Comments (standard) 5053, Internet Engineering Task Force (IETF).

MBMS 2003 (2003) *Multimedia Broadcast/Multicast Service (MBMS); UTRAN/GERAN Requirements.*

MPG 1993 (1993) *Coding of Moving Pictures and Associated Audio for Digital Storage Media at up to about 1.5 Mbit/s, Part 2: Video.*

MPG 1994 (1994) *Generic Coding of Moving Pictures and Associated Audio Information, Part 2: Video.*

Ostermann, J., Bormans, J., List, P., Marpe, D., Narroschke, M., Pereira, F., Stockhammer, T. and Wedi, T. (2004) Video coding with H.264/AVC: tools, performance and complexity. *IEEE Circuits and Systems Magazine*, **4**(1), 7–28, Apr.

Paila, T., Luby, M., Lehtonen, R., Roca, V. and Walsh, R. (2004) *FLUTE – File Delivery over Unidirectional Transport*, Request for Comments (standard) 5053, Internet Engineering Task Force (IETF), Oct.

Postel, J. (1980a) *DoD standard transmission control protocol, Request for Comments (standard) 761*, Internet Engineering Task Force (IETF).

Postel, J. (1980b) *User datagram protocol, Request for Comments (standard) 768*, Internet Engineering Task Force (IETF).

Postel, J. (1981) *Internet protocol, Request for Comments (standard) 791*, Internet Engineering Task Force (IETF).

Ribas-Corbera, J., Chou, P. and Regunathan, S. (2003) A generalized hypothetical reference decoder for H.264/AVC. *IEEE Trans. on Circuits and Systems for Video Technology* **13**(7), 674–687.

Roth, G., Sjöberg, R., Liebl, G., Stockhammer, T., Varsa, V. and Karczewicz, M. (2001) *Common test conditions for RTP/IP over 3GPP/3GPP2*, Doc. VCEG-N80, ITU-T SG16/Q6 Video Coding Experts Group (VCEG), Santa Barbara, CA, USA.

Schierl, T., Kampmann, M. and Wiegand, T. (2005) 3GPP Compliant Adaptive Wireless Video Streaming Using H.264/AVC. In *Proc. of IEEE Int. Conf. on Image Processing (ICIP05)*, Genoa, Italy, Sept.

Schulzrinne, H., Rao, A. and Lanphier, R. (1998) *Real-time streaming protocol (RTSP), Request for Comments (standard) 2326*, Internet Engineering Task Force (IETF).

Schulzrinne, H., Casner, S., Frederick, R. and Jacobson, V. (2003) *RTP: A transport protocol for real-time applications, Request for Comments (standard) 3550*, Internet Engineering Task Force (IETF).

Shokrollahi, A. (2006) Raptor Codes. *IEEE Trans. on Information Theory*, **52**(6), 2551–2567, June.

Singer, D., Färber, N., Fisher, Y. and Fleury, J. (2006) Fast channel changing in RTP. Technical Document ISMA TD00096, Internet Streaming Media Alliance, 2006.
http://www.isma.tv/technology/TD00096-fast-rtp.pdf

Stockhammer, T. (2006) Robust System and Cross-Layer Design for H.264/AVC-based Wireless Video Applications. *Special Issue on Video Analysis and Coding for Robust Transmission, EURASIP Journal on Appl. Signal Processing*, March 2006.

Stockhammer, T., Wenger, S. and Hannuksela, M. (2002) H.26L/JVT coding network abstraction layer and IP-based transport. In *Proceedings of IEEE International Conference on Image Processing*, Rochester, USA.

Stockhammer, T., Hannuksela, M. and Wiegand, T. (2003) H.264/AVC in wireless environments. *IEEE Trans. on Circuits and Systems for Video Technology*, **13**(7), 657–673.

Stockhammer, T., Shokrollahi, A., Watson, M., Luby, M. and Gasiba, T. (2008) *Application Layer Forward Error Correction for Mobile Multimedia Broadcasting*, CRC Press, Taylor & Francis Group.

Wenger, S., Hannuksela, M., Stockhammer, T., Westerlund, M. and Singer, D. (2005) *RTP payload format for H.264 video, Request for Comments 3984*, Internet Engineering Task Force (IETF).

Wiegand, T., Sullivan, G., Bjøntegaard, G. and Luthra, A. (2003) Overview of the H.264/AVC video coding standard. *IEEE Trans. on Circuits and Systems for Video Technology*, **13**(7), 560–576.

8

Cross-layer Error Resilience Mechanisms

Olivia Nemethova, Wolfgang Karner and
Claudio Weidmann

Real-time video services use the Real-time Transport Protocol (RTP) to encapsulate the data stream. The header of the RTP (Schulzrinne *et al.* 2003) is 12 bytes long. Each RTP packet is encapsulated into a User Datagram Protocol (UDP) packet, which adds a header with 8 bytes to the RTP packet. UDP (Postel 1980) is a transport layer protocol that does not provide any retransmission mechanism. It contains 8 bytes, one of them being the checksum for error detection. A UDP packet is further encapsulated into an Internet Protocol (IP) packet. The IP header has a size of 20 bytes for IPv4 and a size of 40 bytes for IPv6. The IP packet enters the UMTS network and is further processed as described below.

If the UDP checksum fails, the whole IP packet is typically discarded at the receiver. However, video applications can benefit from obtaining damaged data rather than from letting them discard by the network. In order to facilitate forwarding of damaged data to higher layers, UDP-Lite (Larzon *et al.* 2004) has been proposed. UDP-Lite provides a checksum with an optional partial coverage. When using this option, a packet is divided into a sensitive part that is covered by the checksum and an insensitive part that is not covered by the checksum. Errors in the insensitive part will not cause the packet to be discarded by the transport layer at the receiving end host. When the checksum covers the entire packet, which should be the default, UDP-Lite is semantically identical to UDP. Compared to UDP, the UDP-Lite partial checksum provides extra flexibility for applications capable of defining the payload as at least partially insensitive to bit errors.

In this chapter, possibilities of how to exploit the link layer information at the application layer and vice versa are discussed. Various cross-layer mechanisms for error detection and scheduling of video packets of a single user in UMTS are proposed and evaluated.

The investigated scheduling methods utilize the predictability of the errors at the UMTS radio link layer. Moreover, the semantics and content of the video are taken into account to determine the priority of the packets. Note that the scheduling of the information belonging to one stream of a single user, as discussed here, can be equally considered as a particular form of unequal error protection.

8.1   Link Layer Aware Error Detection

The smallest unit in which an error can be detected, without additional mechanisms assuming UMTS as the underlying system, is the Radio Link Control (RLC) Protocol Data Unit (PDU). The size of an RLC PDU is typically 320 bits. To enable the usage of RLC CRC information, this information needs to be passed from the RLC layer to the application layer, that is, to a video decoder. The application of UDP-Lite can ensure that the content of damaged IP packets is passed to the application layer. The passing of the CRC information can be performed as an implementation-specific change for the entities that implement the entire protocol stack (for example in mobile phones). An alternative was proposed in Zheng and Boyce (2001), where a new protocol called 'Complete UDP' (CUDP) is developed, enabling the link CRC information to be forwarded. This protocol, however, has not been standardized yet. In the following, we will analyse the benefit provided by the availability of the RLC CRC at the video decoder.

8.1.1   Error Detection at RLC Layer

Having the RLC CRC information at the video decoder, the position of the first erroneous RLC PDU within the slice can be specified. All correctly received RLC PDUs before the first erroneous one can be decoded successfully. This adds no computational complexity and there is no rate increase necessary. The RLC CRC is calculated over a fixed number of bits. Thus, the size of the picture area damaged as a consequence of an error in the RLC packet depends strongly on the content and on the video compression settings.

The benefit provided by detection of errors at the RLC layer is closely related to the distribution of errors at the UMTS link layer, more specifically, to the number of erroneous RLC PDUs within one IP packet. The size of IP packets may be variable or fixed and may typically be configured by the application layer. Choosing the size of the IP packets taking into account the channel configuration in lower layers, in particular further segmentation and/or reassembly, may also help to optimize the end-to-end performance. An example of distribution of the index of the first lost RLC PDU in a slice, illustrated in Figure 8.1, has been obtained by mapping the measured RLC error traces (compare to section 2.2) on the IP packets that contain the H.264/AVC encoded 'foreman' video sequence. The so-mapped RTP video stream was compressed using QP = 28; the number of bytes per slice was limited to 700 bytes at most.

The peak at the first position, which is approximately 27% for a static and 45% for a dynamic scenario, denotes the cases where the detection at the RLC layer does not provide any improvement because the error was detected immediately in the first RLC PDU. In the remaining cases the method is beneficial at no additional costs, that is, in rate or complexity. Even for the dynamic scenario, this method guarantees an improvement in more than 50% of the situations.

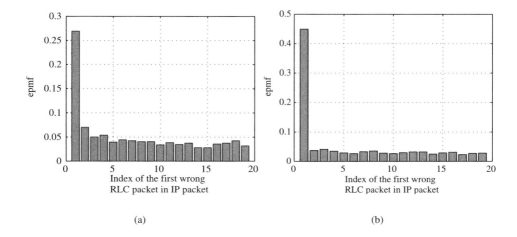

Figure 8.1 Distribution of the index of the first lost RLC PDU in a video slice for a (a) static and (b) dynamic mobility scenario.

8.1.2 RLC PDU Based VLC Resynchronization

Due to the desynchronization of the VLC after the first error within the slice, it is no longer possible to use successive RLC PDUs, although some of them might have been received correctly. The start of the next VLC codeword within the segment of the bitstream, corresponding to such RLC PDU payload, is not known. To use all correctly received RLC packets, adaptation of the slice size to the RLC packet size is an option. In particular, the size of the slice can be selected to correspond to the RLC packet size. Nevertheless, such a step is connected with a rate increase caused by the slice/NAL header and above all by the header of RTP/UDP/IP, which would thus be equally long as the slice itself (40 bytes if the typical size of 320 bits for RLC PDU is assumed). In section 6.4 a method of sending the macroblock position indicator every M macroblocks was introduced. This method allowed for resynchronization of the VLC stream at the position of the first VLC codeword that is entirely included within M macroblocks. To utilize all correctly received RLC PDUs, analogically, side information has to be sent for each RLC PDU to signalize the position of the first VLC codeword in the first macroblock/block that starts inside as introduced by Nemethova *et al.* (2005). If there is no Macro Block (MB) starting within the RLC PDU, a special codeword is sent. The isolated macroblocks, separated from the beginning *and* the end of the IP packet by erroneous RLC PDUs, are placed within the picture using a boundary or block matching algorithm.

The number of cases where this method leads to an improvement is shown in Figure 8.2, where the distribution of the number of erroneous RLC PDUs per slice is illustrated for the same video sequence and encoder settings. Only if all the RLC PDUs within an IP packet are erroneous this method does not succeed. The probability of not succeeding corresponds to the last column of the empirical probability mass function (epmf). A gain is possible in almost 99% of the situations for the static case. It is worsening for the dynamic scenario,

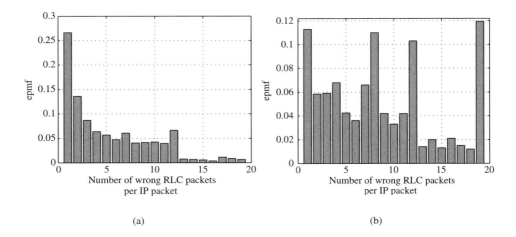

(a)   (b)

Figure 8.2 Distribution of the number of erroneous RLC PDUs per slice for a (a) static and (b) dynamic UE mobility scenario.

where the probability of all RLC PDUs within an IP packet being erroneous grows to almost 12%. Nevertheless, in about 88% of the cases an improvement can be gained by RLC layer detection combined with VLC resynchronization. The peaks of the epmf in Figure 8.2 are caused by the mapping between the TTI and an IP packet; an IP packet can be transported over more TTIs. Moreover, in the dynamic case, a bit stream switching causes a variable number of RLC PDUs per TTI. The peaks at 8 and 12 correspond mainly to the cases where all RLC PDUs within the TTI were damaged.

Side Information Encoding

The resynchronization information can be advantageously represented as the position of the start of the 'first MB starting in the RLC PDU'. The maximum length l_{max} of such Position Indicator (PI) within a q bit large RLC PDU is

$$l_{max} = \lceil \log_2 q \rceil \quad \text{[bit]}, \tag{8.1}$$

corresponding to a Fixed Length Code (FLC). For $q = 320$ bits long RLC PDUs the length of the fixed length PI codeword is $l = l_{max} = 9$. Fixed length coding has an advantage of robustness compared to variable length coding. The compression gain of a VLC depends on the distribution of the source symbols – here, the source symbols are the values of the PIs.

The epmf of macroblock sizes for I and P macroblocks (for QP = 30 and QP = 20), obtained by encoding the 'soccer match' video sequence, is presented in Figure 6.15. The distribution of MB sizes corresponds to the distribution of Length Indicators (LIs) in section 6.4. However, it does not correspond to the distribution of the PI values. The position of the first MB starting in the RLC PDU is in general uniformly distributed between zero (if an MB begins at the beginning of the RLC PDU) and the size of the previous MB minus

Table 8.1 Calculated average codeword length for chosen universal codes, calculated based on the distribution of MBPIs estimated from the 'soccer match' video sequence.

Code	$\bar{l}_I$, QP $= 20$	$\bar{l}_P$, QP $= 20$	$\bar{l}_I$, QP $= 30$	$\bar{l}_P$, QP $= 30$
Unary	78.38	90.40	96.24	43.90
Golomb (ideal)	7.55	7.89	8.03	6.89
Elias-γ	8.46	10.43	11.14	8.44
Elias-δ	7.76	9.64	10.33	8.02
Elias-ω	8.39	10.54	11.28	8.91
Entropy	6.07	7.50	7.80	6.61

one. Nevertheless, the shapes of the LI and PI distributions are similar. Moreover, there are only $q + 1$ values to be signalized for PI – q positions of MBs that begin within one RLC PDU, and one symbol signalizing that there is no MB starting within the q bits of the RLC PDU. Table 8.1 presents the mean codeword length of PIs encoded by VLCs introduced in section 6.4.

The ideal Golomb code is obtained for $g_{I20} = 63$, $g_{P20} = 62$, $g_{I30} = 63$ and $g_{P30} = 30$. Note that the higher lossless compression gain for I frame PIs than for P frame PIs and QP $= 20$ is caused by the high probability of the I MB size greater than q. The gain from using a VLC code is rather small. The Golomb code with an ideal g is nearest to the entropy but the gain is not more than 1 bit per codeword. Since an FLC is more robust against errors, it is in this case preferable to encode the PIs.

8.1.3   Error Detection and VLC Resynchronization Efficiency

To evaluate the efficiency of the presented methods the JM reference software (section 5.1.5) has been used. Missing parts of the video were concealed by the zero-motion method as explained in section 5.3.2. In JM, the RLC segmentation and PI generation were implemented. JM was further adapted to accept link error traces as an input. The baseline profile encoder settings, provided by the configuration file `encoder_baseline.cfg` included in JM, was used with the following changes:

- The slicing with number of bytes per slice was limited to 800.
- The period of I frames (Group of Pictures (GOP) size) was set to 50.
- The experiments were performed with QP $= 20, 22, \ldots, 30$ set equally for both I and P frames.
- The output file mode was set to RTP and RDO as well as rate control were switched off (low-complexity mode).

As a test sequence a 'soccer match' sequence was encoded with a frame rate of 10 f/s, which resulted in a duration of 45 min.

Figure 8.3 shows an empirical cumulative distribution function (ecdf) of the frame Y-PSNR for different error-handling methods for a 'soccer match' video sequence compressed with QP of 26. The two error detection methods (with and without VLC resynchronization, using TB CRC) are compared to error-free decoding and to an approach where the whole

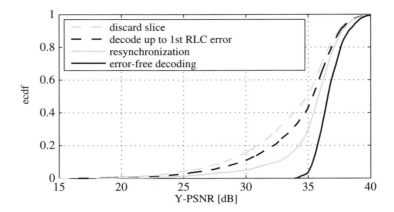

Figure 8.3 Quality of video reconstruction at the receiver for different error-handling mechanisms.

UDP packet is discarded if there was an error detected by UDP checksum. Note that the ecdf contains Y-PSNR of all frames – erroneously received, error-free received but containing impairments due to error propagation and error-free with compression impairments only. To introduce errors, the RLC error traces obtained from measurements in a live UMTS network were used (section 2.2). The overall probability of a TB error was thus 0.266% and the resulting UDP/IP packet error rate was 0.888%. This low error rate is also the reason for the predominantly high Y-PSNR values. The quality improvement achieved by decoding until the first error occurrence is in 20% of cases greater than 2 dB. The gain achievable by the VLC resynchronization information compared to the decoding until the first error occurrence is still more than 5 dB in Y-PSNR for some erroneous frames which is significant from the user experienced quality point of view.

On the other hand, the side information introduces redundancy causing a rate increase even in the cases where no errors occur. Therefore, in the next section, a selective scheduling for sending the PI packets is analysed, that only sends them if the link quality sinks.

8.2 Link Error Prediction Based Redundancy Control

Since the side information for VLC resynchronization is only to be employed if an error occurs, it does not make sense to transmit the PI packets if the channel quality is high. However, in wireless mobile networks the channel quality varies very quickly. In this section the predictability of the UMTS dedicated channel (DCH) errors is investigated. Moreover, a mechanism is proposed that transmits the side information only if the predicted error probability exceeds a certain threshold γ.

8.2.1 Redundancy Control

As shown in Chapter 4, a link layer error model can be used for the prediction of errors. Since the PI information is utilized by the decoder only if an error at the radio interface occurs, the

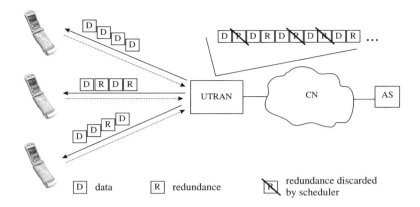

Figure 8.4 The principle of the redundancy control performed for UMTS DCH.

rate can be saved by sending the side information packets only if a higher error probability is predicted. Figure 8.4 illustrates the principle of the redundancy control mechanism if applied to UMTS DCH.

Assume an encoder located on the Application Server (AS) for video streaming; the video stream is encoded including the PI side information packets for each video frame and transmitted over the core network to the UMTS Terrestrial Radio Access Network (UTRAN). In UTRAN, scheduling for the radio interface is performed. The information about the current link quality is only available in UTRAN, the AS has no access to such data. Based on the link quality feedback, the error probability is estimated. If the probability of error is low, the side information packets are discarded in UTRAN, whereas if the probability of error is high, the side information packets are sent over the radio interface. Note that the side information can be optionally sent, protected by additional encoding, or using the acknowledged mode that allows for retransmissions limited by a discard timer. The decision about low/high error probability can be based on a single threshold γ, or alternatively on an adaptive threshold, considering, for example, expected distortion.

For the experiments, again, the 'soccer match' video sequence was used, encoded in the same way as in section 8.1.3. The same error traces were used for the comparison. To predict the errors the feedback delay $d_{FB} = 36$ TBs was assumed, corresponding to three TTIs of 10 ms each. The quality improvement achievable by VLC resynchronization using PIs encoded using a 9-bit FLC and encapsulated into a H.264/AVC NALU and the UDP/IP protocol is illustrated in Figure 8.5 for all investigated thresholds with values: $\gamma_1 = 1 \times 10^{-4}$, $\gamma_2 = 2 \times 10^{-4}$, $\gamma_3 = 3 \times 10^{-4}$, $\gamma_4 = 5 \times 10^{-4}$, $\gamma_5 = 7 \times 10^{-4}$ and $\gamma_6 = 1 \times 10^{-3}$.

The improvement was calculated as a difference between the frame Y-PSNR of the method using resynchronization with threshold γ and the frame Y-PSNR of the method decoding up to the first TB error. Note that this ecdf is derived by considering all investigated QPs (unlike Figure 8.3). In this representation one can observe that, for example, with γ_2, in more than 7% of the cases the improvement in quality is higher than 5 dB. Moreover, in 40% of the cases the quality improvement is more than 1.5 dB. The improved frames comprise the frames where an error occurred as well as frames to which the error propagated due to temporal prediction.

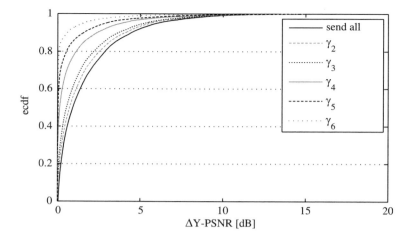

Figure 8.5   Quality improvement resulting from resynchronization of VLC, compared to decoding up to the first error occurrence for $\gamma_{2-6} = \{2 \times 10^{-4}, 3 \times 10^{-4}, 5 \times 10^{-4}, 7 \times 10^{-4}, 1 \times 10^{-3}\}$.

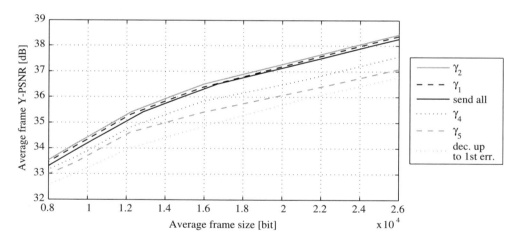

Figure 8.6   Average Y-PSNR over the rate. The rate is expressed as average size of frame and side information in bits ($\gamma_1 = 1 \times 10^{-4}$, $\gamma_2 = 2 \times 10^{-4}$, $\gamma_4 = 5 \times 10^{-4}$, $\gamma_5 = 7 \times 10^{-4}$).

The improvement for the 'send all' strategy is only slightly higher, although it requires a higher rate.

To look at the Y-PSNR without considering the additional rate is not fair. Therefore, Figure 8.6 illustrates the average quality at the decoder normalized by the required rate.

As a rate, the average number of bits per frame was taken, which already also comprises the resynchronization information and its header. The averaging was performed over the

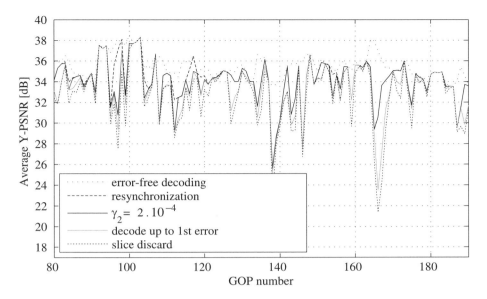

Figure 8.7 Average Y-PSNR per GOP over time (GOP number).

entire video sequence, containing also the error-free packets. Clearly, the threshold $\gamma_2 = 2 \times 10^{-4}$ provides the best quality for the given rate in the whole investigated range. The improvement of approximately 0.25 dB can be achieved by using the error prediction when comparing to the full VLC resynchronization (sending all PIs). This is a considerable improvement since the error rate was small and the increase of the rate caused by signalling side information was taken into account. Furthermore, it can be seen that the resynchronization alone brings approximately 1.5 dB of average quality improvement for the given rate. This also demonstrates that it is possible to reduce the rate at the radio interface significantly while keeping the same quality at the decoder if VLC resynchronization information is used.

Figure 8.7 illustrates the Y-PSNR per GOP for the same simulation scenario. This representation provides an example of what quality the user perceives over time (average per GOP corresponding to approximately 5 s of video in our case). The difference between the full resynchronization and resynchronization using prediction (with the value γ_2 that provides the best quality for the given rate) is in most cases zero. Some small differences are caused by the wrong error predictor decision (false detection or missed detection). It can be seen that resynchronization improves the quality at the receiver considerably, even for a small packet loss rate. The difference from the typically used slice discard method is up to 10 dB per GOP on average; a slightly smaller gain (up to 8 dB) can be seen compared to the decoding up to the first error. These differences are significant from the user perception point of view. Note that the size of a picture area (in MBs) corresponding to one TB depends on the compression ratio of this area; the compression ratio depends on the content of the video sequence (amount and type of the movement, amount of edges) and on the QP. To limit the erroneous picture area rather than the number of bits, signalization of the LIs at each M macroblocks would be needed as already treated in section 6.4.

8.3   Semantics-aware Scheduling

Scheduling is a key task of the radio resource management, especially for wireless mobile networks. In Fattah and Leung (2002), a general overview of scheduling algorithms for various services of more users transmitting over wireless multimedia networks is presented. In this section, the scheduling of video packets over UMTS is proposed, the focus being on the scheduling of video packets from a single stream of a single user.

Already in Cianca *et al.* (2004) a truncated power control is introduced to improve the video quality and for efficient transmission of the video stream. In Kang and Zakhor (2002) and Tupelly *et al.* (2003) opportunistic scheduling algorithms are presented which make use of the characteristics of the streamed video data, whereas in Kang and Zakhor (2002) the more important parts of the video stream are transmitted prior to the less important ones in order to ensure more opportunities for retransmissions in case of error. In Tupelly *et al.* (2003) the priority-based scheduling exploits the diversity gains embedded in the channel variations when having more than one stream. Another approach is shown in Koutsakis (2005), where a prediction of the link errors is used in connection with call admission control and scheduling algorithms for avoiding the system being overloaded and thus improving the quality of the services with higher priority. A semantics-aware link layer scheduling for video streams of more users over wireless systems was proposed in Klaue *et al.* (2003). It adapts the dropping deadline on the type of video frame transmitted – the intra frames have the longest delay, inter frames a smaller one. The perceptual importance of the video packets is utilized in Bucciol *et al.* (2004) for scheduling of retransmissions over 3G networks. An RD-optimized scheduling framework is proposed in Kalman *et al.* (2003) with a simple distortion model; in Chou and Miao (2006) the framework is enhanced.

In this section a scheduling method will be introduced that makes use of the error prediction mechanism presented in Chapter 4. Unlike in Cianca *et al.* (2004), where transmission is stopped in time intervals where the channel quality is poor, the method presented here makes use of all the available bandwidth but delays the packets with higher importance to a position where smallest error probability is predicted. Unlike Klaue *et al.* (2003), the semantically important parts are postponed rather than limiting the dropping deadlines. For details, see also Nemethova *et al.* (2007).

8.3.1   Scheduling Mechanism

In a video stream, the I frames are more important than the P and B frames as they refresh the stream. If an error occurs in an I frame, it propagates in the worst case over the whole group of pictures up to the next I frame. Knowing the predicted probability of the error, the transmission of the I frames can be postponed to the intervals where this error probability is low, as shown in Karner *et al.* (2005). With respect to the RLC layer, an IP packet entering the scheduler corresponds to an RLC Service Data Unit (SDU). The RLC SDU is further segmented to smaller units to which are added a header and output as RLC PDUs. A schematic illustration of the proposed scheduling algorithm is shown in Figure 8.8 where in the upper part the time series of the received TBs can be seen, with the erroneous TBs marked.

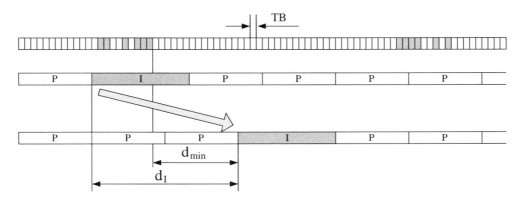

Figure 8.8 Schematic illustration of the proposed semantics-aware scheduling algorithm.

The I packets are the RLC SDUs containing parts of I frames. The P packets contain parts of P frames. The general goal of the scheduling algorithm is to transmit the highest priority packets at a time instant where the smallest error probability is predicted. Here, the packets containing I frames are considered more important, having higher priority than packets containing P/B frames. However, another subdivision of priorities is also possible. The simple semantics-based two-level priority principle has the advantage of being compatible with the standard video codec settings, that is, it does not require any data partitioning, other additional error resilience methods or distortion estimation.

To implement the proposed method in a UMTS scheduler, situated in the RNC at the RLC layer, video semantics awareness is necessary – that is, the RLC layer must be passed the information about the content of the processed IP packets (RLC layer SDUs). To facilitate this, a cross-layer design must be supported by the mobile network. Passing the semantics information can be achieved in several ways. An application-specific way for example, consists in analysing the content of IP packets at the RNC. However, this requires the RNC to know the syntax of a particular codec which is an impractical assumption. Alternatively, the priority can be signalized either in the NRI field of the video NALUs (compare to section 5.1.3) or within the IP header as a DiffServ[1] information. Mapping of the IP packets onto different UTRAN logical channels according to their priority is also an option, although it requires an increased rate. In such a case, the logical channels could be configured for different desired error rates, for instance, via amount of puncturing applied to the turbo code. A differentiation of quality at the level of logical channels is also performed in UMTS for a radio access bearer carrying speech traffic (3GPP TS 34.108 2004).

The proposed semantics-aware scheduling algorithm can thus be applied in UMTS networks without changes of the standards. Additionally to the classical UMTS architecture, the information about the error status of the TBs at the receiving terminal is necessary at the scheduler (transmitting side). This can be achieved by using the RLC AM mode with the maximum number of retransmissions set to zero. Thus, in UMTS uplink (UL) all the

---

[1]DiffServ (Differentiated Services) is a mechanism for assigning different priority levels to the handling of IP packets (Blake *et al.* 1998). The priority is signalized as a 6-bit value encoded into the differentiated services field of the IP header.

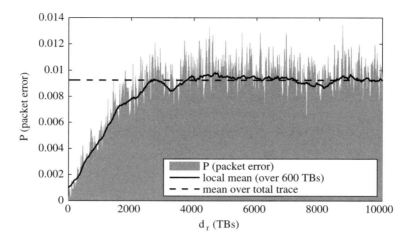

Figure 8.9 Packet error probability versus relative position d_r of a TB to d_{min} (37 TBs) after the last error cluster.

necessary information for error prediction and scheduling is already available at the mobile terminal.

As the intervals with lower error probability can be predicted to occur after a TB error cluster, the scheduling algorithm maps the I packets onto TBs which are to be transmitted within such a predicted interval. This is performed by delaying the I packets by a time d_{min} after the last erroneous TB in the cluster (if there is an erroneous TB within d_{min} the counter for d_{min} is reset) and transmitting the P packets in the meantime. The delay d_{min} has to include the maximum of the short gap lengths. The feedback delay d_{FB} is rather small ($\sim 2\,\mathrm{TTIs}$ if each TB is acknowledged) in comparison to an average gap length and, thus, does not essentially affect the performance of this scheduler. With growing d_{FB}, the effective size of the interval with lower error probability will be reduced from the scheduler point of view. Therefore, possibly lower amounts of I packets will be transmitted without any error.

Figure 8.9 shows the behaviour of the IP packet error probability dependent on the distance to the last TB error cluster. This figure was obtained by simulated transmission IP packets with maximum size of 720 bytes ($= 18$ TBs), with delays to the last erroneous cluster larger than d_{min} of 37 TBs ($\geq 3\,\mathrm{TTIs} = 30\,\mathrm{ms}$). At d_{min} the lowest packet error probability of less than 0.1% is reached, increasing to slightly above the total average (over the entire trace) of 0.925% at a distance of 3000 TBs ($\sim 2.5\,\mathrm{s}$) after d_{min}. To obtain Figure 8.9, simulations were performed by using a pseudo-random trace generated by the two-layer Karner model as explained in Chapter 3.

The proposed method causes an additional transmission delay for I packets and thus also for I frames. However, such delay would not cause any deterioration of the quality as long as it remains within the d_{max} which should be determined according to the application requirements and constraints (playout buffer). In Figure 8.10 the resulting I frame transmission delay in number of TBs ($12\,\mathrm{TB} = 10\,\mathrm{ms}$) is shown for a 'soccer match' video streaming sequence, simulated for the measured trace. Without any limitation, the maximum resulting transmission delay for the I frames is about 7000 TBs ($\sim 6\,\mathrm{s}$). Thus, $d_{max} = 6\,\mathrm{s}$

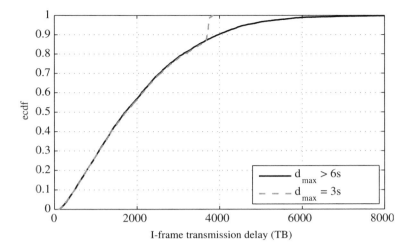

Figure 8.10 Resulting I frame transmission delay in number of TBs for the 'soccer match' video sequence.

– which is a common value for video streaming services – would be sufficient for a full utilization of the proposed method. According to Kang and Zakhor (2002), current video streaming applications usually use a 5–20 s large jitter/playout buffer to cope with the channel fluctuations and the inherent variable-bit-rate nature of coded video sequences. For mobile terminals, the buffer size may be even smaller.

8.3.2 Performance Evaluation

To test the efficiency of the proposed method, the transmission of a video over a UMTS network was simulated. The UMTS network was emulated by means of error traces obtained from measurements. Again, the baseline profile of H.264/AVC is assumed for the experiments, excluding B frames. For these experiments, the JM H.264/AVC encoder and decoder were used as well as a MATLAB program for mapping of the measured RLC trace on the encoded video IP packet trace as shown in Figure 3.21. The JM H.264/AVC encoder and decoder were adapted by introducing the following additions:

- The H.264/AVC encoder outputs the IP packet trace. The IP trace captures for each IP packet its size, the type of the encapsulated slice and the error flag is set to zero (no error).

- The H.264/AVC decoder uses an IP packet trace with modified error flags as input. It decodes the error-free slices and conceals the slices corresponding to the erroneous IP packets (error flag set to one).

- The copy-paste error concealment was used for the inter-predicted packets and spatial weighted averaging error concealment was used for the intra-predicted packets.

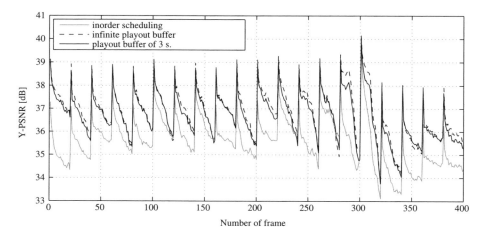

Figure 8.11   Average Y-PSNR over the frame number of the 'foreman' sequence.

The simulation starts by encoding the video sequence. The sizes and types of the IP packets together with the IP error trace are fed into the proposed cross-layer scheduler. The scheduler performs the rescheduling of the IP packets based on the predicted probability of the error and constrained by the size of d_{max}. It outputs two IP packet traces with their error flags set to one in the case of an error. The first trace is the one corresponding to the in-order scheduling of IP packets; the second one corresponds to the semantics-aware scheduling. The decoding is performed separately for these two traces to enable the comparison.

Two different video sequences with QCIF resolution were chosen: the well-known 'foreman' video sequence (400 frames) and the 'soccer match' sequence (11 000 frames). The slicing with number of bytes per slice limited to 700 was chosen with a frame rate of 15 f/s. The quantization parameter was set to QP = 25 so that finally a video stream with an average bit rate of 300 kbit/s was obtained. No DP and no rate control were used.

In order to obtain reliable results, the video was decoded several times, reusing the whole measured trace ten times resulting in approximately 10 hours of video streaming. In Figure 8.11, the average of Y-PSNR per frame of the 'foreman' sequence is shown, corresponding to quality variations perceived by the user.

For the encoding of the 'foreman' sequence the I frame frequency was set to 20. Even with the small IP packet loss probability of 0.88%, the average Y-PSNR improvement of more than 1 dB per frame without limiting d_{max} is obtained. Note that the averaging was also performed over the correct frames. The effect of the proposed cross-layer scheduler on the distribution of the non-averaged frame Y-PSNR is better visualized in Figure 8.12 for the same 'foreman' sequence. The histogram demonstrates that the number of frames with lower Y-PSNRs is reduced whereas the number of frames with higher Y-PSNRs increases. Note that the figure only contains the lower Y-PSNR range. In the higher range, there is a high peak for the Y-PSNRs corresponding to the error-free frames.

The 'soccer match' sequence was encoded with an I frame frequency of 75. The lower I frame frequency better matches to the link error periodicity. Thus, the lower amount of I frames lowers the probability of having an error in such I frame (after scheduling has been

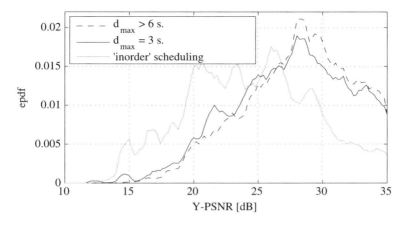

Figure 8.12 Empirical pdf of lower Y-PSNRs per frame for the 'foreman' sequence.

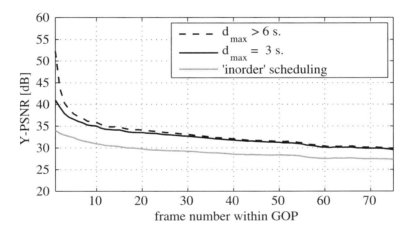

Figure 8.13 Error propagation: Average Y-PSNR over the 75 frames long GOP of the 'soccer match' sequence. The averaging was only performed over the erroneous GOPs.

applied) due to the fact that efficient scheduling of all packets belonging to one I frame can be performed. In Figure 8.13 the average Y-PSNR over the GOP is depicted for the erroneous GOPs only. The error propagation is considerably reduced already for the semantics-aware scheduling with $d_{max} = 3$ s. The Y-PSNR results improve even more if the encoder is set to insert an I frame at every scene cut. The temporal error concealment in the P frames and scene cut situations causes visible artefacts. Another possibility is to assign a higher priority also to the large P frames as it is probable that they contain a new scene.

Table 8.2   Frame error improvement for the 'foreman' and the 'soccer match' video sequences and different d_{max} settings.

Video sequence	Scheduling method	ΔD_I (%)	ΔD_P (%)
'Foreman'	$d_{max} \geq 6\,s$	75.8	3.90
'Foreman'	$d_{max} = 3\,s$	39.5	2.43
'Soccer match'	$d_{max} \geq 6\,s$	83.4	4.20
'Soccer match'	$d_{max} = 5\,s$	81.8	4.09
'Soccer match'	$d_{max} = 3\,s$	54.0	2.62

To demonstrate the benefit of the proposed scheduling algorithm without assuming a particular error concealment method, the frame error improvement ΔD is defined as

$$\Delta D = \frac{N_{err} - N_{err}^{(new)}}{N_{err}} \cdot 100 \quad (\%), \tag{8.2}$$

where $N_{err}^{(new)}$ is the number of erroneous frames after applying the improved scheduling and N_{err} is the number of erroneous frames in the case of 'in-order' scheduling. Here, as an erroneous frame, each frame with PSNR differing from the PSNR of the compressed sequence is considered. In Table 8.2, the frame error improvement can be seen separately for I and P frames, for both the 'foreman' and the 'soccer match' sequences.

The difference between the 'foreman' and the 'soccer match' sequence is caused by the lower key frame rate in the 'soccer match' case which allows for a more efficient scheduling. In the case of $d_{max} = 3\,s$, the I packets exceeding the maximum transmission delay have to be transmitted immediately and thus cannot utilize the predicted low-error-probability intervals – they experience the total mean error probability.

8.4   Distortion-aware Scheduling

A natural enhancement of the semantics-aware scheduling presented in the previous section is a refinement allowing more important and less important parts of the video to be distinguished more finely. The importance of a packet containing a part of the video stream corresponds to the distortion caused by its loss. Thus, the finer the distortion modelled, the more efficient the scheduling will be. In this section, a novel distortion estimation mechanism is proposed and evaluated. This mechanism is used further to refine the scheduling method proposed in the previous section as described in the following.

8.4.1   Scheduling Mechanism

The proposed scheduling mechanism that makes use of both the link error prediction and the estimated distortion is illustrated in Figure 8.14.

Based on the RLC feedback from the receiver (available at the transmitter after the feedback delay d_{FB}), the link error model block identifies the transmission intervals with low error probability and delivers this information to the scheduler. The probability of error is estimated as described in Chapter 4. The importance of packets is a real-valued number,

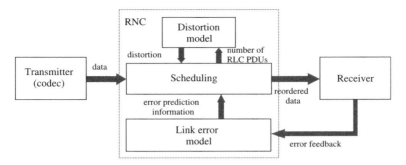

Figure 8.14 Functional scheme of the proposed distortion-aware scheduling mechanism.

given by the distortion that their loss would cause. In section 8.3, only two priority levels were considered and, therefore, the I packets were simply postponed until the start of the interval with lower error probability and then sent. If the number of postponed I packets is greater than the length of the low-probability interval L_i, the remaining I packets suffer the overall mean error probability P_e. Thus, the knowledge of the interval length L_i is not important. For the distortion-aware scheduling proposed in this section, however, this length has to be estimated. The performance of the scheduler can be controlled by the scheduling buffer length $L_b \geq L_i$ and the maximum acceptable delay d_{max}. The delay d_{max} depends on the requirements of the service; it should be chosen to be smaller than the application's playout buffer length. Moreover, the maximum gaplength provides a guiding value for determining the size of d_{max}.

While the link error model predicts transmission intervals with low error probability, the distortion model provides an estimation of the *cumulative distortion* $\widehat{D}_\Sigma$ that would be caused by the loss of a particular RLC PDU. This is calculated for every RLC PDU in the scheduling buffer as

$$\widehat{D}_\Sigma = \sum_{k=n}^{m} \widehat{D}_{k-n},$$ (8.3)

where n denotes the number of the frame within the particular GOP where the RLC PDU under consideration is located and m is the number of frames in that GOP. Consequently, $\widehat{D}_i$ denotes the estimated distortion in the ith frame counted from the frame where the distortion first occurred and for which $i = 0$.

At each step, if an RLC PDU remained in the scheduling buffer longer than d_{max}, then it is scheduled immediately. Otherwise, at each step within the time interval L_i, the RLC PDU associated with the highest estimated cumulative distortion $\widehat{D}_\Sigma$ is chosen from the scheduling buffer. Outside the interval L_i, the RLC PDU with the lowest $\widehat{D}_\Sigma$ is scheduled. If several RLC PDUs have the same (lowest or highest) value of $\widehat{D}_\Sigma$, the oldest among them is scheduled.

8.4.2   Distortion Estimation

Various distortion models for video streaming over lossy packet networks have been proposed to date. In Bergeron and Lamy-Bergot (2006) a semi-analytical model is proposed for H.264/AVC, where the authors assume that impairments due to loss or damage of the frame are caused while decoding the stream. They further assume that no error concealment is applied in such situations. Another approach can be found in Kanumuri *et al.* (2006), where the visibility of packet loss is modelled depending on a set of motion and spatial video parameters. The method proposed in De Vito *et al.* (2004) combines the analysis-by-synthesis approach with a model-based estimation, which requires decoding of all possible loss patterns to obtain the resulting distortion in the first frame. In Stuhlmüller *et al.* (2000), the distortion over the whole sequence is estimated for both the compression and the packet loss impairments.

The proposed scheduling mechanism is based on RLC PDUs. Thus, a distortion model is needed with RLC PDU granularity, instead of the more common IP/RTP/NALU granularity. The task of designing such a model is somewhat simplified by the fact that the decoder will not be able to recover information in RLC PDUs after the first lost PDU of a NALU. Thus, if a NALU is made up of k RLC PDUs, only k cases need to be tested. The impairments caused by the loss of an RLC PDU will propagate from the frame of origin to all frames that reference it. Thus, it makes sense to subdivide the distortion estimation into two steps:

1. Determine the 'primary distortion', that is, the distortion within the frame caused by a lost PDU.

2. Determine the 'distortion propagation', that is, the distortion in the following frames until the end of the GOP.

Note that the distortion is estimated after error concealment, and therefore depends on the performance of the latter. The estimation considers the position and amount of lost data, as well as the size of the corresponding picture area. The number of lost/damaged MBs required by the model can be obtained without fully decoding the video as a proportion of the MBs per NALU packet (if it is fixed) corresponding to the proportion of lost RLC PDUs, by reading of the entropy encoded stream (without decoding the video) or by extra signalling.

To investigate the behaviour of the distortion with respect to the number of missing bits and macroblocks, various experiments were carried out as described in the following.

Dataset Acquisition

In order to obtain a dataset consisting of measured distortions, several assumptions about channel and encoder settings are necessary. In the considered UMTS architecture, the RLC PDU of size $q = 320$ bits is the smallest data unit allowing CRC error detection. Thus, we chose to measure the position and amount of lost data in multiples of q. The size of the missing picture area is measured in multiples of macroblocks (MBs). For temporal error concealment, the zero motion method is taken, as it represents the worst case. Spatial error concealment is used only for the first frame in the sequence. The experiments were all performed using the H.264/AVC JM modified to support RLC segmentation. A training set of five QCIF video sequences with various contents and with a frame rate reduced to 7.5 f/s was used to obtain the distortion model: 'foreman', 'akiyo', 'videoclip', 'soccer' and 'limbach'. All these sequences were encoded using slices of length limited to 650 bytes, QP $= 26$, one reference frame only, no RDO, no rate control, no DP, CAVLC entropy coding, and other settings corresponding to the baseline profile. Three more video sequences were used to verify the prediction performance of the proposed distortion model: 'silent', 'glasgow' and 'squash'.

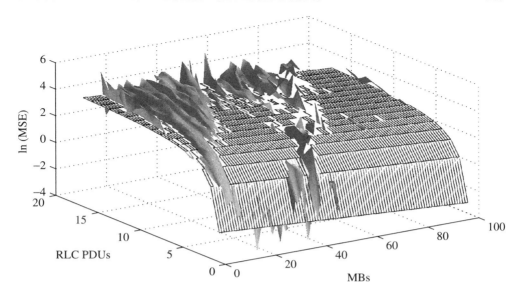

Figure 8.15 Fitting of the measured data by the proposed model.

The mean square error was used as a distortion measure. For each video sequence, each possible position of an RLC PDU error was simulated. There was always only one RLC PDU error per GOP in order to follow the error propagation caused by each error separately. Figure 8.15 visualizes the dataset obtained by these simulations for P frames and a best fit for it, as presented later on.

The PSNR is not a totally smooth function of the number of MBs and RLC PDUs. In particular, the left part of the plot and the part with more than 50 MBs show several points of high distortion. These correspond to intra-coded MBs (I MBs). Unfortunately, this effect on distortion cannot be suppressed or compensated for, since the number of I MBs within the lost part of a slice is not known and their frequency depends on the encoder implementation. Thus, the I MBs inserted into P slices will degrade the performance of the distortion estimation.

Empirical Distortion Modelling

In order to build a model based on the obtained dataset, the estimated primary distortion $\widehat{D}_0$ has to be represented as a function of the number of missing RLC packets (PDUs) N and the corresponding number of missing MBs M:

$$\widehat{D}_0 = f(M, N). \tag{8.4}$$

The task is to find an appropriate model function $f(\cdot)$ and its best fitting parameters for the given scenario, that is, encoder settings and error concealment assumptions. The distortion grows monotonically with increasing number of missing RLC PDUs. We observed that this growth is better described by a rational rather than an exponential function. With increasing number of missing MBs, the distortion decreases. The following model approximates well

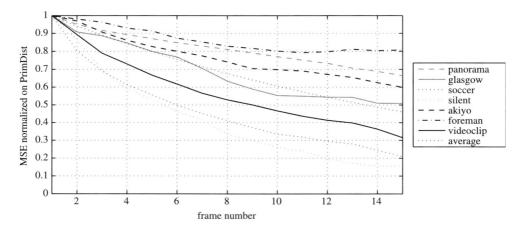

Figure 8.16 Error propagation in the frames of the same GOP for various video sequences (PrimDist = primary distortion).

the measured distortion:

$$\widehat{D}_0 = e^{a \cdot M + (b/N) + c}, \tag{8.5}$$

the parameters a, b and c determined by least square fitting. For the model dataset, the following values were obtained: $a = -0.041$, $b = -6.371$ and $c = 4.482$.

In order to model the error propagation in a GOP, all obtained distortion sequences $\{D_k\}_{k=1,2...}$ where k is the frame index, were normalized by the corresponding primary distortion D_0. On average, an exponential decrease was observed:

$$\widehat{D}_k = \widehat{D}_0 \cdot e^{-s \cdot k}, \tag{8.6}$$

where s determines the speed of decay which depends strongly on the nature of the motion in the sequence. Since we assume that only M and N are available, the error propagation has to be estimated by an average over all sequences in the model dataset. This degrades the performance of the estimator, since the error propagation differs considerably for different sequences as shown in Figure 8.16. The so-obtained decay is $s = 0.08$. To obtain a more exact model, more information about the video sequence is necessary. The propagation of errors depends on the movement in the sequence, which also influences the encoding process, for example, by skipped and intra macroblocks.

The estimation performance of the proposed method was tested for all video sequences in the training set, as well as for the three additional video sequences from the test set. Table 8.3 summarizes the results in terms of the Pearson correlation coefficient ρ for both primary distortion and error propagation models, obtained by decoding of all possible RLC PDU loss error patterns. The comparison was performed with the estimator based on the whole training set, not per video sequence. The performance of the distortion estimation varies for different video sequences. In particular, scene cuts and frames with a significant amount of I-MBs cause lower correlation. However, in the scheduling application absolute distortion values are not as important as the relation between the distortions caused by different error patterns.

Table 8.3 Performance of the distortion estimation for video sequences in the model dataset (1–5) and new sequences (6–8).

Video sequence	$\rho(D_0)$	$\rho(D_k)$
'foreman'	0.86	0.84
'soccer'	0.88	0.81
'akiyo'	0.85	0.78
'limbach'	0.68	0.80
'videoclip'	0.84	0.77
'silent'	0.87	0.85
'glasgow'	0.79	0.67
'squash'	0.83	0.73

8.4.3 Performance Evaluation

For all experiments, JM H.264/AVC was used. The video sequence 'videoclip' containing a music video clip was chosen as a test sequence, since it contains a variety of different scenes separated by scene cuts and gradual transitions. The sequence was encoded with the settings corresponding to those used for obtaining the distortion model. It was decoded 1500 times for each of the considered methods to obtain sufficient statistics. The CRC of the RLC PDUs ($q = 320$ bits) was used to detect the errors and the decoding of a video slice was performed up to the first erroneous RLC PDU within the corresponding packet as proposed in Nemethova et al. (2005). The following methods are compared:

- **In-order scheduling**: decoding using unchanged link error traces obtained from measurements in a live UMTS network.

- **Scheduling limit**: theoretically achievable limit under assumptions that the position of errors is known and both $d_{\max}$ and L_b are larger than one entire video sequence.

- **Semantics-aware scheduling**: prioritized I frame scheduling as proposed in Karner et al. (2005) but assuming the usage of the correctly received data from NALU up to the first missing RLC PDU.

- **Distortion-aware scheduling**: configurations of the proposed method for different $d_{\max}$ and L_b.

Here, simulation results for the following combinations are presented: $d_{\max} = 8400\,\text{TBs}$, $L_b = 3600\,\text{TBs}$ and $d_{\max} = 3600\,\text{TBs}$, $L_b = 1200\,\text{TBs}$ (denoted further as 8400/3600 and 3600/1200), which corresponds to the maximum delay of 7 and 3 s, respectively.

The empirical cdfs of the frame PSNR are exhibited in Figure 8.17 for the presented methods. The difference between the semantics-aware scheduling and the distortion-aware scheduling method is well illustrated. Even if not shown, all cdfs approach one at the PSNR $= \infty$ (error-free frames). The semantics-aware scheduling method globally reduces the number of erroneous frames since it moves the errors from the first frame in the GOP to the later ones, making the error propagation chain shorter. The proposed distortion-aware scheduling reduces the number of frames with higher distortion and consequently increases the number

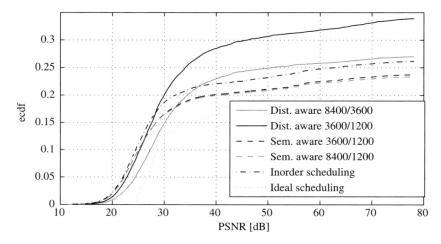

Figure 8.17 Achieved ecdf of the frame PSNR by the presented scheduling methods.

of frames with lower distortion. It can even happen that the total number of frames containing an error increases after applying the distortion-aware scheduling. This is caused by burst errors possibly affecting multiple frames, since the combinations of errors in time are not considered; that is, if an error occurs, followed by another error in the next frame at the same position, then the cumulative distortion is smaller or equal to the sum of the two distortions. However, as there is no information about the motion between the two frames, the additional distortion caused by another error cannot be determined reliably. Considering all combinations of two and more errors would increase the complexity of the scheduler considerably.

Figure 8.18 depicts the PSNR over the frames of the tested 'videoclip' sequence. Again, the difference between the semantics-aware scheduling and the distortion-aware scheduling can be seen. The PSNR at the end of the sequence decreases for the distortion-aware scheduling. This effect is caused by the high predicted distortions for a loss of the first frame. In our simulation we encoded the video sequence 1500 times, thus the first frame follows after the last one again. Since the first frame is concealed only by spatial error concealment, the corresponding distortion caused by a loss of its parts is higher. In Table 8.4, the proportion of error-free frames is summarized together with the mean distortion for all presented methods. The semantics-aware scheduling is not as sensitive to smaller d_{max} as the distortion-aware scheduling.

The reason why semantics-aware scheduling performs for the 'videoclip' test sequence slightly worse than for the 'soccer match' and 'foreman' sequences is the fact that the 'videoclip' test sequence contains several inter-encoded scene cuts. Clearly, the quality of the I frames is improved considerably (see frames 1, 40 and 80 in Figure 8.18). The only case where the quality after scheduling worsens, compared to the quality without scheduling, is the P frames containing a scene cut, for example 18, 75. This problem can be solved easily by inserting an I frame in the case of a scene cut. Nevertheless, semantics-aware scheduling still gains about 1 dB on average.

The proposed distortion-aware scheduling method gains on average 2 dB compared to the common 'in-order' scheduling. The gain achieved by using the distortion model rather than

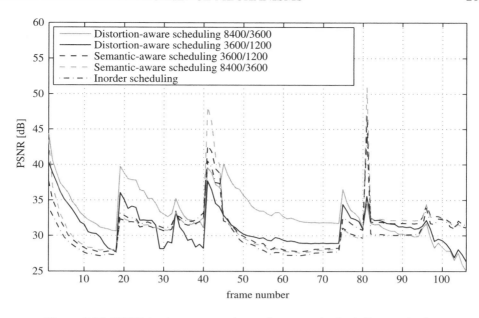

Figure 8.18 PSNR in time: comparison of presented scheduling methods.

Table 8.4 Number of erroneous frames and mean PSNR for the presented scheduling methods.

Method	Error-free frames (%)	Mean PSNR (dB)
In-order	73.86	29.29
Semantics-aware ($d_{max} = 3600$TBs)	76.27	30.26
Semantics-aware ($d_{max} = 8400$TBs)	76.63	30.32
Distortion-aware (3600/1200)	66.07	30.44
Distortion-aware (8400/3600)	73.01	32.03
Scheduling limit	82.09	35.53

the semantics information is about 1 dB. These gains are significant from the point of view of user perception.

References

Blake, S., Black, D., Carlson, M., Davies, E., Wang, Z. and Weiss, W. (1998) *An Architecture for Differentiated Services,* RFC 2475, IETF.

Bergeron, C. and Lamy-Bergot, C. (2006) Modelling H.264/AVC sensitivity for error protection in wireless transmissions. In *Proceedings of IEEE Workshop on Multimedia Signal Processing (MMSP)*, Victoria, BC, Canada.

Bucciol, P., Masala, E. and DeMartin, J.C. (2004) Perceptual ARQ for H.264 Video Streaming over 3G Wireless Networks. In *Proceedings of International Conference on Communications (ICC)*, vol. 3, pp. 1288–1292, Paris, France.

DeVito, F., Quaglia, D., DeMartin, J.C. (2004) Model-based Distortion Estimation for Perceptual Classification of Video Packets. In *Proceedings of IEEE Workshop on Multimedia Signal Processing (MMSP)*, Siena, Italy.

Chou, P. and Miao, Z. (2006) Rate-distortion Optimized Streaming of Packetized Media. *IEEE Trans. on Multimedia*, **8**(2), 390–404.

Cianca, E., Fitzek, F.H.P., DeSanctis, M., Bonanno, M., Prasad, R. and Ruggieri, M. (2004) Improving Performance for Streaming Video Services over CDMA-Based Wireless Networks. In *Proceedings of International Symposium on Wireless Personal Multimedia Communications (WPMC)*, Honolulu, Hawaii.

Fattah, H. and Leung, C. (2002) An Overview of Scheduling Algorithms in Wireless Multimedia Networks. *IEEE Wireless Communications*, **10**, 76–85.

Kalman, M., Ramanathan, P. and Girod, B. (2003) Rate-distortion Optimized Video Streaming with Multiple Deadlines. In *Proceedings of IEEE International Conference in Image Processing (ICIP)*, Barcelona, Spain.

Kang, S.H. and Zakhor, A. (2002) Packet Scheduling Algorithm for Wireless Video Streaming. In *Proc. of 12th Int. Packetvideo Workshop (PV)*, Pittsburgh PA, USA.

Kanumuri, S., Cosman, P.C. Reibman, A.R. and Vayshampayan, V.A. (2006) Modeling Packet-Loss Visibility in MPEG-2 Video. *IEEE Trans. on Multimedia*, **8**(2), 341–355.

Karner, W., Nemethova, O. and Rupp, M. (2005) Link Error Prediction Based Cross-layer Scheduling for Video Streaming over UMTS. In *Proceedings of 15th IST Mobile and Wireless Communications Summit*, Myconos, Greece.

Klaue, J., Gross, J., Karl, H. and Wolisz, A. (2003) Semantics-aware Link Layer Scheduling of MPEG-4 Video Streams in Wireless Systems. In *Proceedings of Applications and Services in Wireless Networks (AWSN)*, Bern, Switzerland.

Koutsakis, P. (2005) Scheduling and Call Admission Control for Burst-error Wireless Channels. In *Proceedings of 10th IEEE Symposium on Computers and Communications (ISCC)*, La Manga del Mar Menor, Cartagena, Spain.

Larzon, L.A., Degermark, M., Pink, S., Johnsson, L.E. and Fairhurst, G. (2004) *The Lightweight User Datagram Protocol (UDP-Lite)*, RFC 3828, IETF.

Nemethova, O., Karner, W. Al-Moghrabi, A. and Rupp, M. (2005) Cross-layer Error Detection for H.264 Video over UMTS. In *Proceedings of Wireless Personal Mobile Communications, International Wireless Summit*, Aalborg, Denmark.

Nemethova, O., Karner, W., Weidmann, C. and Rupp, M. (2007) Distortion-minimizing network-aware scheduling for UMTS video streaming. In *Proceedings of European Signal Processing Conference (EUSIPCO)*, Poznan, Poland.

Postel, J. (1980) *User datagram protocol, Request for Comments (standard) 768*, Internet Engineering Task Force (IETF).

Schulzrinne, H., Casner, S., Frederick, R. and Jacobson, V. (2003) *RTP: A transport protocol for real-time applications, Request for Comments (standard) 3550*, Internet Engineering Task Force (IETF).

Stuhlmüller, K., Färber, N., Link, M. and Girod, B. (2000) Analysis of Video Transmission over Lossy Channels. *IEEE Journal on Selected Areas in Communications*, **18**(6), 1012–1032.

3GPP TSG TS 34.108 2004 (2004) *Common Test Environments for User Equipment (UE) Conformance Testing*, ver. 4.11.0, Jun.

Tupelly, R.S., Zhang, J. and Chong, E.K.P. (2003) Opportunistic Scheduling for Streaming Video in Wireless Networks. In *Proceedings of the Conference on Information Sciences and Systems*, Johns Hopkins University, Baltimore, MD.

Zheng, H. and Boyce, J. (2001) An Improved UDP Protocol for Video Transmission over Internet-to-Wireless Networks. *IEEE Trans. on Multimedia*, **3**(3), 356–365.

Part V

Monitoring and QoS Measurement

Introduction

Network and traffic monitoring plays a central role within the larger process of network operation. Part V of the book reports in Chapter 9 on the project METAWIN and its successor project DARWIN (www.ftw.at) that started as experimental research at the Forschungszentrum Telekommunikation Wien (ftw.) for monitoring a wireless core network on its Gi and Gn interface and is now being implemented as a commercial tool. Much monitoring experience was gained during these four project years being reported here. In particular the problems of huge data amounts to be recorded, stored and made visible are addressed but also entirely different problems such as how to anonymize data for future use. Once such a monitoring system is at hand, detection algorithms are required to handle troubleshooting, detect bottlenecks and anomalies, evaluate the performance of a wireless network and finally optimize its behaviour. Important features are the network-centric as well as the user-centric points of view that finally facilitate the definition of 'quality of service' and 'quality of experience'.

Chapter 10 reports on many typical features that this monitoring system offers, such as finding bottlenecks in a network or diagnosing network states based on aggregated statistical moments. Once the monitoring system was operational the newly running UMTS network in Austria was analysed. As there were hardly any serious customers at that time, we were very surprised at some high loads of traffic. The analysing tool revealed that the main traffic was generated by worms and viruses. One had to ask, what came first, the worm or the Internet?

As it transpired, simple statistical measures allowed many important questions in monitoring to be answered. Surprisingly, simple moments such as the mean, the variance and the kurtosis provide sufficient ways to detect most anomalies. Furthermore, an alternative method of exploring one-way delays is explained and many results using such methods are described. It was found that the delay at a UMTS-SGSN is moderately influenced by user mobility, while flow control and user mobility considerably impact the delay process at a GPRS-SGSN. Simple summary indicators can be extracted from the delay statistics, as a combination of percentiles and threshold-crossing probabilities. Such indicators can be used for the purpose of detecting abnormal delay deviations, pointing to problems in the network equipment.

While measuring quality of service for speech services and high resolution television signals is a standard task now, for video services with low resolution, as is common for mobile terminals, it is a rather new field. In a first step Chapter 11 provides an overview of existing techniques and expands into new ones. Evaluation methods based on many test runs with individual persons to design subjective quality metrics are presented. Such metrics are to replace time-consuming tests by rather simple, low-complex algorithms in a second step. The work reported here by Michal Ries can be found in more detail in his thesis *Video Quality Estimation for Mobile Video Streaming*, available at http://publik.tuwien.ac.at/files/PubDat_170043.pdf as well as in numerous publications cited within the chapter.

9

Traffic and Performance Monitoring in a Real UMTS Network

Fabio Ricciato

Network and traffic monitoring plays a central role within the larger process of network operation. The aim of this chapter is to discuss the technical aspects involved in the monitoring of the 3rd Generation (3G) network infrastructure and the associated benefits. Recent years have seen a surge of studies and research activities involving passive monitoring of Internet network traffic. Several methodologies for the analysis of the collected data are being developed, importing concepts and tools from different areas such as signal processing, data mining and so on. Some of these techniques can provide powerful support for the operation and engineering of real networks, with recognizable benefits in terms of revenue protection and/or cost saving.

Despite the extensive research carried out recently on this topic, the practice of network measurement and traffic monitoring remains a sort of 'esoteric art' to outsider engineers. The content of this and the following chapter is based on the know-how matured during four years of applied research work, dealing with extensive traffic monitoring in a large commercial 3G network (METAWIN and DARWIN projects (DARWIN)).

9.1 Introduction to Traffic Monitoring

With the term Traffic Monitoring and Analysis (TMA, for short) we refer to the task of passively monitoring the traffic traversing the network at a very fine granularity, typically at the packet level, and process it in order to gain insight into the status and performance of the network and into the behaviour of the user population. Traffic monitoring is typically achieved by inserting passive wiretaps on selected network links so as to 'sniff' transit traffic and store locally a copy of each frame. Depending on the intended application, the frame can be recorded integrally or partially – in many monitoring applications only the headers and the initial bytes of the payload are stored. Moreover, the recorded frame can be labelled with external additional information such as time-stamps or Layer-2 identifiers. The resulting data, called 'packet-level traces' or simply 'traces', represents a highly accurate representation of 'what happened' in the network. The traces can be post processed in various ways depending on the particular application. Often, the first elaboration step is the extraction of specific embedded data. For example, from packet-level traces it is possible to reconstruct flow-level traces – by whatever definition of flow, for example L4 connections and traffic between host pairs – by listing the set of observed flows along with their attributes (start/end time, volume, end-point identifiers and so on). For some applications it is useful to extract packet-level sub-traces constituted by all packets matching some specific criteria, for example TCP SYN towards port 80. Counting specific occurrences of certain events in fixed time-bins will deliver discrete time series: simple examples are the number of packets transmitted in each direction, or the number of distinct hosts seen in each bin. Note that the same methods can be applied to the user-plane as well as control-plane traffic at any protocol layer. For instance, in the context of the 3G network a wealth of information can be extracted from the monitoring of control information at the 3rd Generation Partnership Project (3GPP) layers below (IP).

The above few examples illustrate the broad range of intermediate data that can be extracted from packet-level traces to serve different applications. This is just the first step in the post processing chain that will eventually lead to a human action, such as, for example, the reconfiguration of some network element, as illustrated in Figure 9.1. The following stage takes as input the intermediate data and applies processing methods that are highly specific to the particular application. Some prominent examples are:

1. basic statistical analysis to extract synthetic indicators such as averages and percentiles, which can be immediately interpreted by the network staff;

2. direct visualization and multi-dimensional exploration by network experts;

3. application of advanced ad-hoc algorithms and techniques, including data-mining and signal-processing techniques;

4. feeding the intermediate data into another tool performing more elaborate tasks, such as, for example, emulation, trace-driven simulations or 'what-if' analysis (discussed later).

9.2   Network Monitoring via Traffic Monitoring: the Present and the Vision

The classical approach to network monitoring for the purposes of network management relies largely on routine collection of data delivered by the network equipments themselves: built-in counters, logs, Simple Network Management Protocols (SNMP), Management Information Bases (MIBs) and so on. This approach has some limitations. First, the quality of the

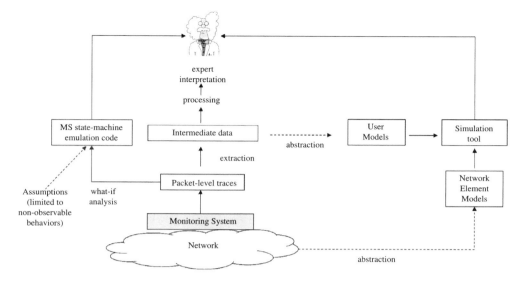

Figure 9.1 Conceptual paths for exploitation of traffic measurements.

available data is not always adequate for the intended task: the time granularity and the aggregation level are coarse, the data semantic is limited, and in some special cases even their reliability can be questioned, for example in the case of overload or equipment malfunctioning. Secondly, the process of extracting, gathering and correlating such data involves considerable costs given the broad heterogeneity of equipment types, vendors, software releases, data formats and data semantic. Moreover, every change in the network, including regular interventions such as replacement of equipment or software upgrades, is likely to require changes in the monitoring infrastructure also. Above all is a general fundamental problem with such an approach, namely the 'lack of decoupling' between the monitored system and the monitoring tool, that produces obvious ambiguity problems and makes the latter unreliable in the case of equipment malfunctioning.

To complement the routine large-scale data collection from network elements, sporadic fine-grain measurements are performed with small-size network protocol analysers. These measurement interventions are generally limited in time and space (one or few interfaces) and are often used for troubleshooting actions, after a problem has been detected by external means – for example, customer complaints.

In summary, the current status of the network monitoring practice is often a combination of large-scale routine data collection from network elements plus sporadic and local fine-grain measurement actions. The future vision of an advanced network monitoring foresees monitoring tools that (a) are capable of performing routine collection of fine-grain measurements in the large scale, (b) are completely decoupled from the production network and (c) can cleverly process the recently collected data and deliver reports and alarms proactively. This is exactly the vision of a large-scale TMA infrastructure based on passive wiretapping at key network links, with null or very limited interaction with the production equipments. The scheme of a monitoring system for the 3G Core Network (CN),

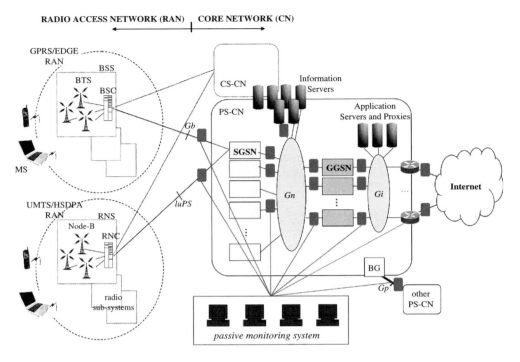

Figure 9.2 Scheme of passive monitoring system for the PS-CN.

like that which we have been developing for our research (DARWIN; Ricciato *et al.* 2006c), is outlined in Figure 9.2. Such a global system would be intrinsically multi-purpose. Observing the traffic allows for the derivation of network-related as well as user-related data. While the former is used by technical departments (operation and maintenance, planning, optimization and so on) the latter can be useful for marketing purposes.

Large-scale monitoring systems are available on the market and have been used in the past few years in the core network of second-generation systems such as the Global System for Mobile communications (GSM). With the deployment of 3G a novel Packet Switched (PS) domain has been added to the legacy Circuit Switched (CS) one, and some monitoring systems were extended to cope with the packet-switched domain. However, the current generation of commercial systems falls short of exploiting the full potential of passive monitoring of a packet network. To some extent this is perhaps due to the fact that they were conceived and developed as mere extensions of the legacy system for the CS domain. The PS section was implicitly considered just as a set of additional protocols to be parsed. Instead, the differences between the two domains are deeper: the PS protocols have completely different dynamics, some of which can be exploited to the operator advantage: for example, the closed-loop dynamics of the Transmission Control Protocol (TCP) can be exploited to infer delays and packet loss, as discussed later in Chapter 10. Furthermore, the PS section supports a completely different usage environment (user populations, terminal types, applications, services and so on) that is infinitely more heterogeneous and complex than the CS telephony.

Accordingly, the choice of meaningful and convenient Key Performance/Quality Indicator (KPI/KQI) for the PS cannot be reduced to a variation of those successfully adopted in the CS. Instead, it requires the application of sound 'dual expertise' in both fields of mobile cellular and packet networking. The latter might not necessarily be present in legacy development groups from the CS world.

9.3   A Monitoring Framework for 3G Networks

The prerequisite to exploit fully the potential of TMA in the context of an operational 3G cellular network is the availability of an advanced monitoring infrastructure. In the METAWIN project (DARWIN) we have developed a working prototype of a large-scale monitoring system covering selected links on all logical interfaces of the Core Network. In this section we list the most important features of the monitoring system that enable the analysis tasks and applications discussed later in the remaining of the paper.

Complete capture. The typical rates of 3G traffic on the CN links are such that all frames can be captured with standard hardware equipments, with no need to resort to packet sampling. This is a major simplification compared to TMA in backbone networks with multi-Gbps traffic rates. Also, the current volumes of daily traffic allow weeks-long storage at accessible costs with Commercial Off The Shelf (COTS) storage equipment.

Joint User- and Control-plane capture. The system is able to capture and parse the complete frame, including the 3GPP headers below the user-IP layer. Along with user-data packets, it captures and parses signalling frames at each layer.

Multi-interface capture. The system covers multiple interfaces and is able to perform cross-link analysis. A prominent example is the computation of one-way delay between two interfaces (see section 10.2.2).

Anonymization. For privacy reasons all subscriber-related fields, most prominently the International Mobile Subscriber Identity (IMSI) and Mobile Subscriber ISDN Number (MSISDN), are hashed with a non-invertible function. The resulting string univocally distinguishes the Mobile Station (MS) but cannot be referred to the user identity. For simplicity we will maintain the term 'IMSI' to refer to the hashed string. This approach preserves packet-to-MS associations, that is, the possibility to discriminate the packets associated to a specific MS, while at the same time protecting the identity of the subscriber. Also, to preserve content privacy, the user payload above the TCP/IP layer can be hashed or simply stripped away.

Stateful tracking of associations. For many applications it is highly desirable to label each packet with certain information associated to the MS it was generated by or directed to. The most important are the MS identifier (hashed IMSI) and the current location (cell or Routing Area (RA)). For other applications it might be useful also to know the MS type (handset, laptop card, PDA and so on) and the equipment capabilities. The former can be directly retrieved by the Type Approval Code (TAC) included in the International Mobile Equipment Identity (IMEI), while the latter are usually advertised by the MS during the Attach request. Similarly, it is often useful to associate individual packets to their Packet Data Protocol (PDP) context and hence to their attributes, for example, the assigned IP address. The relevance of such associations will be highlighted in the next sections. A generic frame crossing the network does not include all such information in its fields, but a passive monitoring system can dynamically reconstruct these associations by continuously tracking the message exchange between the MS and the network. More specifically, it is required to inspect signalling procedures and certain fields of the lower-layer control information (TLLI, T-IMSI) and to maintain for each entity – be it an MS or a PDP-context – a dynamic record of associations. We skip the technical details here. The point is that in general, any attribute that is exchanged between the MS and a generic network element can be captured and later associated to future packets. The associations between packets, PDP-contexts and (hashed) IMSI can be extracted on any interface between the MS and the Gateway GPRS Supporting Node (GGSN), for instance Gn. The localization of the terminal can be achieved in GPRS Enhanced Data rates for GSM Evolution (EDGE) by sniffing the Gb interface (for a detailed description of IMSI-to-cell tracking on Gb see (Borsos *et al.* 2006), Sec. IV.C). For UMTS/High Speed Downlink Packet Access (HSDPA), sniffing on IuPS would allow the localization of the MS only at the granularity of RAs, as intra-RA cell changes are not reported to the SGSN. Exact cell-level localization for UMTS/HSDPA would require monitoring of the Iub interface between the MS and the Radio Network Controller (RNC).

The prototype of a monitoring system fulfilling such requirements was developed as a research tool in the METAWIN project (DARWIN), completely on a Linux platform. We used Endace DAG acquisition cards (Endace) and high-end standard PCs equipped with RAID storage. The system is deployed and currently operational in the GPRS/UMTS network of a major mobile operator in the EU. It covers all CN interfaces including Gi, Gn, IuPS, Gb, Gp and Gr (see also Figure 1.1 of Chapter 1).

9.4   Examples of Network-centric Applications

9.4.1   Optimization in the Core Network Design

An optimization problem found in the engineering of the GPRS CN addresses the wiring of Gb links, that is, the optimal association of Base Station Controllers (BSCs) to SGSNs. Given a certain setting of the Radio Access Network (RAN), and specifically a given grouping of cells into RAs, the problem is then to optimize the assignments of BSC to SGSNs, that is, the Gb wiring. There are multiple concurrent optimization goals to be pursued: minimize the monetary cost of the link distance – which typically increases with the distance, balance the load among the set SGSN – which is typically measured in terms of the peak number of attached users, and minimize the frequency of inter-SGSN routing area updates (RAUs). The latter criterion is motivated by the fact that such a procedure is signalling-intensive and involves four network elements: two SGSN, HLR and GGSN. As a partial sub-problem, one might be interested in optimizing only the wiring between a set of SGSN co-located in the same physical site. In this case only the last two minimization objectives apply. This task was considered in Ricciato *et al.* (2006) where it was formulated as a Mixed Integer Linear

Programming (MILP) problem. The input data for the optimization are two discrete time series:

1. number of attached MSs in each RA i at time t_k;

2. number of MSs passing from RA i to j in the interval (t_{k-1}, t_k).

Both time series can be extracted directly from high-quality traces taken on Gb/IuPS and fed into an appropriate code, as detailed in Ricciato *et al.* (2006). The basic idea of such a code is to reproduce the behaviour of each MS with a simplified state-machine which includes only the states relevant to the specific problem at hand, using the messages found in the traces to trigger state transitions. In our case only two states are sufficient: 'Attached' (A) and 'Detached' (D), the former including the current RA identifier as an internal variable. For example, a successfully completed Attach request in RA i will trigger a transition $D \rightarrow A(i)$, RAUs mark transitions $A(i) \rightarrow A(j)$. The transition $A \rightarrow D$ can be triggered by an explicit Detach procedure, which can be observed in the control-plane traces, or alternatively by a sufficiently long period of silence on the data-plane. In fact, the MS is considered as 'Implicitly Detached' from the network after a certain timeout from the last transmitted frame. We have implemented such a code and run over Gb and IuPS traces for the whole network. Our experience tells that a carefully optimized code can straightforwardly track the state of all the MSs in the whole network in real-time on high-end COTS hardware. The optimization of BSC-to-SGSN assignment provides just one illustrative example of how such data can be exploited for the engineering of the network infrastructure. The results from a case study based on a real dataset can be found in Ricciato *et al.* (2006). In principle, similar data can be used to optimize the design of the Radio Access Network (RAN), for example, to optimize the organization of cells into RAs. Notably, such a measurement-based approach avoids the need to resort to abstract – and often arbitrary – models of user mobility and activity patterns.

9.4.2   Parameter Optimization

In the previous section we discussed the method of 'reading' the traces by means of simplified state-machines in order to reconstruct the internal behaviour of each MS, that is, their internal state. In certain cases this approach can be extended to reconstruct the 'hypothetical' behaviour that would have taken place under different network conditions, for example with different parameter settings.

Consider the following simple example. The Dedicated CHannel (DCH) in UMTS is assigned and released dynamically to each MS, and its bandwidth is adapted dynamically, depending on the actual usage patterns. The algorithms for dynamic resource allocation are implemented in the RNC and are often vendor proprietary, with several parameters tuneable by the operator. In the simplest case, the DCH is assigned when the first packet is transmitted or received, and released after a timeout T_H from the last transmitted packet. The optimal value of the DCH holding timeout T_H must trade off between two competing goals: long values cause resource wastage, while short values involve frequent release/reassign cycles. The latter has a negative impact on the user experience, as each reassignment procedure introduces an additional delay on the arriving packet, and increases the signalling load on the radio link. When the DCH is in place, its assigned bandwidth can be dynamically increased/decreased (by multiples of 64 kbit/s) based on thresholds on the traffic rate measured on a certain time window.

The optimal value of T_H as well as of the other involved parameters (rate thresholds, window length and so on) can be obtained by mathematical analysis or via simulations. In both cases the input is a certain model of user behaviour. If high-level traces from the real network are available, the model can be fitted to the data. However, the process of distilling a model out of empirical data is complex and cannot avoid introducing severe simplifications. This approach can be successful in those cases where (i) the user population is highly homogeneous and (ii) their behaviour involves few dimensions. This is typically the case with voice calls in traditional telephony, with the additional advantage of good mathematical tractability of the resulting models. However, in multi-service networks such as UMTS this is no longer the case: the behaviour of the user population is marked by high heterogeneity, huge disparity (also known as the 'elephants and mice' phenomenon, see also section 13.3.5) and large dimensionality. As a result, the real usefulness of abstract models for any practical engineering task is often questionable. Returning to our example, we recognize that the availability of high-quality traces offers the possibility of estimating what would have been the DCH occupancy pattern of each MS for different values of T_H by adopting the simplified state-machine approach. We can build up a simplified state-machine for each MS, with states corresponding to the different values of DCH bandwidth plus the 'non-assigned' state when the MS is logically placed on the Forward Access Common Channel (FACCH), and state transition rules according to the bandwidth assignment algorithms – typically, release the channel after T_H from the last packet. The recorded traces were captured for a single value of the timer, call it $T_{H,r}$ (real). By feeding such traces into state-machines with a different parameter setting, say $T_{H,v}$ (virtual), we can reconstruct the DCH occupancy profile for each MS in the different network conditions – $T_{H,v}$ instead of $T_{H,r}$. This approach, outlined in Figure 9.3, is similar to trace-driven simulations. It is important to note that the reconstruction is not exact: it implicitly assumes that the user behaviour is independent from the network setting (in our case, the value of the DCH timer), so that the packet pattern observed for $T_{H,r}$ would be maintained for $T_{H,v}$ also. In other words, it neglects the impact of the parameter setting onto the traffic pattern. Such an approach is sometimes referred to as 'open-loop': the user behaviour influences the network performance, but is not influenced by the network setting. This is clearly a simplifying assumption that introduces an estimation error. In fact, the dynamics of the real system are 'closed-loop', that is, there is a mutual (bi-directional) interaction between the user and the network, at least to a certain extent. However, the estimation error introduced by the open-loop approximation is certainly smaller than any model-based simulation, wherein the lack of observability of the would-be user behaviour for $T_{H,v}$ is cumulated with abstractions and simplifications of the actually observed behaviour for $T_{H,r}$. Therefore, measurement-based design, better if corrected by 'educated guess' about the evolution of the user behaviour, is generally more accurate than any model-based optimization.

9.4.3   What-if Analysis

In some cases the trace-driven approach can be extended beyond parameter tuning. There are a number of engineering questions that can be fruitfully supported by a smart 'replay' of high-quality traces. As a first illustrative example, consider the case in which the operator

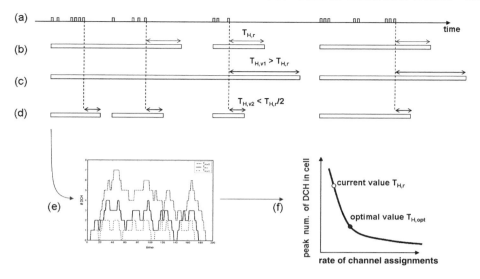

Figure 9.3 Example of 'What-if' analysis for the optimization of DCH release timer.

is planning the introduction of a new service to all users of Access Point Name (APN) X. The new service requires the mobile clients with an active PDP-context in APN X to send periodically a message to a central server every period T. Using trace-driven analysis it is possible to predict rather accurately the additional message load, not only on the server but also 'in each individual cell', for different values of T. As another example, it is possible to predict the network load caused by macroscopic pathologic events, for example a large-scale infection by scanning worms affecting a certain share of terminals, or the impact of Denial-of-Service (DoS) attacks. Regarding the latter, there is a growing awareness within the research community that 3G networks are exposed to novel forms of DoS attack that are launched from the user plane but can have potential impact on the control plane: a number of theoretical attack models specific to 3G were presented in Lee *et al.* (2007), Serror *et al.* (2006), Traynor *et al.* (2007) and Yang *et al.* (2006) and collectively reviewed in Ricciato *et al.* (2006b). More concretely, one could ask questions such as:

1. How many DCH channels would remain allocated in cell Z at time t if $Y\%$ of the MS would be scanning the local address space at rate R?, or

2. How many paging requests would be caused by an external source contacting $Y\%$ of the MS in APN X?,

and similar questions. In other words, the availability of accurate packet-level traces would offer an important basis to carry out quantitative risk assessment for attacks and to predict future bottlenecks under projected variations of current usage scenarios and network setting.

9.4.4 Detecting Anomalies

One of the most common applications of the monitoring tool in real networks deals with the detection of anomalies and troubleshooting. There is a growing scientific literature in the Internet measurement community about algorithms and analysis techniques oriented at detecting anomalous and malicious behaviour. Most if not all such techniques can be applied to 3G networks to reveal anomalies on the user plane; but, in addition, operators of 3G cellular networks must cope with anomalies on the control plane too.

Monitoring the traffic on the control plane allows for the detection of problems and misbehaviours in the 3GPP layers. The most common causes of such events are configuration errors and software bugs in the terminals. In some cases we have observed also that the poor implementation of TCP/IP stack and application-layer modules on the user terminal can cause an undesired overload on the control plane – this is particularly risky when applications designed for wired hosts are blindly ported onto 3G mobile terminals – for an example, see Ricciato *et al.* (2006c). Other error sources include misconfigurations and bugs in the network elements (local or foreign), hardware or software failures, macroscopic infections (Ricciato 2006b) and deliberate attacks.

Recognizing such events requires the ability to inspect signalling traffic at a high level of accuracy and across different interfaces. Moreover, in some cases it is important to correlate observations from the control plane and user plane in order to identify the problem.

The detection of anomalies is relatively easy in the control plane compared with the data plane. In fact many kinds of problem are explicitly encoded into specific error messages. The monitoring system is able to identify the MS associated to each signalling frame, and to count the occurrence of erroneous messages and/or procedures for each MS. In general, more errors are found on those interfaces with a higher level of signalling interactions, for example Gb.

By analysis of several sample traces, we found that most of the error messages are concentrated on a few recurrent error types, and that the vast majority of them are generated by a relatively small portion of MSs, typically due to a software bug in the terminal. Simple thresholds on the absolute number of signalling messages per-MS are often sufficient to identify buggy terminals, as the frequency of error messages for a normal terminal is typically null or extremely sporadic, while buggy terminals tend to iterate erroneous procedures continuously. Therefore, the 'signalling activity' of buggy terminals is typically several orders of magnitude higher than the rest of the population. In other cases we were able to spot a bug in some network element simply by noting a higher-than-usual frequency of error messages and/or incomplete procedures. In general, detecting anomalies on the control plane is a relatively easy task – provided that the monitoring system is able to monitor completely and accurately the whole signalling traffic – as the normal and abnormal behaviours are clearly well separable, also in statistical terms. This is not surprising as the control plane procedures and message semantics are defined by the specifications. Instead, the user plane is a rich and evolving ecosystem of diverse applications and user patterns, without any 'a-priori' definition of what is to be considered 'normal' and what is not.

9.5   Examples of User-centric Applications

The above applications have focused on the network infrastructure: planning, optimization, troubleshooting. On the other hand, the network traffic carries a wealth of important information about the users that have ultimately generated it. In principle, the analysis of their traffic can say much about how the network users behave: which types of terminals they are using, what are their preferred applications, what is the level of quality they are experiencing. Such information is of paramount importance for those operator departments that are in charge of designing marketing strategies, billing schemes and, ultimately, new services.

9.5.1 Traffic Classification

The most fundamental information about the users is: what types of application do they prefer? Or in other words: what do they use the network for? Answering such simple questions is becoming increasingly difficult due to the proliferation of new applications based on the peer-to-peer paradigm, and to a general trend of modern applications towards the adoption of obfuscation and encryption mechanisms. At the start of the commercial Internet area, the most popular applications – email, FTP and web browsing above all – were following standard specifications (IETF RFC), were based on the client–server model and could be easily recognized from passive traces simply by looking at the well-known L4 port number (Svoboda *et al.* 2006). Conversely, in the modern Internet a large number of popular applications are based on peer-to-peer paradigms. Most of them are proprietary or anyway non-standard, and are not bound to any specific port. Furthermore, they tend to adopt encryption and obfuscation mechanisms to escape third-party identification. In other words, more and more such applications are designed to be stealthy or, at least, difficult to identify. This is indeed the boldest outcome of the worldwide efforts set in the last decade to contrast some classes of application, most prominently music/video file-sharing and Voice over IP (VoIP), in order to safeguard the traditional streams of revenues associated to the distribution of copyrighted material and to the circuit-switched telephone service.

The term 'traffic classification' refers to the problem of identifying the type of application used by the user based on the passive observation of the network traffic. Some such techniques rely on the reconstruction and analysis of the application layer data, an approach which is also called 'Deep Packet Inspection' (DPI). Other techniques follow the so-called statistical approach: they aim to infer the application layer class from the statistical patterns of the packet stream. For example, VoIP traffic is characterized by a prevalence of short packet size (below 100 bytes) and a relatively regular packet rate. Peer-to-peer file-sharing applications tend to generate a huge number of parallel L4 connections toward many different peers. In principle such patterns can be recognized without the need to inspect each individual packet content, with obvious advantages in terms of processing overheads. There has been a great deal of research on traffic classification mechanisms and several commercial tools are available on the market. Generally speaking, such tools must implement dedicated modules with specific identification logic for each specific application – Skype, BitTorrent and so on. As applications are in constant evolution, with the continuous release of new versions and patches, any classification tool must be continuously adapted to track their changes, in a way that is reminiscent of the regular updates of antivirus systems. As stated, one of the primary evolutionary trends is toward increased obfuscation, and for that reason it is unclear whether classification tools can continue to maintain an acceptable level of accuracy in the future.

Within the context of a cellular network the task of profiling users' behaviour acquires two additional dimensions: localization and mobility. The output of traffic classification performed at the IP layer can be coupled with user location information available from lower 3GPP layers. In this way it is possible for the operator to explore how the usage patterns correlate with user location, for example to tell which applications are used in different geographical areas, and mobility.

Most commercial tools for traffic classification are designed for wired networks and cannot handle location information, let alone extract it. One possible workaround is to collect separately the location information (cell identifier) with another tool and then combine the two datasets in postprocessing. In practice, however, it is preferable to use a unique monitoring system able to tap directly those 3GPP interfaces where location information is available (Gb, IuPS, Iub) and at the same time run the traffic classification logic at the IP layer.

9.5.2   QoS and QoE monitoring

The notion of 'Quality of Service' (QoS) is network-centric: it tells how well the network performs its task, that is, transporting packets between the users' end-points, in terms of objective quantities such as delay, loss, throughput and similar. On the other hand, the notion of 'Quality of Experience' (QoE) is exquisitely user-centric: it tells how well the users perceive the service offered by the network. Although there is a clear correlation between QoE and QoS values, the mapping between the two is generally non-trivial.

An advanced TMA system can include QoS/QoE estimation features from the traffic observed passively through the network. For example, standard methods have been defined to map the quality of a lossy voice or video stream into a perceptual quality scale. Other types of QoE indicator can be defined for translational applications, for example gaming and browsing, based on statistics of response delays and download times.

In a cellular network QoS/QoE estimation can be coupled with spatial information and mobility data. As discussed above, this requires a unified TMA system that is able to process jointly the 3GPP layers (especially on the control-plane) and the user-plane at the IP layer. The quality indicators can be extracted separately for different areas (cells, RA, RNC areas and so on), in order to identify spots of low quality and trigger a revision of the radio network configuration. Also, based on the cell information it is possible to single out the quality indicators for moving users during handovers. The TMA system can differentiate quality indicators for different types of terminal – this information is encoded in the TAC part of the IMEI, which is normally signalled during the attachment of the terminal to the network. Such combined data can be used to identify terminal types affected by recurrently low quality, which indicates a suboptimal configuration and/or implementation of the protocol stack.

9.6   Summary

A packet-level passive monitoring system able to monitor and correlate IP and 3GPP data can serve multiple purposes. The present chapter has presented a set of possible analysis tasks that, although not complete, is illustrative of the breadth and diversity of the possible applications of traffic monitoring within an operational UMTS network. The information contained in packet-level traffic data enables a wide range of analysis and exploration tasks, both in the direction of the network infrastructure and of the user populations. The multi-purpose nature of the system compensates for the cost and complexity of deploying and operating a pervasive monitoring infrastructure along with the production network. It is important that the monitoring system is versatile and flexible enough to be evolved quickly with new processing logic, so as to respond readily to the future analysis needs encountered by the operator. The independence from the underlying network equipment is another important feature: it ensures correctness of the reported data also during anomalous conditions, that is, when they are most valuable to detect and pinpoint the problem, and at the same time avoids the burden of collecting and mediating among the highly heterogeneous data formats and semantics produced by the network equipments.

References

Borsos, T., Szabo, I., Wieland, J. and Zarandi, P. (2006) A Measurement Based Solution for Service Quality Assurance in Operational GPRS Networks. In *Proceedings of IEEE INFOCOM*, Barcelona, April 2006.

DARWIN and METAWIN, home page available at http://userver.ftw.at/~ricciato/darwin/.

Endace DAG, Endace Measurement Systems. [Online.] Available at: http://www.endace.com.

Lee, P., Bu, T. and Woo, T. (2007) On the Detection of Signaling DoS Attacks on 3G Wireless Networks. In *Proceedings of IEEE INFOCOM 2007*, Phoenix, USA, April.

Ricciato, F. (2006a) Traffic monitoring and analysis for the optimization of a 3G network. *IEEE Wireless Communications – Special Issue on 3G/4G/WLAN/WMAN Planning*, **13**(6), 42–29, Dec.

Ricciato, F. (2006b) Unwanted Traffic in 3G Networks. *ACM Computer Communication Review*, **36**(2), April.

Ricciato, F., Pilz, R. and Hasenleithner, E. (2006) Measurement-based Optimization of a 3G Core Network: a Case Study. In *Proceedings of 6th International Conference on Next Generation Teletraffic and Wired/Wireless Advanced Networking (NEW2AN'06)*, St Petersburg, Russia, May.

Ricciato, F., Coluccia, A. and D'Alconzo, A. (2006b) DoS attack models for 3G cellular networks: a survey and a system-design reading, *submitted*, 2009.

Ricciato, F., Svoboda, P., Motz, J., Fleischer, W., Sedlak, M., Karner, M., Pilz, R., Romirer-Maierhofer, P., Hasenleithner, E., Jäger, W., Krüger, P., Vacirca, F. and Rupp, M. (2006c) Traffic monitoring and analysis in 3G networks: lessons learned from the METAWIN project, *Elektrotechnik und Informationstechnik*, **123**(7/8), 288–296.

Serror, J., Zang, H. and Bolot, J.C. (2006) Impact of Paging Channel Overloads or Attacks on a Cellular Network. In *Proceedings of ACM WiSe'06*, Los Angeles, USA, Sep.

Svoboda, P., Ricciato, F., Hasenleithner, E. and Pilz, R. (2006) Composition of GPRS/UMTS traffic: snapshots from a live network. In *Proceedings of 4th International Workshop on Internet Performance, Simulation, Monitoring and Measurement (IPS-MOME'06)*, Salzburg, Austria, pp. 27–28, Feb.

Traynor, P., McDaniel, P. and La Porta, T. (2007) On Attack Causality in Internet-connected Cellular Networks. *Proceedings of 16th USENIX Security Symposium (SECURITY)*, Boston, MA, Aug.

Yang, H., Ricciato, F., Lu, S. and Zhang, L. (2006) Securing a Wireless World. *Proceedings of the IEEE*, **94**(2), 442–454, Feb.

10

Traffic Analysis for UMTS Network Validation and Troubleshooting

Fabio Ricciato and Peter Romirer-Maierhofer

Chapter 9 has addressed the methodological aspects of network monitoring, providing a high-level overview of the potential applications. In this chapter we deal with a primary class of applications of network monitoring, namely the validation and troubleshooting of the network infrastructure. The goal is to describe how the *data* collected with the measurement techniques described in Chapter 9 can be turned into *information* which is useful to support some important tasks in the network operation process. As with the previous one, the content of the present chapter is also based on the know-how gained during a four-year research activity involving extensive analysis of the real traffic from a large commercial 3rd Generation (3G) network.[1] Part of the material presented in this chapter appeared in Ricciato (2006), Ricciato *et al.* (2007, 2008), Romirer-Maierhofer and Ricciato (2008) and Romirer-Maierhofer *et al.* (2008).

10.1 Case study: Bottleneck Detection

10.1.1 Motivations and Problem Statement

The 3G environment is continuously under evolution: the subscriber population and the traffic volume are in a growing phase; the relative distribution of terminal types and their capabilities is changing quickly; the portfolio of services that are offered by the operators evolves rapidly;

---

[1] The METAWIN and DARWIN projects: http://userver.ftw.at/ ricciato/darwin/.

Video and Multimedia Transmissions over Cellular Networks   Edited by Markus Rupp
© 2009 John Wiley & Sons, Ltd

prospective changes to the network infrastructure are in the agenda of many operators, including technology upgrades such as High Speed Downlink Packet Access (HSDPA) and IP Multimedia Subsystem (IMS) (Bannister *et al.* 2004). All these aspects build a potential for changes in the traffic pattern that can occur at the macroscopic scale (network-wide) and in a relatively short time frame. In such a scenario, the ability to accurately and extensively monitor the actual network state and to early detect global performance drifts or localized anomalies is fundamental to the network operation and engineering processes. Among other problems it is important to early detect emerging bottlenecks, that is, network elements which are under-provisioned and therefore cause performance degradation to some traffic aggregate as its volume increases. A bottleneck in the Radio Access Network (RAN) can be a geographical area that is frequently overloaded because the provisioned radio capacity does not match the peak-hour traffic demand. A bottleneck in the Core Network (CN) is often a link or other network element with insufficient capacity to carry peak-hour traffic. A capacity bottleneck always impacts a certain traffic aggregate rather than isolated flows – for example, all the traffic directed to a certain radio area, or routed over a certain network element.

The classical approach to reveal capacity bottlenecks is to compare coarse measurements of the actual traffic volume, typically obtained by Simple Network Management Protocol (SNMP) counters, with the 'nominal' provisioned capacity. This approach requires access to configuration parameters such as provisioned link bandwidth, logs and built-in counters from several network elements, and considerable costs, complexity and complications are found in practice where it comes to extraction, gathering and correlation of such heterogeneous data – in the format and semantics – from different elements, with different software platforms and from different vendors. In other words, deploying and maintaining a monitoring infrastructure with the same capillarity of the production network might be very expensive. On the other hand, the structure of a 3G mobile network is highly hierarchical and centralized: the whole traffic is concentrated in the CN and there are only a few Gateway General Packet Radio Service (GPRS) Supporting Nodes (GGSN) that connect it to external networks such as the Internet, therefore the whole traffic can be captured at a few monitored links.

The aim of this case study is to develop methods to infer the presence of performance bottlenecks in the network from the analysis of the traffic captured at a few centralized monitoring points, without direct access to all network links. A possible approach is to observe the Transmission Control Protocol (TCP) behaviour: since TCP is closed-loop controlled, its dynamics and performances are dependent on the state of the whole end-to-end flow path. In principle it should be possible to infer the presence of performance bottlenecks by looking at the evolution of the TCP aggregate and/or to individual connections at any point along the path. We devised two possible strategies to diagnose the presence of a bottleneck on some network element from the analysis of the traffic aggregate routed through it but observed at a different point:

- from the analysis of the aggregate traffic rate;

- from the analysis of TCP performance indicators such as frequency of retransmissions and/or round trip times.

In this case study we explore both approaches. This study is based on several weeks of packet traces collected by passive monitoring the core network links of a large operational 3G network during 2006. In the middle of the measurement period a bottleneck link within

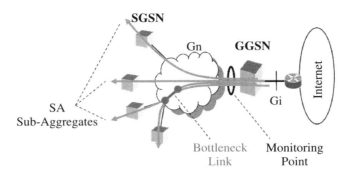

Figure 10.1 Reference network scenario (top) and example of bottleneck on Gn (bottom).

the Universal Mobile Telecommunications System (UMTS) Core Network was detected and removed by a capacity upgrade. Since the traces are complete – that is, all packet headers were captured and time stamped – it is possible to analyse and compare several aspects of the traffic dynamics in the two network conditions: with and without the bottleneck. The aim is to identify those patterns that more clearly discriminate between the two conditions and can be taken as 'signatures' of the presence of a bottleneck. This would allow the implementation of intelligent agents to be integrated directly into the online monitoring system, watching for future occurrences of similar patterns and reporting early warning about emerging bottlenecks. This approach is similar to the so-called 'post-mortem' analysis that is commonly performed in the area of network security in order to learn about the dynamics of past attacks so as to devise appropriate detection mechanisms.

The reference network scenario is depicted in Figure 10.1. The 3G network has a tree-like deployment: the Mobile Stations (MSs) and the Base Transceiver Stations (BTSs) are geographically distributed (nationwide), but the level of concentration increases when moving towards the boundary of the 3G CN towards the Internet. There are several Serving GPRS Support Nodes (SGSNs) and a few GGSNs and the traffic is concentrated on a small number of Gn/Gi links near the GGSNs. Therefore, with few probes on these interfaces one can capture the whole traffic. The problem addressed here is how to infer the presence of a bottleneck in the 3G core network from the passive observation of the traffic at a single monitoring point – on Gn in our case. Remarkably, we assume only minimal information about the structure and settings of the whole network: for instance, *the bandwidth provisioned at each link is not known* to the monitoring agent, nor is the detailed network structure known. This allows a dramatic reduction in the complexity and maintenance effort of the monitoring system itself, as discussed later.

Definition of Sub Aggregate (SA). The only required information is that enabling the discrimination of different sub-aggregate components inside the total traffic. An SA defines the portion of the overall traffic observed at the monitoring point that is routed through a specific part of the network. An SA can be associated to each network element, typically an SGSN, Radio Network Controller (RNC) or cell. For example, one could define an SA for a single SGSN x, meaning that all the traffic routed through SGSN x can be separated from the rest and examined separately. At a coarser granularity one could consider the SA towards a cluster of SGSNs y co-located at a single physical site and sharing the same access bandwidth to the site. At a finer granularity SAs can be associated to individual RNCs (say z). Due to the typical tree-like structure of the 3G network SAs are hierarchically nested ($z \subset y \subset x$), and the analysis of SAs at different levels might help to individuate the position of a bottleneck, an approach that is in principle similar to network tomography (Castro *et al.* 2004). In the simplest approach, the analysis of each SA is performed in isolation. In other cases the comparison between different SAs at the same hierarchical level will help to pinpoint an abnormal behaviour. If one is able to extract summary indicators that are invariant to different load conditions, then the direct comparison between SAs would provide a precious contribution to the task of bottleneck detection. The identification of synthetic and invariant indicators is one of the ultimate goals of this case study.

SA discrimination. Sub-aggregate analysis requires sub-aggregate discrimination, that is, the ability to map each traffic unit (a packet or a connection) to a specific SA. In practice the way that this association is implemented depends on several technical details that are to some extent dependent on the specific configuration of the network. In this respect, it is preferable to set the monitoring point on Gn rather than Gi, since the lower layers of the Gn protocol stack include useful information for this purpose. For instance, the Gn network is Internet Packet (IP) based: after GPRS Tunnelling Protocol (GTP) encapsulation user packets are encapsulated into User Datagram Protocol (UDP)/IP packets carrying the IP address of the destination SGSN/GGSN: this allows direct per-SGSN and per-GGSN discrimination. Note that the detailed tracking mechanism is different for GPRS and UMTS.

As a working hypothesis we assume that the monitoring system is capable of capturing and discriminating all packets belonging to a certain SA and the problem is to diagnose the presence of a bottleneck affecting such specific SA. This should trigger an alarm to the network staff, which should then start an in-depth inspection of the network and specifically of the SA path. The fact that the alarm is raised for a specific SA gives a useful initial hint about the possible location of the bottleneck but it is not sufficient pointedly to localize it, nor to decide whether it is located upstream or downstream of the monitoring point. However, restricting the attention within the scope of a single SA path can lead in practice to a dramatic speed-up of the troubleshooting process.

Obliviousness to the provisioned capacity. We assume that the capacity provisioned on each link is not known to the monitoring system, nor is the *detailed* network structure. This eliminates the burden of maintaining the monitoring system synchronized with the configuration of the network elements, with clear savings in the complexity and maintenance efforts of the whole monitoring system. To appreciate such gain, consider that a real network might have a very complex deployment. For example, the connectivity between GGSN x and SGSN y (Gn interfaceGn interface) might involve a combination of physical circuits, Asynchronous Transfer Mode (ATM) virtual circuits and Local Area Network (LAN) segments internal to each site. Therefore, the path from x to y might include multiple links of different nature, dedicated or shared with other paths, and nodal equipments spanning multiple technologies. Our monitoring system is oblivious to the *detailed* deployment of the Gn links: we only look at the SA traffic flowing between x and y, regardless of its path. A conceptually simple approach to detect congestion would be to compare the measured traffic volume with the provisioned path capacity, that is, the minimum net capacity between x and y. However, there are several complications associated to this approach:

1. Operational complexity: in order to derive the exact capacity of the path $x - y$, it might be necessary to access the configuration files of several nodes along the path.

2. Deviations from nominal behaviour: in the case of malfunctioning of some network equipment, the actual capacity of the link might depart from the nominal one.

3. L2 provisioned capacity: in the case of rate-limited virtual channels (typically ATM) the net capacity 'visible' at the IP layer depends on the actual distribution of IP packet size. The larger the fraction of short packets, the smaller the net available capacity. In this scenario it is possible that a shortage of capacity emerges due to macroscopic changes in the statistics of the packet size, for example following the spreading of new popular applications, or multiplication of undesired traffic resulting from worm infections (Pang *et al.* 2004; Ricciato 2006b; Zou *et al.* 2002).

By taking an oblivious approach about the provisioned capacity, the above complications are avoided. Furthermore, a number of additional advantages emerge. First, being decoupled from the actual configuration of the network equipments, the monitoring system does not need to be updated upon each reconfiguration or re-provisioning event, which saves costs in terms of system maintenance and communication overhead within the staff. Secondly, the resulting diagnosis method is more robust, since it will detect shortage of the 'actual' bandwidth rather than of the 'planned' one. Possible causes of mismatching between the two include human mistakes in the configuration of the network elements, equipment malfunctioning, mismatching between L3 and L2 bandwidth and others. After the signature of a capacity bottleneck has been recognized, the network staff has to locate it exactly, which requires knowledge of the detailed network deployment and eventually inspection of several elements during the troubleshooting process. However, being able to trigger an early alarm without relying on external information about the underlying network configuration greatly improves the robustness of the overall process.

10.1.2 Input Traces

For this case study we collected four weeks of traces during December 2004. A capacity bottleneck was in place in the network, affecting a certain portion of the UMTS traffic, and was removed during the measurement period by a capacity upgrade. The bottleneck link was found on an IP-over-ATM virtual circuit with a cell-level rate-limiter based on the standard General Cell Rate Algorithm (GCRA). The link was serving a physical site hosting multiple SGSNs. We were able to discriminate the SA component crossing the bottleneck element out of the total Gn trace and to analyse it separately. In the rest of the section we will therefore refer exclusively to the analysis of this sub-trace. For proprietary reasons we cannot disclose several absolute quantitative values such as traffic volumes, number of users, number of Gn links and so on. For the same reason we provide only relative values, that is, fractions, or rescaled values instead of absolute values.

As discussed above we look at the TCP behaviour because of the end-to-end nature of its dynamics. We found a large prevalence of TCP traffic in the 3G network (for more details on the observed traffic mix given later in Chapter 12, see also Svoboda *et al.* (2006)). Furthermore, it is likely that a large part of the traffic seen as UDP packets in the CN is indeed TCP-controlled: for instance TCP connections tunnelled into IPsec VPNs are seen as UDP packets on the Gn monitoring point.

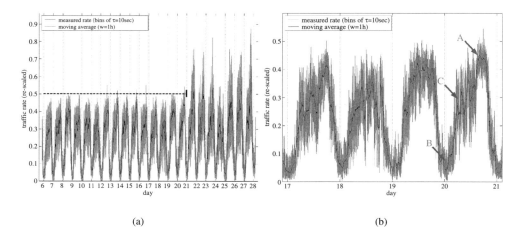

(a)   (b)

Figure 10.2   SA total rate for (a) the monitored period and (b) zoomed in days 17–20. Aggregation bins of 10 s, rescaled values.

10.1.3   Diagnosis based on Aggregate Traffic Rate Moments

Exploration of Data

Here we focus on some basic statistical properties of the 'instantaneous rate' for the traffic aggregate. The rate is measured in time bins of fixed length τ. The value of τ must be larger than a few times the typical Round Trip Time (RTT), as for smaller values the notion of 'rate' for TCP-controlled flows vanishes. On the other hand, too large values for τ will average out short-term fluctuations, which is against the aim of measuring the 'instantaneous' rate. Given that the typical RTT measured in the operational UMTS network is in the order of half a second as shown in Vacirca *et al.* (2005) we used $\tau = 10$ s. The results given below in this section will confirm the wisdom of such a choice for our purposes. Only the packets belonging to the specific SA under study are considered here in the traffic count.

The measured traffic is highly asymmetric, with a large prevalence of web: the ratio between uplink/downlink (from/to the MS) traffic was approximately 1:3. Only the downlink traffic reached the bottleneck capacity; therefore we will not consider here the uplink rate. Note that all user data packets are considered in the byte count, regardless of the specific payload protocol. However, due to the strong prevalence of TCP traffic, the overall SA rate is shaped by the TCP dynamics.

The time series counting the total number of bytes of downlink traffic in time bins of length s_τ will be denoted by $s_\tau(t)$. For proprietary reasons the whole time series has been rescaled by an arbitrary factor denoted by u, therefore all rate values in the rest of this section will be expressed in terms of this arbitrary unit (note that u is *not* related to the link capacity, therefore a rate value of 0.6 does *not* refer to a link load of 60%). Figure 10.2 plots the signal $s_\tau(t)$ with time bins of duration $\tau = 10$ s. Figure 10.2(b) refers to the entire monitoring period. A bottleneck with bandwidth limit $0.5u$ was in place during the first part of the period, and was removed in the night between days 20 and 21. Figure 10.2(b) zooms in the last few days before the removal. The presence of the bottleneck can be diagnosed by noting that the variability range in the peak hour as indicated by point 'A' in Figure 10.2 was approximately the same as during the night time 'B'. Nevertheless, the variability range at intermediate traffic levels 'C' was broader. We deemed it strange: we would have expected higher variability corresponding to higher traffic level. Based on such qualitative observation, we made the hypothesis that some capacity restriction along the path of this SA was acting as a capping barrier for the TCP-controlled aggregate, preventing the maximum downlink rate from exceeding the provisioned bandwidth 'which was unknown to us'. Triggered by our report, a quick investigation by the network staff resulted in the identification of a capacity shortage along the path of this SA that was removed within hours by a capacity upgrade. More precisely, the bottleneck was found on an IP-over-ATM link with a cell-based rate-limiter. Inspired by this event, we recognized the possibility to implement an automatic method for detecting future occurrences of similar events, by translating the qualitative observation based on the visual inspection into a quantitative indicator.

We analysed the evolution of some basic properties of the marginal distribution of s_τ – namely variability and symmetry – at different traffic loads. To do so, we considered a sliding window of length w and analysed the empirical distribution of the samples in $D_{\tau,w}(t) = \{s_\tau(\theta) : t - w/2 \le \theta \le t + w/2\}$ for different values of t. The value of w should be short enough such that the signal s_τ can be considered stationary within this window; on the other hand, too short values for w will include too few signal samples to have a robust estimation of the moments. We set $w = 1$ h as a compromise between two concurrent needs. We extracted the first moments of the empirical rate distribution $D_{\tau,w}(t)$, specifically the mean $m(t)$, the variance $\sigma^2(t)$ and the skewness $\gamma(t) = E\left((s_\tau - m(t))^3\right)/\sigma^3(t)$. Recall that the skewness (third-order central moment) is a simple measure of the asymmetry of the distribution: positive values $\gamma(t) > 0$ indicate a right-skewed distribution, negative values $\gamma(t) < 0$ indicate left-skewness, symmetric distributions hold $\gamma(t) = 0$. The values of these empirical moments depend on the time-scale parameters τ and w, but for the sake of simplicity we will omit them in the notation. In addition to the moments, we also computed the median $\mu(t)$ and percentiles $p_{10}(t)$, $p_{90}(t)$ of $D_{\tau,w}(t)$.

Before computing the moments, we pre-filtered the data from the outliers. This prevents a few large samples from biassing the estimated moments in their neighbourhood. An outlier can be caused, for example, by a large burst of packets coming from the Internet, produced, for example, by a scanning source as discussed later in section 10.2.3 – recall that the aggregate rate is monitored upstream of the bottleneck. A sample value $s_\tau(t)$ is classified as outlier if it falls outside the range $\mu(t) \pm 3\sigma(t)$, with median and standard deviations computed locally in the window $D_{\tau,w}(t)$ centred around t. This resulted in fewer than 200 outliers out of 60 480 values (0.3%). Generally speaking, for some applications sporadic large values carry useful information about the phenomenon under study and should be considered as an important part of the traffic process rather than being classified as outliers and filtered out. However, this is not the case here, since the proposed scheme for bottleneck detection relies on the analysis of the mass and near-tails of the rate distribution.

We are interested in looking at the evolution of the moments and percentiles of the rate distribution with the mean traffic rate. For every instant t, we measure the variance and mean value of the rate samples in the observation window. We project such data into bi-dimensional space, where each point represents a pair $\{\sigma^2(t), m(t)\}$ for a specific value

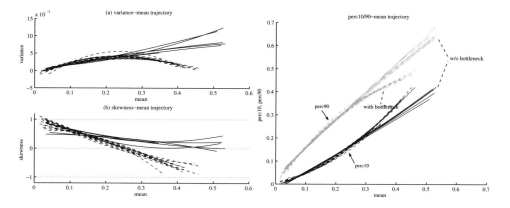

Figure 10.3   Left: variance-mean and skewness-mean trajectories. Right: 90- and 10-percentile-mean trajectories. Dashed line: days 6–21, with bottleneck. Solid line: days 21–27, without bottleneck.

of t. On one day we obtain a total of 8640 data points in a bi-dimensional space (recall that we are using 10 s time bin). In order to summarize this data, we fit the 8640 data points representing a single-day trajectory with a quadratic curve, which is a degree-two polynomial. In addition to the variance, the same procedure is applied also to the skewness, so as to observe the skewness-mean trajectory. The resulting curves are given in the left-hand plot of Figure 10.3 for 22 days. The dashed and continuous lines refer respectively to days 6–20 (with the bottleneck) and days 21–27 (after the bottleneck removal). The variance-mean trajectory is such that when the bottleneck is in place the variance reaches a maximum when the mean is approximately half of the bottleneck capacity, after which it decreases again. Instead, without the bottleneck the variance keeps increasing steadily with the mean traffic intensity. The presence of the bottleneck also impacts the skewness of the distribution: in the high traffic region the bottleneck clearly induces left-skewness ($\gamma < 0$), while after the bottleneck removal the distributions tend to be more symmetric, with small positive values of γ. We applied a similar method to the percentiles $p_{10}(t)$ and $p_{90}(t)$. The resulting regressed trajectories are shown in the right-hand plot of Figure 10.3, from which it can be easily seen that the trajectory of the inter-percentile distance $\{p_{90}(t) - p_{10}(t), m(t)\}$ displays a concave behaviour similar to the variance.

Interpretation of the Results

All the above observations are consistent with the following interpretation. The empirical rate distribution evolves with the mean traffic intensity – therefore with the time-of-day – as outlined in Figure 10.4. For low traffic intensity (night and early morning) the distribution is right-skewed: since rates cannot be negative only positive fluctuations are possible. For higher traffic intensity (morning and afternoon) the rate distribution becomes more dispersed (higher variance) and more symmetric (lower skewness). In fact, the rate of individual connections is limited by the bandwidth of the radio link, but such a limit caps individual flows (per-MS channel) or at most on a bundle of a few flows belonging to the active MSs in the same

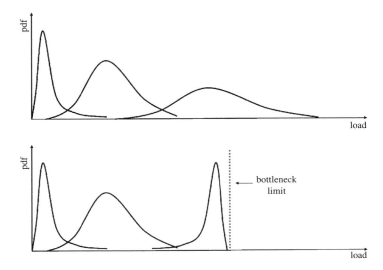

Figure 10.4 Evolution of the marginal rate distribution.

cell (per-cell bandwidth). As a result, in the total SA under analysis – covering hundreds of parallel connections spanning several cells – the correlation between the rates of individual connections is very low, and the shape of the total traffic distribution is dominated by the arrival/departure process of the connections, assumed independent at a first approximation. Since higher traffic during the day means essentially more active MSs and more parallel connections, this translates into a marginal rate distribution with increasing mean, variance and symmetry, as is typical with the superposition of many independent On/Off sources. Without any common bottleneck, such a trend would be maintained also during the peak hour (evening and early night), with more variance and more symmetry as in the top graph of Figure 10.4. Instead, when the bottleneck is in place (bottom graph of Figure 10.4) it introduces an additional limit on the total rate of the whole SA whose effect becomes stronger as the total traffic approaches the bottleneck capacity. At this point two phenomena take place. First, due to the TCP congestion control mechanisms, the bottleneck introduces correlation between the rates of the individual connections. Secondly, an additional control-loop emerges from the user behaviour: users will delay the next download until the current one is completed, and arguably some users will abandon the network if the perceived quality degrades below what they consider to be the 'acceptable' level. As a result, the user behaviour contributes to modulation of the number of parallel connections during the congestion period. The joint effect of these two control loops (TCP congestion control and user behaviour) is that the total marginal rate distribution becomes compressed towards the capacity limit: consequently, its total variance diminishes, and the hard-limit on the upper tail – but not on the lower one – leaves the distribution left-skewed.

Diagnosis Method

The basic idea of our work is that by simply observing the evolution of the marginal rate distribution during one day cycle, it is possible to infer the presence of a bottleneck, which

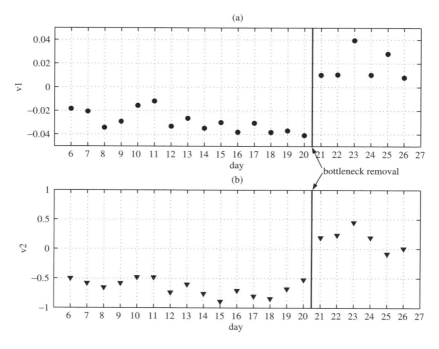

Figure 10.5 Measured values of summary indicators v_1, v_2.

is a compressing barrier on the aggregate rate, *without knowing the value of the bottleneck limit*. The goal is to develop one or a few empirical indicators telling whether the marginal distribution of the traffic evolves 'freely' with the mean traffic intensity or, rather, displays signs of strain. Our first proposal is to derive simple indicators based on the coefficients of the regressed variance-mean and skewness-mean trajectories along one-day cycles, as explained in the following. For a generic day we denote by a_j and b_j the coefficients of the regressed polynomial for the variance-mean and skewness-mean trajectory, respectively, and by $\overline{m}$ the maximum value of the mean traffic registered during the same day. Denote by $v_1 = 2a_0\overline{m} + a_1$ the slope of the variance-mean trajectory at the extreme point $\overline{m}$. Similarly, denote by $v_2 = b_0\overline{m}^2 + b_1\overline{m} + b_2$ the extreme value of the regressed skewness. In Figure 10.5 we plot the measured values for v_1 and v_2 for the whole monitored period. There is an evident correlation between negative values for both indicators and the presence of the bottleneck. This suggests that a preliminary alarm can be triggered whenever any such indicators fall below zero, while persistent negative values for a few days should trigger a check intervention by the network staff. In this specific case, by setting the alarm region at $v_1 < 0$ or $v_2 < 0.25$ such simple indicators would have revealed the presence of the bottleneck several days in advance. In practice, the alarm region should be identified based on the historical time series for the same SA under known bottleneck-free conditions.

In summary, the proposed algorithm would include the following steps, to be run regularly each day for each SA:

1. Collect rate measurements at small timescale ($\tau = 10$ s).

2. Remove outlier samples falling outside the range $\mu(t) \pm 3\sigma(t)$, computed in a sliding window of length w (we used $w = 1$ h).

3. Compute running moments (mean, variance, skewness) and percentiles ($\mu(t)$, $p_{10}(t)$, $p_{90}(t)$) in sliding window of length w.

4. Apply quadratic fitting to the X-mean trajectories (where X stands for variance, skewness, inter-percentile distance, etc.) to reduce the data to a few coefficients.

5. From these coefficients derive a synthetic vector indicator **V**.

6. Collect historic records and identify the 'typical' region for **V**.

7. Raise an alarm whenever **V** persistently falls outside such region, for example for two to three consecutive days.

Each step is quite easy to implement and requires only a limited amount of memory and processing resources. The overall scheme is ultimately a collection of simple pieces of time-series analysis, typically available as standard routines in any mathematical library. We note that the quadratic regression (Step 4) is a key stage: by collapsing the set of observations along one day into a few parameters, namely the coefficients of the polynomials, it increases the stability of the indicators and hence the robustness of the whole scheme. We have seen that the presence of the bottleneck has a direct impact on these trajectories. For instance, the variance-mean trajectory monotonically increases under normal 'bottleneck-free' conditions, and becomes folded and non-monotonic when the bottleneck is in place. Such patterns can be taken as a signature for the presence of the bottleneck. The reason for choosing quadratic fitting is that degree-two polynomials are the simplest family of curves able to capture non-monotonic behaviours with a local maximum. Polynomials with degrees higher than two would not be more helpful for our purpose. On the contrary they might hamper the robustness of the overall scheme by offering more degrees of freedom to reflect secondary details and/or random fluctuations.

10.1.4 Diagnosis based on TCP Performance Indicators

Methodology

In parallel to the rate analysis, we also considered several TCP performance parameters as candidate indicators for the presence of a bottleneck. The first group of potential indicators refers to the frequency of retransmission events, discriminated into the following categories:

- **FRTX** (Fast Retransmit Retransmissions): the number of TCP packet retransmissions triggered by duplicate ACKnowledgments (ACK).

- **LRTO** (Loss-induced Retransmission Time Outs): TCP retransmission events triggered by the expiration of the TCP Retransmission Time Out (RTO) caused by packet loss.

- **SRTO** (Spurious Retransmission Time Outs): TCP retransmission events due to RTO expiration caused by a large delay, without packet loss.

- **AMB** (AMBiguous retransmission timeouts): TCP packet retransmissions that we are not able to classify into one of the previous categories. Note that AMB events are always associated to RTO expiration.

All these events were inferred with the procedure proposed in our previous work (Vacirca *et al.* 2008). There the focus was on the estimation of the SRTO events, which in turn required the discrimination of LRTO, FRTX and AMB. We have implemented the estimation algorithms in a modified version of the `tcptrace` tool,[2] which was run over the whole trace.[3] For each type of event we measured the relative frequency into each time bin as follows. For each active MS_i we counted the number of occurrences n_i of the specific event, as, for example, LRTO, and the total number of DATA packets N_i. The global frequency of such events is then defined as the ratio $f_{tot} = \sum_i n_i / \sum_i N_i$.

In general it is desirable to have exact MS discrimination, that is, to identify the source/destination MS for each packet. This requires tracking of the signalling messages associated to each PDP-context and reconstruction of all state transitions, so that each packet can be labelled with a unique and non-ambiguous identifier associated to the MS. Such a feature was not yet in place in the dataset from December 2004; therefore, we resorted to (approximate) MS discrimination based on the IP address. The results presented below indicate that this approach does not limit the capability of the proposed method to expose the presence of the bottleneck from TCP indicators. This is good news, as it suggests that the proposed method can be applied also on 'simple' IP-level Gn traces.

In addition to the above events we also considered the RTT values. A first exploratory analysis of TCP RTT in UMTS and a comparison with GPRS was reported in our previous work (Vacirca *et al.* 2005). Instead of the end-to-end RTT, we considered the semi-RTT between the monitored Gn interface Gn and the Mobile Station. For the sake of simplicity, in the rest of this work we will refer to the Gn-MS-Gn semi-RTT simply as 'RTT'. An RTT sample is defined as the elapsed time $t_{data} - t_{ack}$, where t_{data} and t_{ack} are the timestamps, respectively, of a TCP DATA packet arriving from the Internet and of the associated ACK from the MS as observed at the monitoring point on Gn. The RTT defined in this way includes three components: the downlink delay (Gn→MS), the uplink delay (MS→Gn) and delay component internal to the MS, which includes processing time and I/O buffering. Only the first two components are accountable as network dependent, while the latter depends exclusively on the terminal. However, the TCP dynamics, and ultimately the user-perceived performances, will be impacted by the cumulative RTTs. Note that only non-ambiguous DATA-ACK pairs are considered to produce a valid RTT sample: the acknowledgment number of ACK must be at least one byte greater than the last sequence number of the DATA packet; furthermore, it is required that the packet being acknowledged was not retransmitted, and that no packets that came before it in the sequence space were retransmitted after t_{data}. The former condition invalidates RTT samples due to the retransmission ambiguity problem. This is the same procedure that TCP utilizes to estimate the RTT and to set the value of the Retransmission Timeout (Karn's algorithm; see Stevens 1994). The latter condition invalidates RTT samples since the ACK could acknowledge cumulatively the retransmitted packet rather than the original DATA packet.

---

[2]Tcptrace 6.6.1, available at http://www.tcptrace.org.

[3]The modified `tcptrace` version can be downloaded from http://userver.ftw.at/ vacirca.

Before the analysis we filtered out all packets on ports tcp:4662 and tcp:445/135. The former is used by popular peer-to-peer file sharing applications. Typically, such applications run many parallel TCP connections; therefore, they are likely to induce self-congestion on the radio channel and/or on the terminal internal resources. This would result in poor TCP performances that are application-specific rather than network-dependent, therefore do not carry information about the network state. Additionally, during the exploratory analysis we found the presence of a large number of packets directed to ports tcp:445/135, mainly TCP SYN in the uplink direction. This is probably due to some self-propagating worms attached to infected 3G terminals. The presence of such unwanted traffic was expected since laptops with 3G data cards – often equipped with popular operating systems – populate the 3G networks nowadays along with handsets and smart phones. It is well known that unwanted traffic has been a steady component of the traffic in the wired networks for years (Pang *et al.* 2004). For a discussion on the potential impact of unwanted traffic onto a 3G cellular network see Ricciato (2006b). The point here is that most such packets did not bring any valid contribution to the problem of bottleneck detection while consuming resources in the analysis software; therefore, filtering them out speeds up the analysis process.

Besides pre-filtering the bulk of unwanted traffic and file-sharing applications based on port filtering, we adopted a filtering procedure to prevent the problem of bias by dominant MSs. The problem arises whenever the distribution of data samples across users is such that a few MSs can dominate the global average. In fact, a handful of MSs generating large traffic volumes – hence contributing with many samples within a measurement time bin – and large RTT values can influence the measured RTT distribution and bias statistical indicators such as average and percentiles. Similarly, they can inflate indicators representing the global frequency of retransmission events, such as SRTO. In other words, high values of RTT and/or SRTO indicators may result from the sporadic presence of a few 'bad-and-big' sources, rather than from network-wide conditions. This is more likely to happen when the number of active MSs is low, that is, at night. In order to prevent false alarms, it is desirable to eliminate or at least mitigate the bias introduced by such sources, for example filtering out in each time bin the few 'worst' MSs with the strongest impact on the global statistics. A possible heuristic procedure to achieve this goal is presented in Ricciato *et al.* (2007).

Results

In Figure 10.6 we plotted the measured values for each parameter in time bins of 1 h. The top graph shows the cumulative number of TCP DATA packets ($\sum_i N_i$), normalized to the peak value during the entire monitoring period in order not to disclose the absolute value. The second graph reports the average and several percentiles (5%, 50%, 95%) of the RTT samples extracted in each time bin. The remaining graphs report the measured frequency of FRTX, LRTO, SRTO and AMB events, respectively. In each time bin, the samples of the few 'worst' MSs were filtered out to avoid bias. Recall that the bottleneck was removed in the night between days 20 and 21. All the frequencies of retransmission events present a clear change-point at this time. As expected, the intensity of retransmissions climbs to high values during the peak hour when the bottleneck is in place, while after the bottleneck removal they never exceed a physiological level, which is reasonably low. The change-point is so clear that one can immediately pinpoint the bottleneck removal time. Note also that under 'normal'

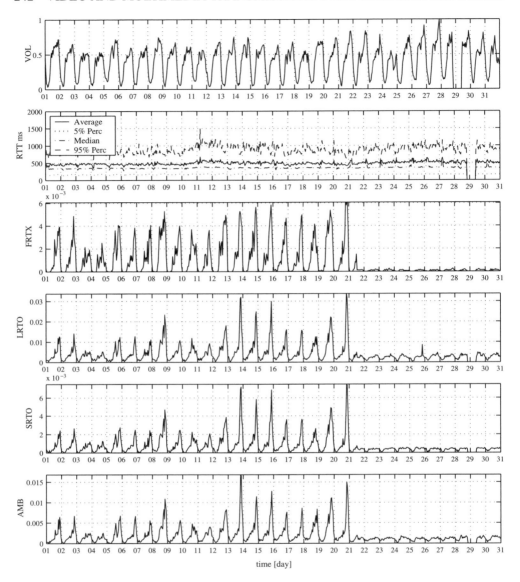

Figure 10.6 TCP performance indicators after filtering the worst 10 MSs in each 1 h bin.

operating conditions (without bottleneck) the SRTO frequency stays at a 'physiological' level below 0.1% that appears to be independent of the time of day, hence of the network load.

Regarding the RTT statistics, the filtering process had a dramatic effect and almost completely cancelled the fluctuations of the average – now firmly anchored to the level of 500 ms – and of the lower percentiles that were present before the filtering (compare Figures 10 and 11 in Ricciato *et al.* (2007)). However, there is no evidence of correlation between the

RTT values shown in Figure 10.6 and the presence of the bottleneck. The conclusion is that the RTT process estimated with the methodology described above is not a good indicator for this type of bottleneck. A likely explanation is that the RTT estimation process only considers selected DATA-ACK pairs that do not hold any ambiguity in the RTT estimation. This method filters away 'invalid DATA-ACK pairs' that typically emerge in the neighbourhood of events such as packet losses, retransmissions and timeouts. These are exactly the events that are generated by the bottleneck. In other words, RTT statistics were intrinsically 'cleaned-up' due to the way RTT samples are extracted, which explains why they did not react to this type of bottleneck. Still, it can be argued that different types of bottleneck, more oriented to buffer excess packets, might be better captured by indicators associated to the packet delay rather than to retransmissions or timeouts.

Such results show that it is possible to build powerful bottleneck indicators from performance parameters estimated by passive monitoring TCP traffic. How good an indicator is depends on its '**predictability**' under nominal operational conditions, which in turn requires good stability, and on its '**responsiveness**' to abnormal conditions. Once a 'good' indicator of the network state is defined, statistical testing methods can be applied to provide automatic alarms when an abnormal condition occurs. From a general point of view, the 'better' the indicator signal in terms of predictability and responsiveness, the simpler the statistical testing technique. In the specific case considered here, all the proposed indicators associated to retransmission events are so effective that even the simplest test, that is, an appropriate fixed threshold set – would have provided an early warning about the presence of the bottleneck several days in advance. However, this is true only for indicators that have been adequately filtered to counteract the instabilities and bias due to a sporadic 'big-and-bad' MSs.

Another important performance metric for the practical applicability of the proposed method is its scalability with the traffic volume – number of packets and number of MSs – and the required processing/memory resources. While a formal analysis is beyond the scope of this work, we can report that the overall algorithm could be run in real-time for the total traffic aggregate in the whole network (absolute values cannot be disclosed). The modified version of `tcptrace` that we have used for this work, which includes a patch for optimized memory management, is publicly available from DARWIN.

10.2 Case Study: Analysis of One-way Delays

10.2.1 Motivations

The availability of synchronized packet-level traces captured at different links allows the extraction of one-way delays for the network section in between. Delay statistics can be used as quality indicators to validate the health of the network and to detect global performance drifts and/or localized problems. This is particularly important in the context of 3G cellular networks, given their complexity and relatively recent deployment.

The aim of this case study is to show how the analysis of the one-way delay process can effectively contribute to the task of revealing network and equipment problems. In the following sections, we present four concrete examples where the investigation of the one-way delay provided useful partial insight into the dynamics of the monitored 3G core network. By relying on packet delay statistics, we detected micro congestion caused by sequential

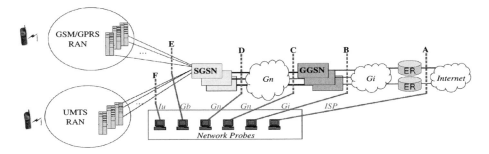

Figure 10.7 Monitoring setup.

scanners as well as configuration errors of network equipment. Additionally, current radio conditions in the RAN could be inferred. We conclude this case study by elaborating on the possibility of leveraging one-way delay measurements for an autonomous and online detection of abnormal deviations in the delay process.

10.2.2   Measurement Methodology

Monitoring Setup

The present work is based on the online monitoring of packet-level traces captured in the operational network of a mobile provider in Austria. The monitoring system was developed in a past project (DARWIN). The monitoring setup used for the present study is depicted in Figure 10.7. The one-way packet delay measurements presented in this study were performed using the network probes monitoring the Gb (Section E), IuPS (F), the Gn interfaces of a SGSN (D) and a GGSN (C), the Gi interface (B) as well as the ISP links (A) to the public Internet. Each network probe is equipped with at least one Endace DAG [4] capture card. These capture cards allow GPS-synchronized packet capturing, offering a time-stamp accuracy of ± 100 ns or better (Donnelly 2002, pp. 97–98). The single network probes are interconnected via a separate IP network. This network is used for measurement tasks requiring information from more than one interface (for example, one-way delay computation) and for maintaining the monitoring system itself.

Delay Computation

The one-way delay was extracted in post-processing with a similar methodology to Papagiannaki *et al.* (2003). In our work, each IP packet is hashed into a string of length $N = 128$ via the MD5 function[5] – note that CRC-32 was used instead in Papagiannaki *et al.* (2003). This guarantees a negligible collision probability. A few selected header fields were excluded by the hashing, for instance the time-to-live value which is changed at each hop. For each processed IP packet, the hashed string, the arrival time stamp and both IP addresses are written to a binary file. The same procedure is repeated for the traces captured at every

[4]Endace Measurement Systems. [Online.] Available at: http://www.endace.com.
[5]The MD5 Message-Digest Algorithm, *RFC 1321*, April 1992.

interface. In order to extract the delay samples from A to B, we look for string matches in the two corresponding binary files, and take the difference of the respective timestamps. In order to save processing power and computation time we can resort to packet sampling, and compute the delays only on a fraction of the total traffic. This strategy is particularly important for online delay computation on highly loaded links in order to meet the real-time constraint. Denote the sampling rate by r_s ($0 < r_s \leq 1$). Our monitoring system is capable of running online sampling and delay computation with a tuneable value of r_s. However, for off-line analysis we often resort to complete delay computation ($r_s = 1$). The delay samples are then aggregated into $N = 300$ exponentially spaced bins between 0 and 60 s in order to summarize their distribution. The bin boundary is given by the following formula $b(n) = e^{(n-\delta)\cdot\alpha}$; $n = 0, \ldots, N - 1$, with the parameter setting $\alpha = 0.07$ and $\delta = 240$. While this scheme is conceptually simple, a number of practical issues must be taken into account.

Multiple matches. In some cases, a single hashed value observed at A has multiple matches at B. This may happen due to a retransmission of IP packets at lower protocol layers. However, only the first observation holds a meaningful delay sample, hence any duplicated samples must be filtered out. Also, routing loops may cause multiple matches; a concrete example was reported in Ricciato *et al.* (2008).

Missing matching. A hashed value observed at, for instance, the ISP interface might have no match at the Gn-GGSN interface. This might indicate that the packet was lost on its way from the ISP link to the GGSN, or that the packet arrives at a path that does not include the ISP interface. The latter case includes all traffic generated by a source internal to the Gi network, for example a Domain Name System (DNS) reply, or a Wireless Application Protocol (WAP) connection. In the network under study, part of the web browsing traffic is handled by a transparent proxy that modifies the TCP/IP headers, hence the related hash value. In this case no match can be found and no delay sample is produced. This reduces the actual rate of valid delay samples.

Tuning the sampling rate. In the present case study, the distribution of the delay samples is summarized into $N = 300$ exponentially spaced bins. Let us denote N_S as the number of samples and N_B the number of bins containing at least one delay sample. For an accurate estimation of the actual delay distribution, the sampling rate r_s should be chosen such that $N_S \gg N_B$. For a given value of r_s, N_S is proportional to the traffic load, which normally varies by one or two orders of magnitude during a daily cycle (time-of-day effect). In order to maintain a constant value of N_S, it would be desirable to vary r_s dynamically, following the changes in network load. Note, however, that the dynamic adaptation of the sampling rate requires synchronization across the capture points.

Packet Fragmentation. In Romirer-Maierhofer and Ricciato (2008), we showed that the arrival time stamp of the 'last' fragment of an IP packet has to be chosen for the delay computation whenever fragmentation occurs at lower layers. In this way the delay components due to upstream fragmentation, which are typically dominated by a remote cause (for example radio link conditions for uplink packets), are excluded from the measurements.

10.2.3 Detecting Micro Congestion Caused by High-rate Scanners

The one-way delays presented in this section were recorded during a full day in September 2006. We start by plotting the delay samples measured within 1 h (3–4 pm) between the peering links and the Gi interface (points 'A' and 'B' in Figure 10.7) in Figure 10.8(a).

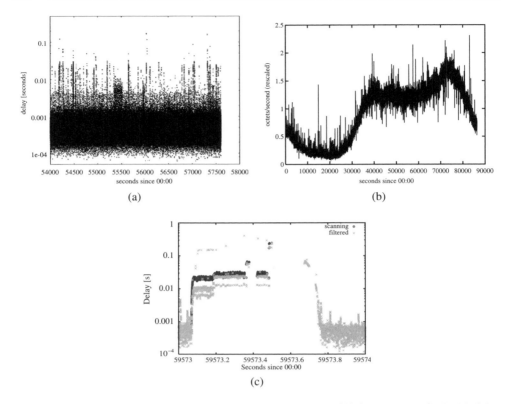

Figure 10.8 Downlink delay between the peering links and Gi links (A–B section): (a) delay versus time, (b) traffic rate versus time at Gi and (c) delay during scanning burst.

As expected, most of the samples take very low values, below 2 ms. However, we also observe large delay values, up to 200 ms. The high delays are not persistent but appear concentrated into vertical lines that are scattered uniformly across the whole day. The simplest hypothesis is that they are associated with traffic bursts. To further investigate this hypothesis, the full-day traffic rate arriving in downlink at the Gi interface is shown in Figure 10.8(b). The rate is measured as byte counts in time bins of 1 s (the graph is arbitrarily rescaled; see note [1]). The time series yields the typical time-of-day behaviour found in any public network: the traffic mean and variance are lower at night, when fewer users are active, and reach their peak in the late evening. In addition, we noticed the presence of *traffic spikes* (positive impulses) scattered across the whole day on Gi. In Figure 10.8(b), we also note the presence of *traffic notches* (negative impulses) on Gi, an aspect that will be discussed below.

High-rate Sequential Scanners

By manual investigation we found that the primary cause of traffic spikes at the Gi links are *scanners*. Consider a host in the Internet that is performing a *sequential scanning* of the full address space by sending some kind of probe packet (for example, ICMP, TCP SYN, UDP)

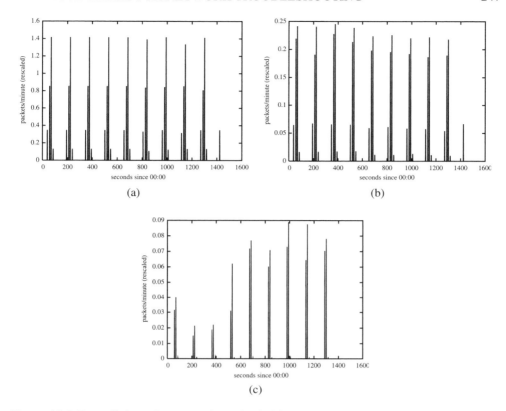

Figure 10.9 Downlink packet count for a single high-rate scanner (S1 source), time bins of 1 min, one full day: (a) peering Links; (b) Gi links; (c) Gn links.

at the rate of R probes/s. At some point it will start spanning the address block assigned to the local network, say of size L (for example, $L = 2^{12}$ for a /12 block). This will result in an impulse of incoming traffic lasting L/R s. If we count the traffic in time bins of length $\Delta > L/R$ and assume that the whole burst falls within a single time bin, the packet count in this bin will jump by L. If the address space allocated to the network consists of several non-contiguous blocks, we will observe a different spike for each block. The height of each spike equals the width of the block, and the separation in time directly relates to the distance in the address space between the blocks. In other words, the probe traffic generated by a sequential high-rate scanning source causes an arrival rate pattern at the peering link that mirrors the address space allocation to the local network. If the scanner source keeps cycling into the address space, such patterns will appear periodically. We observed exactly this phenomenon in the real network: in Figure 10.8(c) we report the arriving packet rate measured at the peering links for the traffic sent by a single external IP address identified as a high-rate sequential scanner (S1 source). In the figure, packets are counted in time bins of 1 min. In practice, the probe packets arrive clustered into bursts of very short duration, corresponding to a relatively high bit rate (undisclosed). We observed several high-rate scanners, periodic and not, using different packet types (for example, TCP SYN, UDP) to different ports. If the scanning occurs on a blocked port (firewalled), the probe packets will be immediately discarded at the edge router. Therefore, no associated traffic spike will be observed at Gi. On the other hand, in the case of non-blocked ports, the scanning burst will penetrate into the Gi network. At the time of measurement, we verified that most of the incoming high-intensity scanning occurred on a single UDP port (undisclosed). Note that some other ports that are commonly used by scanning worms and various malware were already filtered by the firewall.

In Figure 10.9, we report the packet rate from the scanning source S1 as observed at the different links. Comparing the packet rate at the peering links and at the Gi links (note the different vertical scales), we find that the spikes are lower at Gi than on the peering links: at the peak, only approximately 20% of the probing packets reach the GGSN, the rest being lost somewhere along the A–B path. The microscopic analysis of the delay patterns revealed that the loss of probe packets is due to the micro-congestion in the A–B path caused by the scanning traffic itself. Comparing the Gi and the Gn links, we observe that in each time bin only a fraction of the packets observed at Gi passes the GGSN. In fact, the GGSN forwards only the packets directed towards active IP addresses, that is, currently assigned to a MS within an active PDP-context. The number of packets passing the GGSN provides a rough estimate of the number of contemporary active PDP-contexts within each block (in order to hide this information, we had to rescale the graphs in Figure 10.9 by an undisclosed factor). This is lower at night when most MSs are inactive. Therefore, the penetration of scanning traffic is subject to a time-of-day effect, and is maximal at the peak-hour. In the specific case shown in Figure 10.9, we found that a small fraction of MSs responded to each packet with an ICMP 'port unreachable' message, thus causing backscatter traffic in the uplink direction.

We remark that S1 was just one of several sequential scanning sources with patterns similar to Figure 10.9 that were observed in the traces. Moreover, we also found that other sources were performing (pseudo-)random scanning for long periods (whole days). In this case, the visits to each address block are scattered across the whole scanning period instead of being concentrated into distinguishable spikes. Hence, the rate of arriving probes is low and steady. As a result, random scanning is considerably less invasive then sequential scanning from the perspective of the network infrastructure.

Micro-congestion

The simplest explanation for the loss of scanning packets in A–B is that the probe bursts were clipped after hitting the capacity limit of some internal resource, for example, a link or the CPU of some router. To verify this hypothesis, we zoom into a sample scanning burst from S1 and extract a complete ($K = 0$) set of delay measurements in its neighbourhood. We divide the A–B delay samples into two groups: 'scanning' packets, identified by the IP address of source S1, and other traffic ('filtered' time series). The delay samples for both groups are shown in the right graph of Figure 10.8. For each packet arriving at A at time t_A and observed at B at time $t_B > t_A$, we mark a point of coordinates $\langle t_A, t_B - t_A \rangle$. Let us focus on the 'scanning' time series: the delay pattern is consistent with the presence of a buffer that fills up rapidly (initial slope) and then remains persistently saturated (plateau). After the fill-up phase, most arriving packets are lost, while others gain access to the (almost) saturated buffer and experience an approximately constant delay equal to the buffer depletion time, about 20 ms in this case. Such a delay pattern is consistent with the hypothesis that some link is being saturated for a short period by the scanning burst along the A–B path (micro-congestion). The exact peak rate of the scanning packets arriving at Gi (undisclosed in our case) would provide an immediate indication about its bandwidth to the network staff. This can facilitate the localization of the (micro-) congested link.

After this phase, in Figure 10.8 we observe a 'cluster' of considerably larger delay samples, around 200 ms, followed by an empty period of approximately 200 ms where no packets are received at all. This pattern is consistent with the so-called 'coffee-break' event, that is, a temporary interruption of the packet forwarding process at some intermediate router. In our traces such events only occur during large scanning bursts, suggesting that the coffee-breaks observed in this network are *not* due to 'normal' router dynamics as in Papagiannaki *et al.* (2003) but rather a symptom of short-term CPU congestion. Note also that in Figure 10.8, the delay pattern just described for the 'scanning' packets is followed by the remaining traffic as well ('filtered' series). This indicates that the micro-congested resources (buffer, CPU) are shared by all traffic. In other words, the scanning traffic is causing a small impairment to the other legitimate traffic, causing micro-congestion and temporarily high delay and loss. However such events last a very short time, typically a fraction of a second, and remain invisible to individual users, at least as long as they do not occur too frequently.

Having investigated the causes for large impulses in the one-way packet delay, we find that high-rate sequential scanners represent a common source of traffic impulses. In a concrete measurement example from the Gi-section of a 3G core network, we demonstrate that the microscopic analysis of delay and traffic patterns at short timescales can contribute effectively to the task of troubleshooting IP networks. This is particularly important in the context of 3G cellular networks, given their complexity and relatively recent deployment. For more details about high-rate sequential scanning and for a discussion of further potential consequences of such traffic onto the underlying 3G network, refer to Ricciato *et al.* (2008).

10.2.4 Revealing Network Equipment Problems

In a further concrete measurement example, we show that the investigation of one-way packet delays can substantially support the process of detecting network and/or equipment problems in a 3G core network. The one-way packet delay between the Gi and the Gn interface (Sections B and C in Figure 10.7) of two GGSNs denoted by 'GGSN-A' and 'GGSN-B' was recorded during a full day in November 2007. In Figure 10.10(a) we plot the empirical cumulative distribution function (eccdf) along with an empirical histogram for the interval 7–8 pm, representing a period of high load. At GGSN-A 90% of the delay samples take very low values, below 1 ms, while less than 1% of the delay values are higher than 4 ms. The histogram at GGSN-A (Figure 10.10(b)) shows an absolute peak at only around 0.3 ms. However, the delay distribution of GGSN-B tells a different story. A significant fraction, around 70% of all samples at GGSN-B, suffers from delays greater than 30 ms, while only 10% of the delays are lower than 6 ms (see eccdf in Figure 10.10). In Figure 10.10(b), most of the values fall into a histogram bin at around 400 ms. This is clearly an anomalous behaviour indicating congestion. The overload was caused by a hardware problem within the reassembly of IP packets, resulting in substantial packet delays under high traffic load. The problem was readily fixed by reconfiguration. In Figure 10.10 we

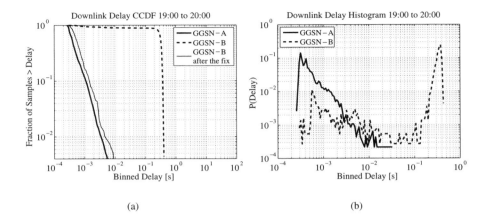

Figure 10.10 Empirical delay distribution at two GGSNs: (a) eccdf and (b) histogram.

also report the eccdf of GGSN-B one day after the fix. Clearly, the delay distribution at GGSN-B looks 'healthy' again. Notably, both eccdfs in Figure 10.10 show a very regular shape (straight line in log-log scale). Having learned that a hardware problem may manifest itself in a change of the delay distribution, one could think of implementing an automatic online agent, continuously measuring the delay distribution and reporting alarms in case of abnormal deviations. Alarming based on a simple threshold on some delay percentile would be sufficient to detect similar congestion events in the GGSN.

In principle, detecting anomalies in the GGSN delay process is a relatively easy task. This is because the delays internal to the GGSN are caused only by queuing. Hence, the typical delay values are very low in a 'healthy' and well-dimensioned GGSN. This is not the case with the SGSN, where packets can be buffered due to *user mobility* and *flow control*. As we will show later, long delays emerge in the SGSN delay process, in addition to the short delays due to queuing (see section 10.2.5).

10.2.5   Exploiting One-way Delays for Online Anomaly Detection

In this section we elaborate on the question of how one-way packet delays may be exploited to develop an autonomous anomaly detection scheme for the packet delay process. This requires stable statistics as input for detecting anomalous delays. Ideally, the input signal is stable enough that simple threshold-based alarms would suffice. In the following section, we show that finding such stable input signals involves two steps: (a) a visual inspection of the packet delay process of the single network sections, and – based on the lessons learned from this inspection – (b) the definition of input signals tailored to the specific network sections under question. For the subsequently presented analysis, we recorded one-way delays during two full days in November 2007 and December 2007. We start presenting the single-hop delay of a GPRS-SGSN and a UMTS-SGSN in uplink. In Figure 10.11(a) we plot the delay histogram for the interval from 8 a.m. to 4 p.m., representing a period of medium load. Since queuing is the only source of delay for packets transmitted in uplink at both SGSNs, the majority of

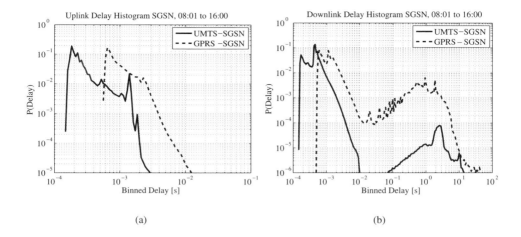

Figure 10.11 Empirical delay histograms: (a) uplink and (b) downlink.

the samples take very low values, below 2 ms. However, the UMTS-SGSN is showing lower delays than the GPRS-SGSN. The empirical histogram at the UMTS-SGSN in Figure 10.11 shows a clear spike at around 1.5 ms. At this point we speculate that this peak is caused by the internal buffer handling of this SGSN.

In Figure 10.11(b), we plot the downlink delay histograms of the two SGSNs. The empirical histogram at the GPRS-SGSN shows a clear bump for delays higher than 20 ms. Also at the UMTS-SGSN, we observe a histogram bump for higher delays. However, the relative share of this bump is lower than at the GPRS-SGSN. As reported in Romirer-Maierhofer and Ricciato (2008), we believe that the delay induced by flow control and mobility is the main cause of the observed histogram bumps. Their different relative shares are due to the differently designed flow control and mobility mechanisms of GPRS and UMTS. A detailed explanation of the differences in these mechanisms between GPRS and UMTS and their impact on the one-way packet delay can be found in Romirer-Maierhofer and Ricciato (2008).

Having visually inspected the delay distribution at two different SGSNs, we have to seek summary indicators that can be used as input signals for the detection of abnormal delay deviations. These indicators must be stable over time, that is, should not present wide variations due to small fluctuations in the underlying delay distributions. The simplest candidate indicators are 'percentiles' and 'threshold-crossing frequencies'. Given a continuous random variable x, the generic pth percentile $d_x(p)$ denotes the value that is exceeded with probability p $(p < 1)$, that is, $d_x(p) : P(x \geq d_x(p)) = p$. Therefore, we fix the value of p and obtain from the measurements an empirical estimation of the $d_x(p)$. Conversely, we can define a fixed threshold h in the domain of the random variable x, and denote by $q_x(h)$ the relative fraction of samples that exceed this value, that is $q_x(h) : P(x \geq h) = q_x(h)$. Such a quantity will be denoted as 'threshold-crossing frequency'. By plotting the empirical ccdf of x, it is immediately clear that the value of $d_x(p)$ relates to the 'horizontal' distance of the

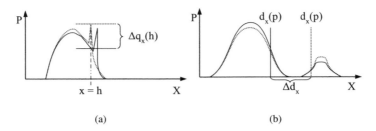

(a)   (b)

Figure 10.12 Possible offsets of indicators based on probability density functions: (a) dual-components distribution and (b) spiky distributions.

ccdf from the vertical axis at the pre-defined height p, while the crossing frequency $q_x(h)$ refers to the 'vertical' distance of the ccdf from the horizontal axis at point h.

It is natural to consider percentiles and crossing frequency as simple candidates to serve as key summary indicators for the entire delay process. The underlying assumption is that 'small' fluctuations of the underlying distribution cause only small variations of such indicators. However, in some particular cases such an assumption does not hold. Consider for example the bi-modal probability density functions (pdfs) given in Figure 10.12(b), consisting of two separate components of mass $1 - p$ and p, respectively. In this case, small random fluctuations may cause large shifts of the p-percentile from the upper tail of the left-hand bell to the lower tail of the right-hand bell. If the distance between the two bells is large, the p-percentile will amplify the random fluctuations, which is an undesirable effect for our purposes. Another notable example is provided by the distribution of Figure 10.12(a), where a large concentration of probability mass (spike) is present around the value $x = h$. Small shifts in the location of the spike would cause large fluctuations in the value of the crossing frequency $q_x(h)$. In summary, the presence of discontinuities in the distribution (that is, 'straight' segments in the ccdf) might cause the percentiles and/or crossing frequencies to vary widely following relatively small random fluctuations that can be considered physiological to the process – that is, 'normal'. Curiously, in our measurements we found occurrences of both the 'pathological' scenarios (compare Figures 10.12 and 10.11).

To illustrate, we plot the downlink delay percentiles along with different crossing frequencies for three days at time bins of 5 min in Figure 10.13(a). In downlink the 0.9 percentile is heavily fluctuating between two orders of magnitude. This is because the empirical histogram in Figure 10.11(b) has exactly the shape as depicted in Figure 10.12(b). Small variations in the distribution modes may cause significant fluctuations in some key percentiles, such that they can hardly be used for the detection of abnormal delays. The crossing frequencies should be preferred in this case. In Figure 10.13(b), we plot crossing frequencies for different threshold values. They appear to be stable over time, showing only small variations during night hours, where the traffic load is low. The crossing frequency for $h = 1$ ms is shaped by the time-of-day profile. This is expected, as such a value falls within the mass of the queuing delay which correlates with the global traffic load.

In Figure 10.14, we plot some key percentiles along with delay crossing frequencies of the uplink delay distribution at the UMTS-SGSN, again for three days and at time bins

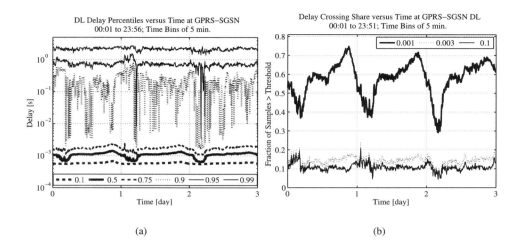

(a) (b)

Figure 10.13 Downlink delay indicators versus time at GPRS-SGSN, 3 days.

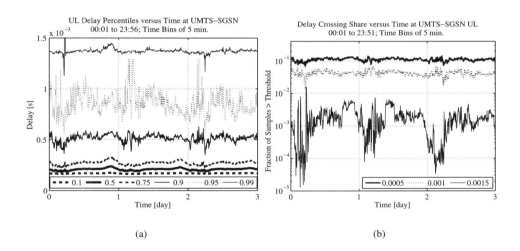

(a) (b)

Figure 10.14 Uplink delay indicators versus time at UMTS-SGSN, 3 days.

of 5 min. The percentiles take very low values and are much more stable than in the case of downlink GPRS traffic. Again, we observe small percentile variations during the night hours. In Figure 10.14(b), we observe that the delay crossing frequency for $h = 1.5$ ms is very unstable, with fluctuations over two orders of magnitude. This is because the delay distribution in Figure 10.11(a) has a shape similar to the distribution sketched in Figure 10.12(a). Small vertical shifts of the spike observed at ≈ 1.5 ms may result in a large horizontal distance of the delay crossing share. In this case the delay percentiles are the preferred input signals for an anomaly (delay) detection scheme. From the above analysis we can draw an important general lesson: the task of identifying simple indicators for spotting anomalies in the delay process within a mobile network cannot be reduced to the blind extraction of percentiles and/or crossing frequencies, which are sometimes misleading. The definition of synthetic indicators for the network health must always rely on a preliminary phase of careful exploration of the process at hand, starting from the understanding of the entire first-order distribution.

References

Bannister, J., Mather, P. and Coope, S. (2004) *Convergence Technologies for 3G Networks*, John Wiley & Sons Ltd., 2004.

Castro, R., Coates, M., Liang, G., Nowak, R. and Yu, B. (2004) Network tomography: Recent developments. *Statistical Science*, **19**(3), 499-517, 2004.

DARWIN and METAWIN home page available at: http://userver.ftw.at/ ricciato/darwin/.

Donnelly, S.F. (2002) *High Precision Timing in Passive Measurements of Data Networks*. PhD Dissertation, CS Dept., Waikato Univ., Hamilton, New Zealand, June.

Pang, R., Yegneswaran, V., Barford, P., Paxson, V. and Peterson, L. (2004) Characteristics of Internet Background Radiation. In *Proceedings of ACM Internet Measurement Conference (IMC'04)*, Taormina, Oct.

Papagiannaki, K., Moon, S., Fraleigh, C., Thiran, P. and Diot, C. (2003) Measurement and analysis of single-hop delay on an IP backbone network. *IEEE Journal on Selected Areas in Communications*, **21**(6), 908–921, Aug.

Ricciato, F. (2006) Traffic Monitoring and Analysis for the Optimization of a 3G network. *IEEE Wireless Communications – Special Issue on 3G/4G/WLAN/WMAN Planning*, **13**(6), 42–29, Dec.

Ricciato, F. (2006b) Unwanted Traffic in 3G Networks. *ACM Computer Communication Review*, **36**(2), April.

Ricciato, F., Vacirca, F. and Svoboda, P. (2007) Diagnosis of capacity bottlenecks via passive monitoring in 3G networks: an empirical analysis. *Computer Networks*, **51**(4), March.

Ricciato, F., Hasenleithner, E. and Romirer-Maierhofer, P. (2008) Traffic analysis at short time-scales: an empirical case study from a 3G cellular network. *IEEE Trans. Network and Service Management*, **5**(1), 11–21, Mar.

Stevens, W.R. (1994) *TCP/IP Illustrated, Volume 1, The Protocols*, Addison-Wesley.

Romirer-Maierhofer, P. and Ricciato, F. (2008) Towards Anomaly Detection in One-way Delay Measurements for 3G Mobile Networks: A Preliminary Study. In *Proceedings of Eighth IEEE International Workshop on IP Operations and Management (IPOM 2008)*, Samos Island, Greece, 24-25 Sept.

Romirer-Maierhofer, P., Ricciato, F. and Coluccia, A. (2008) Explorative Analysis of One-way Delays in a Mobile 3G Network. In *Proceedings of 16th IEEE Workshop on Local and Metropolitan Area Networks (LANMAN 2008)*, Cluj-Napoca, Romania, 3-6 Sept.

Svoboda, P., Ricciato, F., Hasenleithner, E. and Pilz, R. (2006) Composition of GPRS/UMTS Traffic: Snapshots from a Live Network. In *Proceedings of Fourth International Workshop on Internet Performance, Simulation, Monitoring and Measurement, Salzburg (IPS-MOME'06)*, Austria, 27–28 Feb.

Vacirca, F., Ricciato, F. and Pilz, R. (2005) Large-scale RTT Measurements from an Operational UMTS/GPRS Network. In *Proceedings of First International Conference on Wireless Internet (WICON'05)*, Budapest, July.

Vacirca, F., Ziegler, T. and Hasenleithner, E. (2008) Large Scale Estimation of TCP Spurious Timeout Events in Operational GPRS Networks, *COST-279 Technical Document, TD(05)003* (submitted).

Zou, C.C., Gong, W. and Towsley, D. (2002) Code Red Worm Propagation Modeling and Analysis. In *Proceedings of Ninth ACM Conference on Computer and Communications Security (CCS'02)*.

11

End-to-End Video Quality Measurements

Michal Ries

Although perceptual video quality is rather limited for low bit rates, frame rates and resolutions, mobile multimedia streaming applications are becoming more and more popular. It is thus essential to guarantee a required level of customer satisfaction, defined by the perceived video stream quality. It is important to choose the compression parameters as well as the network settings so that they maximize the end-user quality. Due to video compression improvements of the newest video coding standard H.264/AVC (ITU-T H.264, Series H, Mar. 2005), video streaming at low bit and frame rates is possible while preserving its perceptual quality. This is especially suitable for video applications in mobile wireless networks such as 3rd Generation (3G) and Wireless Local Area Network (WLAN).

Video streaming is a one-way quasi real-time data transport where the content is consumed (viewed/heard/read) while it is being delivered. To compensate jitter (variance of the end-to-end delay) at the receiver, a portion of the received data is buffered in a play-out buffer. In the case of video streaming, the video content is rendered on the screen with the signalized frame rate, making the inter-packet arrival time variations invisible for the user. Therefore, the end-user quality for video streaming does not depend on the absolute end-to-end delay as long as it is kept in the order of seconds and video streaming is usually referred to as a quasi real-time service. Moreover, mobile video streaming is characterized by low resolutions and low bit rates. The Universal Mobile Telecommunications System (UMTS) Release 4, implemented by the first UMTS network elements and terminals, provides a maximum data rate of 1920 kbit/s shared by all users in a cell, Release 5 (emerging) offers up to 14.4 Mbit/s in Downlink (DL) direction for High Speed Downlink Packet Access (HSDPA). The availability of such data rates initiated the launch of new services, out of which real-time services are the most challenging from the provider point of view. Commonly used resolutions are Quarter

Video and Multimedia Transmissions over Cellular Networks   Edited by Markus Rupp
© 2009 John Wiley & Sons, Ltd

Common Intermediate Format (QCIF) for cell phones, Common Intermediate Format (CIF) and Standard Interchange Format (SIF) for data-cards and palmtops (PDAs). The mandatory codec for UMTS streaming applications is H.263 but the 3GPP Release 6 (3GPP TS 26.234 V6.11.0) already supports a baseline profilebaseline profile of the new H.264/AVC codec (ITU-T H.264, Series H, Mar. 2005). The appropriate encoder settings for UMTS streaming services differ for various streaming content and streaming application settings (resolution, frame and bit rate) as is demonstrated in Nemethova *et al.* (2004), Ries *et al.* (2005) and Koumaras *et al.* (2005).

Conventional subjective quality measurement methods (ITU-T Rec. P.910, Sep. 1999) involve confronting a test subject with one or more video clips. The subject is asked to evaluate the quality of the test clips, for example by recording the perceived degradation of the clip compared with a reference video clip. A typical test requirement is to determine the optimum choice from a set of alternative versions of a video clip, for example versions of the same clip, or versions of the same clip encoded with different codecs, alternative strengths of a post processing filter and alternative trade-offs between encoder settings for a given bit rate. In this type of scenario, each of the alternative versions of a video clip must be viewed and graded separately by the test subjects, so that the time taken to carry out a complete test increases exponentially with N, the number of alternatives to be tested. When choosing a preferred trade-off between frame rate and image quality there is a large number of possible outcomes and the test designer is faced with the choice between running a very large number of tests in order to obtain a fine-grained result or limiting the number of tests at the expense of discretizing the result (Richardson and Kannangara 2004). Moreover, the subjective testing is extremely personal and time consuming.

The subjective video quality reflects the subjective perception of individual viewers. The evaluation is performed by a psycho-visual experiment and therefore is influenced by the following subjective and objective factors: video content, encoding parameters, usage scenario and network performance. Moreover, objective parameters are only poorly correlated with subjective quality. The estimation of subjective video quality for mobile scenarios is a challenge. To date, several methods have been proposed for the estimation of video quality. Such methods can be classified as follows:

- The first quality distortion measures are Mean Square Error (MSE) and Peak Signal to Noise Ratio (PSNR). Both of them poorly reflect the subjective video quality (Nemethova *et al.* 2004; Ries *et al.* 2005). Nevertheless, PSNR is still widely used as a reference method for comparing the performance of video coding algorithms.

- Several metrics based on the Human Visual System (HVS) have been proposed in recent years (Ong *et al.* 2003; Rix *et al.* 1999; Westen *et al.* 1995; Winkler and Dufaux 2003; Zetzsche and Hauske 1989). The usage of a metric based on the HVS is expected to be very general in its nature and applicability. These metrics compute a distance measure based on the outputs of a multiple channel cortical model of the human vision which accounts for known sensitivity variations of the HVS in the primary visual pathway. Moreover, the metrics assume that the multiple channels mediating visual perception are independent of each other. However, recent neuroscience findings and psychophysical experiments (Bonds 1989; Heeger 1992; Teo and Heeger 1994) have established that there is interaction across the channels and that such interactions are important for visual masking. Thus, in the near future even better HVS models for reliable quality evaluation are expected. The main disadvantage of these HVS models is their high computational complexity.

- Metrics based on a set of objective parameters (Kusuma *et al.* 2005; Marziliano *et al.* 2002; Pinson and Wolf 2004; Ries *et al.* 2005, 2007a,b,c) – and ANSI T1.801.03 – provide a good trade-off between accuracy and complexity. The parameter set consists of quality sensitive objective parameters. This approach is very suitable for quality estimation in scenarios with defined usage, content and video service conditions.

A second classification is possible, depending on the required knowledge of the source material:

- Reference-based metrics (Hauske *et al.* 2003; Wang *et al.* 2004): measurements based on the computation of differences between the degraded and the original video sequences. The differences can be used to compute comparative distortion measures, which have a low correlation with the perceived impairment but are easy to extract. The reference is required at the input of the measurement system, strongly restricting their applicability.

- Measurements obtained by computing a set of parameters on the degraded picture and comparing them with the same parameters computed on the reference picture (Ries *et al.* 2007b; Rix *et al.* 1999; Winkler and Dufaux 2003) and ANSI T1.801.03. Quality indications can be obtained by comparing parameters computed separately on the coded pictures and the reference pictures. These parameters can be distributed in the network at low bit rates to be used when the entire reference signal is not available.

- Reference-free metrics (Ries *et al.* 2006, 2007a,b,c) do not require any knowledge of the original video source. These metrics find a basic difficulty in telling apart distortions from regular content which is something that humans do well by experience. Their biggest advantage is their versatility and flexibility.

The complexity of recently proposed metrics is considerable. Moreover, most of them were proposed for broadband broadcasting and Internet video services. In contrast, our approach focuses on quality estimation of low bit rate and low resolution video in mobile environments.

This chapter thus focuses on measures that do not need the original (non-compressed) sequence for the estimation of quality because such reference-free measures reduce the complexity and at the same time broaden the possibilities of the quality prediction deployment. Moreover, the objective measures of video quality should be simple enough to be calculated in real time at the receiver side.

We address the design of reference-free video quality metrics in mobile environments. The whole chain of metric design looks at the definition of mobile streaming scenarios, a subjective test methodology, the selection and statistical evaluation of objective parameters and the estimator design. The proposed video quality metrics are applicable for quality monitoring in mobile networks.

11.1   Test Methodology for Subjective Video Testing

Mobile video streaming scenarios are determined by usage environment, streamed content and the screen size of the mobile terminals (Nemethova *et al.* 2005). Therefore, mobile scenarios are strictly different in comparison with classical TV broadcasting services or broadband IP-TV services. Most of the recommendations for subjective video testing (ITU-T Rec. P.910, Sep. 1999 and ITU-R Rec. BT.500, Jun 2002) are designed for broadband video services with CIF resolution or higher and for static scenarios. Conversely, the mobile scenario is different due to technical conditions and usage.

Therefore, the initial research limit was to focus on the design of a methodology for subjective video testing. The aim of the test methodology was to provide a real world viewing and usage conditions for subjective testing and to make them reproducible. In ITU-T Rec. P.910 even more methodologies are proposed. The difference between such methods is in whether or not explicit references are utilized. The non-reference methods Absolute Category Rating (ACR) and Pair Comparison (PC) do not test video system transparency or fidelity. On the other hand, reference methods should be used when testing transmission fidelity with respect to the source signal. This is frequently an important factor in the evaluation of high-quality systems. The reference method Degradation Category Rating (DCR) has also long been a key method specified in ITU-T Rec. P.930, for the assessment of television pictures whose typical quality represents extreme high-quality levels of video telephony and video conferencing. The specific comments of the DCR scale (imperceptible/perceptible) are valuable when the viewer's detection of impairment is an important factor. Thus, when it is important to check the fidelity with respect to the source signal, the DCR method should be used. DCR should also be applied for high-quality system evaluation in the context of multimedia communications. Discrimination of imperceptible/perceptible impairment in the DCR scale supports this, as well as a comparison with the reference quality.

On the other hand, ACR is easy and fast to implement and the presentation of the stimuli is similar to commonly used systems. Thus, ACR is well suited for qualification tests. The principal merit of the PC method is its high discriminatory power, which is of particular value when several of the test items are nearly equal in quality. When a large number of items have to be evaluated in the same test, the procedure based on the PC method tends to be lengthy. In such a case an ACR or DCR test may be carried out first with a limited number of observers, followed by a PC test solely on those items which have received about the same rating.

To achieve the most effective methodology for subjective testing of wireless video streaming, the following conditions were defined:

- Viewers do not have access to the test sequences in their original uncompressed form. Only encoded sequences are displayed; a reference-free subjective evaluation is the outcome.

- The sequences are presented on a handheld mobile device as shown in Figure 11.1.

- The encoding settings reflect a typical UMTS streaming setup (see Table 11.2).

- The most frequent streaming content types are displayed.

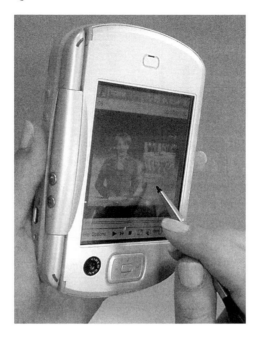

Figure 11.1 PDA with display resolution 480 × 640 pixels.

11.1.1 Video Quality Evaluation

The proposed test methodology is based on ITU-T Rec. P.910 and adapted to our specific purpose and limitations. For this particular application it was considered that the most suitable experimental method, among those proposed in the ITU-T recommendation, is ACR, also called 'single stimulus method'. The ACR method is a category judgement in which the test sequences are presented one at a time and are rated independently on a category scale. Only degraded sequences are displayed, and they are presented in arbitrary order. This method imitates the real-world scenario because the customers of mobile video services do not have access to original videos (high-quality versions). On the other hand, ACR introduces a higher variance in the results, as compared to other methods in which also the original sequence is presented and used as a reference by the test subjects. The results of quality assessments often depend not only on the actual video quality but also on other factors such as the total quality range of the test conditions. A description of reference conditions and procedures to produce them is given in ITU-T Rec. P.930. Note that this ITU-T recommendation proposes LCD monitors for subjective testing. However, the mobile video streaming domain offers a large choice of parameters, and uses a variety of proprietary and standardized decoders, players and UEs as opposed to standard broadband video services such as IP-TV and DVB-T where the system parameters do not vary much. Therefore, it is far more difficult to evaluate the quality of multimedia images than of those in the broadband video services (Kozamernik *et al.* 2005). Experience shows (see Figure 11.2) that this combination strongly influences the final subjectively perceived picture quality. In

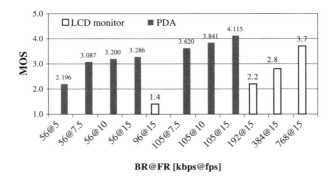

Figure 11.2 Subjective evaluation at LCD monitor and PDA.

Table 11.1 Test environment.

Environment parameter	Value
Viewing distance	20–30 cm
Viewing angle	0°
Room illumination	Low: ≤20 lux at about 30 cm
(ambient light level [lux])	in front of the screen

Figure 11.2 comparisons of subjective evaluations for semantically identical soccer video content are depicted for various combinations of bit rates (BRs) and frame rates (FRs). The sequences are encoded by an H.264/AVC baseline profile codec. The evaluations of LCD monitors were performed within the formal verification tests (JVT 2003) on H.264/AVC defined by Joint Video Team (JVT). Evaluations of PDAs are reported in section 11.2. The results clearly show that test subjects evaluate the sequences at LCD monitors much more critically.

In order to emulate real conditions of the mobile video service, all the sequences were displayed on mobile handheld devices. The viewing distance from the phone is not fixed but selected by the test person. We noticed that users are comfortable to hold UMTS terminals at a distance of 20–30 cm. Our video quality test design follows these experiences in order to reflect better the real world scenarios. The test environment (see Table 11.1 for details) fulfils all criteria defined by ITU-T Rec. P.910. The critical part was to achieve suitable lighting conditions in order to eliminate UE display reflection.

After each presentation the test subjects were asked to evaluate the overall quality of the sequence shown. In order to measure the quality perceived, a subjective scaling method is required. However, whatever the rating method, this measurement will only be meaningful if there actually exists a relation between the characteristics of the video sequence presented and the magnitude and nature of the sensation that it causes on the subject. The existence of this relation is assumed. Test subjects evaluated the video quality after each sequence in a prepared form using a five-grade Mean Opinion Score (MOS) scale: '5 – Excellent', '4 – Good', '3 – Fair', '2 – Poor', '1 – Bad'. Higher discriminative power was not required,

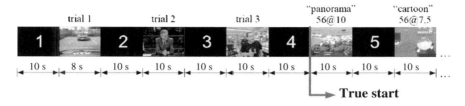

Figure 11.3 Time pattern of the video quality survey.

because our test subjects were used to five-grade MOS scales (school). Furthermore, a five-grade MOS scale offers the best trade-off between the evaluation interval and reliability of the results. Higher discriminative power can introduce higher variations in MOS results.

11.1.2   Subjective Testing

At the beginning of each test round a trial run was presented with three sequences. The subjective quality of these trial sequences varied substantially, in order to offer the test subject initial experience with subjective quality evaluation. The contents and audio or video coding artefacts of these sequences were similar to video sequences. The subjective evaluation of trial run sequences was not taken into account in the statistical analysis of the test results. After this trial run the test sequences were presented in an arbitrary order, the only conditions being that two clips of the same content, although differently degradated, must not appear in succession, and consecutive sequences must not have identical BRs and FRs. If, on the contrary, all the versions of one sequence were displayed in succession, subjects would perform a degradation rating rather than an absolute rating. Since the intention is the subjective evaluation on different sequences of different contents relative to each other, it is important, therefore, to alternate the sequences.

The duration of clips was approximately 10 s. The length was not identical in all the clips, because the sequences were adjusted to a scene cut, in order to keep the contents consistent. The voting time to respond to questions was also set to 10 s. A still image showing the order number of the following sequence – large, white digits on a black background – was displayed during this voting time between sequences, in order to guide the test person through the questionnaire. The succession of clips was presented using a playlist, to ensure that the subject did not have to interact with the device, and could be fully concentrated in his viewing and evaluation task. The use of a playlist assures a level of homogeneity in the viewing conditions for all the viewers, as the presentations cannot be stopped, the voting time is fixed and each test sequence is viewed only once before it is evaluated. The time pattern for the presentation of the clips is illustrated in Figure 11.3, showing that after three initial trials the actual evaluation process starts.

11.1.3   Source Materials

All the original sequences were formatted to CIF or SIF resolutions. Source material in a higher resolution was converted to SIF resolutions. For mobile video streaming the most

Figure 11.4 Snapshot of typical content classes: News (CC1), Soccer (CC2), Cartoon (CC3), Panorama (CC4), Video clip (CC5).

frequent contents with different impacts on the user perception were defined, resulting in the following five Content Classes (CCs), samples of which are displayed in Figure 11.4.

- **Content Class News (CC1)**: The first content class includes sequences with a small moving region of interest (face) on a static background. The movement in the Region of Interests (ROI) is determined mainly by eyes, mouth and face movements. The ROI covers up to approximately 15% of the screen surface.

- **Content Class Soccer (CC2)**: This content class contains wide angle camera sequences with uniform camera movement (panning). The camera is tracking small rapid moving objects (ball, players) on the uniformly coloured (typically green) background.

- **Content Class Cartoon (CC3)**: In this content class object motion is dominant; the background is usually static. The global motion is almost not present owing to its artificial origin (no camera). The movement object has no natural character.

- **Content Class Panorama (CC4)**: Global motion sequences taken with a wide angle panning camera. The camera movement is uniform and in a single direction.

- **Content Class Video Clip (CC5)**: The content class contains a great deal of global and local motion or fast scene changes. Scenes shorter than 3 s are also associated to this content class.

In ITU-T Rec. P.910 a measure of spatial and temporal perceptual information characterizes a video sequence. The Spatial Information (SI) measurement reflects the complexity (amount

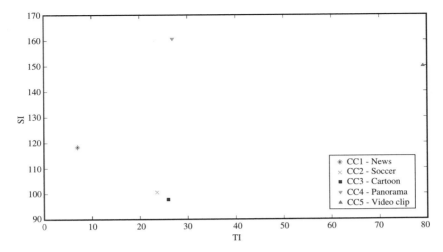

Figure 11.5 Spatial (SI) and temporal (TI) features of the original test sequences.

of edges) of still pictures. The Temporal Information (TI) measurement is based on the motion difference feature. In Figure 11.5, The SI and TI values of the original sequences are depicted. As depicted, the SI and TI features of the content types vary significantly.

11.2   Results of Subjective Quality Tests

This chapter describes the results of subjective video tests. These subjective tests are performed for mobile scenarios but for different video codecs, resolutions and content classes. MOS values were captured by the ACR method for a wide range of video sequences. Their content was selected so as to cover a representative range of coding complexity and content type. The subjective MOS data was obtained according to the subjective test methodology described in section 11.1. This content was then encoded at a variety of BRs, FRs and resolutions to represent high- and low-medium qualities. The obtained subjective data was analysed for each resolution and CC. Furthermore, the data was screened for consistency. The preliminary look at the obtained data shows that the subjective video quality is strongly content dependent, especially for lower BR. Furthermore, the results were the base for the design of subjective quality estimators.

11.2.1   Subjective Quality Tests on SIF Resolution and H.264/AVC Codec

For the tests in SIF resolution all sequences were encoded with the H.264/AVC baseline profile 1b. For subjective quality testing, frame and bit rate combinations for CC1, . . . , CC5 shown in Table 11.2 were used. In total there were 39 combinations.

To obtain MOS values, we worked with 36 test persons for two different sets of test sequences. The first set was used for the design of a metric and the second for the evaluation

Table 11.2 Tested combinations of FRs and BRs for SIF resolution.

FR [frame/s]/ BR [kbit/s]	24	50	56	60	70	80	105
5	CC1, CC3, CC4	CC5	CC1, CC2, CC3, CC4				CC1
7.5	CC1, CC3, CC4		CC1, CC2, CC3, CC4	CC5	CC5		CC1, CC2,CC5
10	CC1, CC3		CC1, CC2, CC3, CC4		CC5	CC5	CC1, CC2, CC5
15	CC1		CC1, CC2			CC5	CC1, CC2, CC5

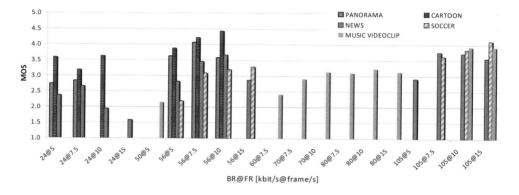

Figure 11.6 MOS for all tested sequences in SIF resolution.

of the metric performance. The training set test was carried out with 26 test persons and the evaluation test set was carried out with 10 test persons. The training and evaluation tests were collected from different sets of the five video classes CC1, . . . , CC5. Our group of test persons were of a range of different ages (between 20 and 30), gender, education and experience with image processing. Two runs of each test were taken. To avoid a learning effect, we made a break of half an hour between the first and the second run. The size (over all tested sequences) of the confidence interval was 0.27 on the five-grade MOS scale.

The most important outcome of our tests is that human visual perception of video content is strongly determined by the character of the observed sequence. As can be seen in Figure 11.6, the measured subjective video quality is strongly content dependent, especially at low BR and resolution. The difference between two contents can result in up to 1.6 MOS grades.

The subjective assessment of the SIF resolution (see Figure 11.6) shows high-quality variation at lower BRs but a quality saturation at BRs over 100 kbit/s can also be observed. For the 'News' sequence, the highest score is obtained by the configuration BR@FR = 105@7.5 kbit/s@frame/s, closely followed by 105@10 kbit/s@frame/s and 56@10 kbit/s@frame/s. Very interesting is the fact that the viewers seem to notice no difference in quality between the combination 56@10 kbit/s@frame/s and 105@10 kbit/s@frame/s, which both receive very positive evaluations. The most dynamic sequence, 'Soccer' received the best evaluation at 105 kbit/s. An increasing frame rate in soccer videos always has a positive effect on the perceived quality, which is in contrast with other content types, especially in the more static 'News' case. In the 'Soccer' sequence viewers prefer smoothness of motion rather than static quality.

Moreover, the video quality survey allows which coding parameters to be used to be estimated, according to the character of the video content, in order to maximize the end user's perceived quality. Furthermore, the results can be used to determine the most suitable trade-off settings in terms of bit rate and subjective quality requirements.

11.3 Video Quality Estimation

The human visual perception of video content is determined by the character of the observed sequence. It is necessary to determine different content characters/classes or content adaptive parameters because the video content itself strongly influences the subjective quality (see also section 11.2.1). The character of a sequence can be described by the amount of edges (SI) in the individual frames and by the type and direction of camera and object movement (TI). The constant BR of the video sequence is shared by the number of frames/s. Higher frame rates at constant BR result in a lower amount of spatial information in individual frames and possibly in some compression artefacts.

In the literature the focus is mainly on the spatial information (Kusuma *et al.* 2005; Marziliano *et al.* 2002). Such approaches originate from the quality estimation of still images (Saha and Vemuri 2000; Wang *et al.* 2004). However, especially in small resolutions and after applying compression, not only the speed of movement (influencing at most the compression rate) but also the character of the movement plays an important role in the user perception (Ries 2008). Therefore, in this section the focus is on the motion features of the video sequences that determine the perceived quality.

Here, the design of content classifiers and video quality estimators for different content classes, codecs and resolutions is described. Since each shot of a sequence can have a different content character, scene-change detection is required as a pre-stage of content classification and quality estimation. Moreover, the quality estimation is based on content-sensitive video parameters.

11.3.1 Temporal Segmentation

The temporal segmentation of a video into its basic temporal units – so-called shots – is of great importance for a number of applications today. Video indexing techniques necessary for video databases rely on it. Segmentation is also necessary for the extraction of high-level semantic features. Moreover, it provides important information for video preprocessing, compression codecs and error concealment techniques. Temporal segmentation is also a prerequisite in the process of video quality estimation. In this context, 'a shot' is a series of consecutive video frames taken by one camera. Two consecutive shots are separated by a shot boundary which can be abrupt or gradual. While an abrupt shot boundary (cut) is generated simply by attaching one shot to another without modifying them as in Figure 11.7, a gradual shot boundary is the result of applying an editing effect to merge two shots. For the purpose of video quality estimation it is sufficient to detect scene cuts because abrupt shot boundaries are the most frequent and gradual shot boundaries (dissolve, fades or wipes) usually not leading to a content class change.

$n-1$ n $n+1$ $n+2$

Figure 11.7   Abrupt scene change between frames n and $n+1$.

For the purpose of quality estimation within two cuts the method based on dynamic threshold boundaries was adopted for temporal segmentation of all content types (Ries 2008; Ries *et al.* 2007b). The threshold function is based on the statistical features of the local sequence.

11.3.2   Video Content Classification

The character of a sequence can be described by the amount of edges in the individual frames, SI, by the type and direction of movement, TI and colour features. Moreover, the video quality estimation can be significantly simplified by content recognition as an independent issue. An automatic content recognition module is able to distinguish between a discrete set of classes and can then be used in combination with an individual metric to estimate the subjective quality of all typical video contents.

11.3.3   Content Sensitive Features

Especially at low resolutions and BRs after applying compression, not only speed and the amount of movement but also the character of movement play an important role in user perception. Moreover, user content perception is also determined by the colour features of the sequence. Therefore, in this section the motion and colour features of the video sequences within one shot that determine the perceived quality are investigated.

Motion Vector Extraction

Block-based motion compensation techniques are commonly used in inter frame video compression which is also supported by H.263 and H.264 in order to reduce temporal redundancy. The difference between consecutive frames in a sequence is predicted from a block of equal size in the previous frame which serves as reference frame. The blocks are not transformed in any way apart from being shifted to the position of the predicted block. This shift is represented by a Motion Vector (MV). This technique was used to analyse the motion characteristics of the video sequences.

The block from the current frame for which a matching block is sought is known as the 'target block'. The relative difference in the locations between the matching block and the target block is called the MV. If the matching block is found at the same location as the target block then the difference is zero, and the corresponding MV is the *zero vector*.

The difference between the target and matching blocks increases approximately linearly with the size of the blocks. Thus, smaller blocks better describe the actual motion in

Figure 11.8 Snapshot of two successive frames – soccer sequence.

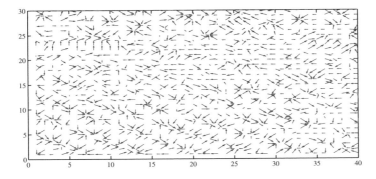

Figure 11.9 MV field obtained with 30×40 MBs of 8×8 pixels without DCT filtering.

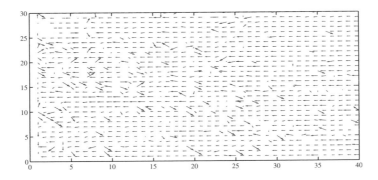

Figure 11.10 MV field obtained with 30×40 MBs of 8×8 pixels with DCT filtering.

the frame. On the other hand, an increase of the objective accuracy does not always imply a better performance. If the blocks are chosen too small, the resulting MVs no longer reflect the motion as it is perceived by a viewer (Ries 2008). Due to the unavoidable presence of noise in video sequences, and the characteristics of the human visual system, movement can be detected although a human observer does not see it. A pixel block size of 8×8 was selected to ensure a good trade-off for Quarter Video Graphics Array (QVGA) resolution sequences. The 320×240 pixels are divided into 30×40 blocks (see Figure 11.9), which gives a total number of 1200 MVs per frame.

The presence of noise, lighting variation and the existence of multiple local minima in the SAD distribution in the video sequences causes the system to detect movement, although a human observer does not perceive it. Such effect introduces significant deviation to further usage of MV features for content classification and quality estimation. The vector fields found after performing the described exhaustive search on raw luminance frames do not always represent the true motion. A good example can be seen at the MV field in Figure 11.9 made out of two consecutive soccer frames as depicted in Figure 11.8. A possible solution to erroneous motion detection is smoothing techniques (Choi and Park 1989).

It was thus necessary to develop a new, low-complex and reference-free method. The most suitable approach was to analyse the AC components of the DCT coefficients and introduce additional low-pass filtering before the block matching (Ries 2008). The matching algorithms are applied using 8×8 pixel blocks and then the first ten coefficients of the inverse DCT are extracted. The comparison of the MV field before filtering in Figure 11.9 and MV after filtering in Figure 11.10 shows considerable improvement.

Extraction of Motion Sequence Parameters

The extracted and filtered MVs allow the further analysis of motion features in the sequence. The static or dynamic character of a sequence is one of the main causes for the differences in perceived quality (Ries *et al.* 2007b, 2008). This investigation leads to a classification not only in terms of 'static sequences' and 'dynamic sequences', but also to deeper understanding of these aspects and to determine typical levels of quantity of movement for every content class. The overall amount of movement or, equivalently, the lack of movement in a frame can be easily estimated from the proportion of blocks with zero vectors, that is, blocks that do not move from one frame to the other. Therefore, the average proportion of static blocks in a sequence of frames is very useful when it comes to distinguishing contents with typical different 'levels' of overall movement.

The length of the MV indicates how far the block has moved from one frame to the next, and its angle tells us in which direction this movement occurred. Therefore, the mean size of the MVs in a frame or sequence of frames is an indicator of how fast the overall movement happens. Moreover, detecting a main direction of movement that corresponds to a big proportion of MVs, pointing in the same direction, is valuable information. Thus, it can be assumed that the analysis of the distribution of sizes and angles of the MVs can provide substantial information about the character of the motion in the sequence (Ries 2008).

A set of statistical MV features were investigated in order to study their level of significance for further content classification. As an initial step, the motion features were analysed throughout the sequence in order to observe their temporal behaviour. The following statistical and resolution-independent features of MVs within one shot (over all the frames of the analysed sequence) were investigated:

1. **Zero MV ratio N_z**
   Percentage of zero MVs in a frame. It is the proportion of the frame that does not change at all (or changes very slightly) between two consecutive frames. It usually corresponds to the background if the camera is static within one shot.

2. **Mean MV size M**
   Proportion of mean size of the MVs within one frame normalized to the screen width, expressed in percentage. This parameter determines the amount of the global motion.

3. **Mean non-zero MV size n**
   Proportion of mean size of the non-zero MVs within one frame normalized to the screen width, expressed in percentage. This parameter determines the amount of the global motion.

4. **Intra frame MV standard deviation F**
   Standard deviation of MV sizes in the frame. MV standard deviation is expressed as a percentage of the MV mean size in the frame. This feature is sensitive at dynamic motion changes.

5. **Intra frame standard deviation of non-static blocks f**
   Standard deviation of the sizes of non-zero MVs, expressed as a percentage of the mean size of these MVs in the frame. It is a measure of the uniformity or dynamic changes of the moving regions.

6. **Uniformity of movement d**
   Percentage of MVs pointing in the dominant direction (the most frequent direction of MVs) in the frame. For this purpose, the granularity of the direction is 10 degrees.

7. **MV direction uniformity of non-static blocks B**
   Percentage of MVs pointing in the main direction in the moving regions (zero MVs are excluded).

A Principal Component Analysis (PCA) was carried out on the obtained dataset. PCA is a suitable method (Krzanowski 1988) for viewing a high-dimensional set of data in considerably fewer dimensions (see Figure 11.11). The first two components proved to be sufficient for an adequate modelling of the variance of the data.

Finally, the biplot technique (Gabriel 1971) was applied for representing the first two principal components by means of parameter vectors. The biplot provides a useful tool of data analysis and allows the visual appraisal of the structure of large data matrices. It is especially revealing in principal component analysis, where the biplot can show inter-unit distances and indicate clustering of units as well as display variances and correlations of the variables.

The visualization of PCA results shows that all parameter vectors have approximately similar influence on the dataset. Therefore, it was not possible to set hard decision criteria for selection of video quality parameters according to PCA results. The chosen parameters have low computational complexity and good distribution over the most significant subspace of PCA within all content classes. According to PCA results and computational complexity, the following three parameters were chosen:

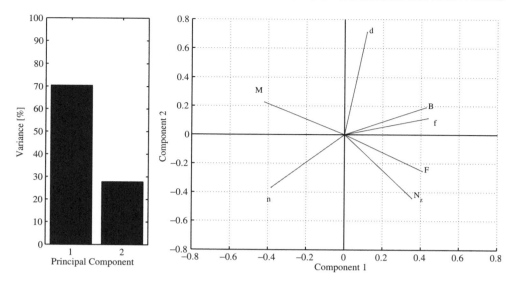

Figure 11.11　Visualization of PCA results for content Soccer (CC2).

- the zero MV ratio N_z;

- the mean non-zero MV size n;

- the uniformity of movement d.

Finally, the selected content-sensitive parameters reflect the static part of the surface, the significance of the movement in the non-static part and the dominant motion direction.

In order to increase the accuracy of the proposed content classifier it was necessary to define content-sensitive features to detect movement in the horizontal direction as well as green colour. Therefore, an additional feature for detecting the amount of horizontal movement was defined.

8. **Horizontalness of movement h**

    Horizontalness is defined as the percentage of MVs pointing in the horizontal direction. Horizontal MVs are from intervals $\langle -10; 10 \rangle$ or $\langle 170; 190 \rangle$ degrees.

This feature allows enhanced detection of the Panorama class and the most accurate detection of Soccer sequences. As can be seen in Figure 11.12 in Panorama and Soccer, the horizontal movement is dominant (Ries 2008).

Soccer sequences, for example, contain many varying green colours while cartoon sequences exhibit discrete saturated colours; therefore, the colour distribution of CCs was investigated. Colour histograms provide additional information about the spatial sequence character because in different types of content, the density and magnitude of colours differ also. This characteristic has important consequences for the compression and transmission artefacts. Therefore, the following parameter was analysed:

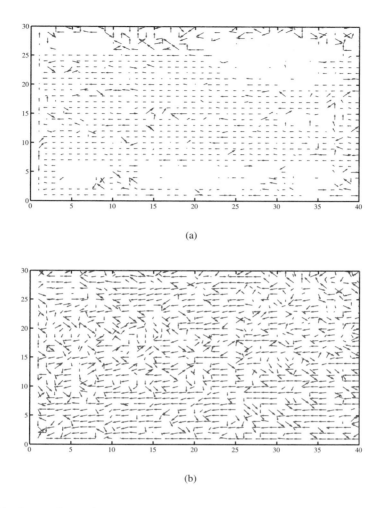

(a)

(b)

Figure 11.12 MV field of typical (a) Panorama and (b) Soccer sequences, 30×40 MBs.

9. **Greenness *g***

   Defines greenness as percentage of green pixels in a frame. For this purpose the RGB colour space was down-sampled to two bits per colour component, resulting in 64 colours. Five out of the 64 colours cover all variations of the colour green.

For calculating colour histograms, sequences were converted to RGB colour space and down-sampled to 64 colours. The colour bins were regularly spaced. As can be seen in Figure 11.13 the five green colour bins proved to be an effective element in the detection of the green sequence.

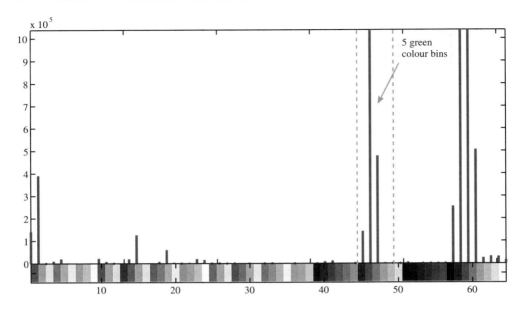

Figure 11.13   Colour histogram of Soccer sequence (displayed in greyscale).

11.3.4   Hypothesis Testing and Content Classification

The content classification of our proposed algorithm is based on the above-defined parameters. Due to the extensive set of objective parameters, a statistical method was used for data analysis and content classification. This excludes content classification based on thresholds, a limited and inaccurate method for evaluating larger datasets.

A statistical method based on hypotheses testing was introduced. Each of the described content classes is determined by unique statistical features of motion and colour parameters (see Figure 11.14). Due to their unique statistical features of well-defined content classes, it is unnecessary to perform M-ary hypothesis testing and sufficient to formulate a null hypothesis (H0) for each content class separately. The obtained empirical cumulative distribution functions (ecdfs) from the typical set of sequences for each content class show substantial differences as depicted in Figure 11.14. It ends up being very difficult to determine a single parametric distribution model representation from the obtained ecdfs alone. For this purpose a hypotheses testing method allowing non-parametric, distribution-free H0 hypotheses to be defined was of interest.

For hypothesis evaluation a method is needed that is capable of working with empirical (sample) distributions. The most suitable for this purpose is the non-parametric Kolmogorov–Smirnov (KS) test (Bosch 1998). The KS test is used to determine whether two underlying probability distributions differ, or whether an underlying probability distribution differs from a hypothesized distribution, in either case based on finite samples. The two-sample KS test is one of the most used and general non-parametric methods for comparing two sets of samples, as it is sensitive to differences in both location and shape of the ecdfs of the two sample sets.

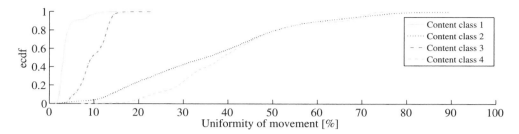

Figure 11.14 Model ecdf for uniformity of movement.

From the typical set of sequences for each content class the ecdfs are obtained. The model ecdfs were derived from a set of 142 typical sequences. Each content class is described with five model ecdfs (zero MV ratio N_z, mean MV size n, uniformity of movement d, horizontalness of movement h, greenness g), which correspond to their H0 hypothesis, respectively. Furthermore, it is necessary to find the maximal deviation ($D_{CC\ max}$) within one content class for all parameters (for each model ecdf). If $Q_n(x)$ is the model ecdf and $Q(x)$ is the ecdf of the investigated sequence, D_n is the maximal difference between $Q_n(x)$ and $Q(x)$:

$$D_n = \max_x |Q_n(x) - Q(x)|. \tag{11.1}$$

The content class estimation is based on a binary hypothesis test within the four content classes (CC1–CC4). With the KS test the ecdfs of the investigated sequence and all model ecdfs of the first four content classes are compared. The KS test compares five ecdfs (of defined MV or colour parameters) of defined content classes specified by the H0 hypothesis with all five ecdfs of the investigated content.

$$D_n \overset{?}{\leq} D_{CC\ max}. \tag{11.2}$$

If the ecdfs of the investigated sequence do not fit with any of the first four content classes, the content classifier decides for the remaining content class number five. The classifier estimates the content at the transmitter side from the original sequence.

The performance of the content classifier was evaluated with two parameters. **False detection** reflects the ratio of improper detection of a content class, when investigated sequences belong to any **other** content class. **Good match** reflects the ratio of successful classification of investigated sequences, when investigated sequences belong to any of the first four classes. Note, the sequences contain almost only cuts and no gradual changes. The scene change detector was sensitive to gradual shot boundaries (dissolve, fades or wipes). Some 786 sequences were tested to evaluate the performance of the content classifier; 98% were classified correctly. The achieved precision of the content classifier is shown in Table 11.3, a satisfying result for further quality estimation.

11.3.5 Video Quality Estimation for SIF-H.264 Resolution

In this section, three methods for quality estimation in SIF resolution are presented. The proposed methods focus on reference-free video quality estimation. The character of motion

Table 11.3 The evaluation results of content classifier.

Content class	False detection [%]	Good match [%]
1	0	97
2	0	100
3	5.6	92
4	0	100

is determined by the amount and direction of the motion between two scene changes. The first method estimates video quality in two steps. First, a content classification with character-sensitive parameters is carried out (Ries *et al.* 2007b). Finally, based on the content class, frame rate and bit rate, the video quality is estimated in a second step. The following method thus presents the design of a quality metric based on content adaptive parameters, allowing for content-dependent video quality estimation. The second method estimates the quality with one single universal metric. In contrast to the first two, the third method exploits the estimation ensemble of models. The performance of the proposed method is evaluated and compared to the ANSI T1.801.03 metric. The results show that the motion based approach provides powerful means of estimating the video quality for low-resolution video streaming services.

11.3.6 Content Based Video Quality Estimation

The estimation is based only on the compressed sequence without the original (uncompressed) sequence. If estimation is performed on the receiver side, the information about the content class needs in parallel to be signalled with the video stream as depicted in Figures 11.15 and 11.16. Such measurement setup allows for continuous real-time video streaming quality measurement on both sides: user and provider. The video quality is estimated after content classification (compare to section 11.3.2 and Figure 11.16 within one cut).

Due to limited processing power of the user equipment it was necessary to identify low-complex objective parameters. In order to keep the complexity as low as possible the most suitable parameters are already provided: FR and BR. These parameters are the codec compression settings and are signalled during the initiation of the streaming session, requiring no computational complexity for estimation as they are known at both transceiver and receiver.

The proposed low-complexity metric is thus based simply on the two objective parameters BR and FR, for each content class:

$$\widehat{MOS} = f(BR, FR, CC). \tag{11.3}$$

A general model is proposed with linear and hyperbolic elements as shown in equation 11.4. The coefficients vary substantially for each content class. Typically, some of them even take on a zero value. On the other hand, rather good correlation was achieved with one offset and two non-zero coefficients (see Table 11.4) for each class.

$$\widehat{MOS} = A_{CC} + B_{CC} \cdot BR + \frac{C_{CC}}{BR} + D_{CC} \cdot FR + \frac{E_{CC}}{FR}. \tag{11.4}$$

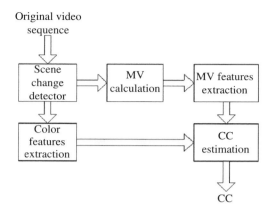

Figure 11.15 Content classifier.

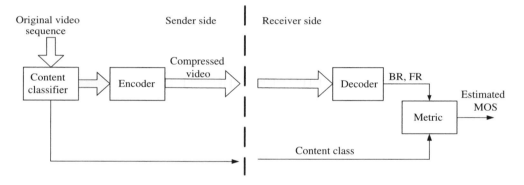

Figure 11.16 Content based video quality estimation.

Table 11.4 Coefficients of metric model for all content classes (CC1–CC5).

Coefficients	CC 1	CC 2	CC 3	CC 4	CC 5
A_{CC}	4.0317	1.3033	4.3118	1.8094	1.0292
B_{CC}	0	0.0157	0	0.0337	0.0290
C_{CC}	−44.9873	0	−31.7755	0	0
D_{CC}	0	0.0828	0.0604	0.0044	0
E_{CC}	−0.5752	0	0	0	−1.6115

The metric coefficients were obtained by a linear regression of the proposed model with the training set (MOS values averaged over two runs of all 26 subjective evaluations for a particular test sequence). The performance of the subjective video quality estimation compared to the subjective quality data is summarized in Tables 11.6 and 11.7 and is

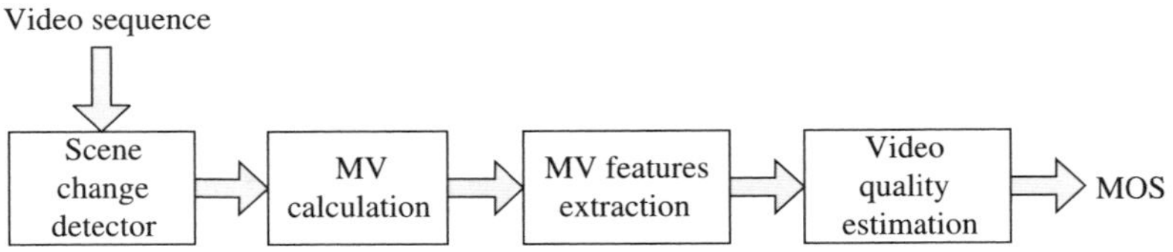

Video sequence

```
  ┌──────────┐   ┌──────────┐   ┌──────────┐   ┌──────────┐
  │  Scene   │   │    MV    │   │MV features│  │  Video   │
  │  change  │⇒ │calculation│⇒│extraction │⇒ │ quality  │⇒ MOS
  │ detector │   │          │   │          │   │estimation│
  └──────────┘   └──────────┘   └──────────┘   └──────────┘
```

Figure 11.17 Video quality estimation based on content adaptive parameters.

shown in Figure 11.19. The obtained correlations with the evaluation set show very good performance of the proposed metric for all content classes except for Cartoon (CC3), containing two- and three-dimensional cartoon movies. The proposed metric exhibits a particular weak estimation performance for the three-dimensional cartoon sequences.

Quality Estimation Based on Content-sensitive Parameters

In this section we focus on the motion features of the video sequences. The motion features can be used directly as an input into the estimation formulas or models as shown in Figure 11.17. Both possibilities were investigated in Ries *et al.* (2007a) and Ries *et al.* (2007c), respectively. The investigated motion features concentrate on the motion vector statistics, including the size distribution and the directional features of the MVs within one sequence of frames between two cuts. Zero MVs allow the size of the still regions in the video pictures to be estimated. That, in turn, allows MV features for the regions with movement to be analysed separately. This particular MV feature makes it possible to distinguish between rapid local movements and global movement.

Extraction of content adaptive features

The aim was to define measures that do not need the original (non-compressed) sequence for quality estimation because this reduces the complexity and at the same time broadens the possibilities of the quality prediction deployment. Furthermore, the size distribution and the directional features of the MVs were analysed within one sequence in between two cuts. The details regarding MV extraction can be found in section 11.3.3. The still and moving regions were analysed separately. The size of the still region was estimated by the amount of zero MV vectors. That allows MV features to be analysed separately for regions with movement. This particular MV feature makes it possible to detect rapid local movements or the character of global movements. The content-sensitive parameters as well as BR were considered. For this purpose motion sequence parameters (as defined in section 11.3.3) were reinvestigated. Furthermore, it was necessary to investigate the influence of these motion parameters and the BR on investigated contents. A PCA was carried out to verify further applicability of the motion characteristics, and BR for metric design. It turned out that the first two PCA components proved to be sufficient for an adequate modelling of the variance of the data. The variance of the first component is 60.19% and the second 18.20%. As Figure 11.18 reveals, the PCA results show sufficient influence of the most significant parameters on the dataset for all content classes.

The following features of MV and BR represent the motion characteristics:

- **Zero MV ratio within one shot** Z

  The percentage of zero MVs is the proportion of the frame that does not change at all (or changes very slightly) between two consecutive frames averaged over all frames in the shot. This feature detects the proportion of a still region. The high proportion of the still region refers to a very static sequence with small significant local movement. The viewer attention is focused mainly on this small moving region. The low proportion of the still region indicates uniform global movement and/or a lot of local movement.

- **Mean MV size within one shot** N

  This is the percentage of mean size of the non-zero MVs normalized to the screen width. This parameter determines the intensity of a movement within a moving region. Low intensity indicates the static sequence. High intensity within a large moving region indicates a rapidly changing scene.

- **Ratio of MV deviation within one shot** S

  Percentage of standard MV deviation to mean MV size within one shot. A high deviation indicates a lot of local movement and a low deviation indicates a global movement.

- **Uniformity of movement within one shot** U

  Percentage of MVs pointing in the dominant direction (the most frequent direction of MVs) within one shot. For this purpose, the resolution of the direction is $10°$. This feature expresses the proportion of uniform and local movement within one sequence.

- **Average $\overline{BR}$**

  This parameter refers to the pure video payload, calculated as an average over the whole stream. Note that the parameter $\overline{BR}$ reflects the inverse of a compression gain in spatial and temporal domain. Moreover, the encoder performance is dependent on the motion characteristics. A $\overline{BR}$ reduction causes a loss of the spatial and temporal information that is usually annoying for viewers.

The perceptual quality reduction in spatial and temporal domain is very sensitive to the chosen motion features, making these very suitable for reference-free quality estimation because a higher compression does not necessarily reduce the subjective video quality.

Direct Motion Based Quality Estimation

The initial approach is to design a universal metric based on content-sensitive parameters (Ries *et al.* 2007a). The subjective video quality is estimated with five objective parameters. Additional investigated objective parameters do not improve the estimation performance. On the other hand, reducing objective parameters significantly decreases the estimation accuracy. The proposed model reflects the relation of objective parameters to the MOS values. Furthermore, the mix-term $S \cdot N$ reflects the dependence of the movement intensity N and its motion character S. Finally, one universal metric is proposed for all contents based on the defined motion parameters $\{Z, S, N, U\}$ and $\overline{BR}$:

$$\widehat{MOS} = a + b \cdot \overline{BR} + c \cdot Z + d \cdot S^e + f \cdot N^2 + g \cdot \ln(U) + h \cdot S \cdot N. \qquad (11.5)$$

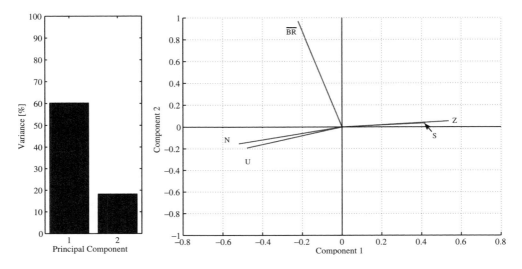

Figure 11.18 Visualization of PCA results.

Table 11.5 Coefficients of estimation model.

Coefficients		Coefficients	
a	4.631	e	0.783
b	8.966×10^{-3}	f	-0.455
c	8.900×10^{-3}	g	-5.272×10^{-2}
d	-5.914×10^{-2}	h	8.441×10^{-3}

The metric coefficients (Table 11.5) were obtained by a regression of the proposed model with the training set (MOS values averaged over two runs of all 26 subjective evaluations for a particular test sequence). To evaluate the quality of the fit of the proposed metrics for the data, a Pearson correlation factor (VQEG 2000) was used. The metric model was evaluated with MOS values from the evaluation set (MOS values averaged over two runs of all ten subjective evaluations for the particular test sequence). The performance of this proposed model was not satisfying (Tables 11.6 and 11.7). In the following, estimates so-obtained were used as input for a further design of an ensemble based video quality estimator.

11.3.7   Ensemble Based Quality Estimation

In order to obtain higher prediction accuracy, ensemble based estimation was investigated. Ensemble based estimators average the outputs of several estimators in order to reduce the risk of unfortunately selecting a poorly performing estimator. The first idea was to use more than one classifier for estimation coming from the neural network community (Dasarathy and Sheela 1979). In the past decade, significant research progress was achieved in the field of individual classifier strategies (Kuncheva 2005) as well as on strategies employing combined classifiers.

The aim is to train a defined ensemble of models with a set of four motion-sensitive objective parameters $\{Z, N, S, U\}$ and $\overline{BR}$. The ensemble consists of different model classes to improve the performance in regression problems (Ries 2008; Ries *et al.* 2007c). The theoretical background (Krogh and Vedelsby 1995) of this approach is that an ensemble of heterogeneous models usually leads to a reduction of the ensemble variance because the cross terms in the variance contribution have a higher ambiguity. This prerequisite was applied to an ensemble of universal models. In order to estimate the generalization error and to select models for the final ensemble a cross-validation scheme for model training (Hastie *et al.* 2001) was used. These algorithms increase the ambiguity and thus improve generalization.

The final step in the design of an ensemble based system is to find a suitable combination of models. Due to outliers and overlapping in data distribution of the dataset, it is impossible to propose a single estimator with perfect generalization performance. Therefore, an ensemble of many classifiers was designed and their outputs were combined such that the combination improves upon the performance of a single classifier. Moreover, classifiers with significantly different decision boundaries from the rest of the ensemble set were chosen. This property of an ensemble set is called diversity. The above-mentioned cross-validation introduces model diversity; the training on slightly different datasets leads to different estimators (classifiers). Additionally, diversity was increased by using two independent models. Furthermore, in cross-validation classifiers with correlation worse than 50% on the second set were automatically excluded.

As the first estimation model, we chose a simple non-parametric method, the k-Nearest Neighbour rule (kNN) with adaptive metric (Hastie *et al.* 2001). This method is very flexible and does not require any preprocessing of the training data. The kNN decision rule assigns to an unclassified sample point the classification of the nearest sample point of a set of previous classified points. Moreover, a locally adaptive form of the kNN was used for classification. The value of k is selected by cross-validation.

As the second method an Artificial Neural Network (ANN) was used. A network with three layers was proposed; input, one hidden and output layer using five objective parameters as an input and estimated MOS as output. Each ANN has 90 neurons in the hidden layer. As a learning method Improved Resilient Propagation (IRPROP+) with back propagation (Igel and Husken 2000) was used. IRPROP+ is a fast and accurate learning method in solving estimation tasks for the dataset. Finally, the ensemble consists of two estimation models kNN and ANN and six estimators, three kNN and three ANN. The performance of the ensemble based estimator is the best out of the proposed ones. The results show a very good agreement between estimated and evaluated MOS values (Tables 11.6 and 11.7).

Video Quality Estimator Performance

To validate the performance of the proposed metric, the Pearson (linear) correlation factor (VQEG 2000) was applied:

$$r = \frac{(\mathbf{x} - \overline{\mathbf{x}})^T (\mathbf{y} - \overline{\mathbf{y}})}{\sqrt{((\mathbf{x} - \overline{\mathbf{x}})^T (\mathbf{x} - \overline{\mathbf{x}}))((\mathbf{y} - \overline{\mathbf{y}})^T (\mathbf{y} - \overline{\mathbf{y}}))}}. \tag{11.6}$$

Table 11.6 Performance on evaluation set by Pearson correlation.

Metric	Pearson correlation
Content based	0.8303
Direct motion based	0.8190
Ensemble based	0.8554
ANSI	0.4173

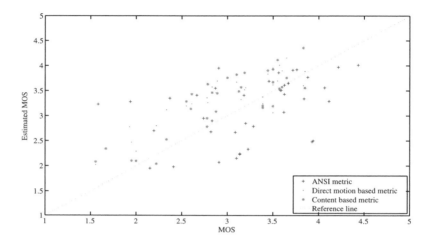

Figure 11.19 Estimated versus subjective MOS results.

Here, the vector **x** corresponds to the 'average' MOS values of the evaluation set (averaged over two runs of all obtained subjective evaluations for a particular test sequence and one encoding setting) for all tested encoded sequences and $\bar{\mathbf{x}}$ corresponds to the average over **x**. Vector **y** corresponds to the prediction made by the proposed metric and $\bar{\mathbf{y}}$ corresponds to the average over **y**. The dimension of **x** and **y** refers to the amount of tested sequences. In order to provide a detailed comparison, the performance of the ensemble based estimator (Ries *et al.* 2007c) was compared with the content class based (Ries *et al.* 2007b) and direct motion based (Ries *et al.* 2007a) estimators as well as the ANSI T1.801.03 metric on the evaluation set. The depicted results for the Pearson correlation factor in Table 11.6 reflect the goodness of fit (cf. Figure 11.19) with the independent evaluation set for all content types together.

This correlation method only assumes a monotone relationship between the two quantities. A virtue of this form of correlation is that it does not require the assumption of any particular functional form in the relationship between data and predictions. The results in Table 11.6 clearly show a good agreement between the obtained and the estimated values for all proposed metrics. In addition, the goodness of the fit on different content classes (see Table 11.7) was investigated. The best performance over all content classes is provided by ensemble and content class based metrics. A fair performance was obtained by the motion based metric and very poor performance by the ANSI metric.

Table 11.7 Individual CC performance on evaluation set by correlation.

Metric/Content type	CC 1	CC 2	CC 3	CC 4	CC 5
Content based	0.93	0.97	0.99	0.90	0.93
Direct motion based	0.85	0.98	1.00	0.71	0.95
Ensemble based	0.93	0.97	0.77	0.91	0.97
ANSI	0.63	0.85	0.95	0.93	0.97

The ensemble based metric shows a performance similar to the content class based metric. The content classification can be understood as a kind of pre-estimation in order to obtain a more homogeneous set of results within one content class, which allows for more accurate quality estimation. This effect was achieved by introducing cross-validation in ensemble based metrics. The direct motion based metric suffers from weak estimation performance for 'News' and 'Cartoon' sequences in comparison to the other content classes.

The weak performance of the ANSI metric shows that it is not suitable for a mobile streaming scenario. The usage of the mobile streaming services influences the subjective evaluation. Therefore, a universal metric such as ANSI is not suitable for the estimation of mobile video quality. Only for higher MOS values which occur at high bit rates ($\geq$ 90 kbit/s) does the ANSI metric perform comparably to the proposed metrics as depicted in Figure 11.19.

References

3GPP TS 26.234.V6.11.0 (2006) Transparent end-to-end Packet-switched Streaming Service (PSS); Protocols and codecs (Rel. 6), June 2006.

ANSI T1.801.03-2003, Digital transport of one-way video signals. Parameters for objective performance assessment. American National Standards Institute, Jul. 2003.

Bonds, A. (1989) Role of inhibition in the specification of orientation selectivity of cells in the cat striate cortex. *Visual Neuroscience*, **2**, 41–55.

Bosch, K. (1998) *Statistik-Taschenbuch*, Oldenbourg Wissensch, Vlg, Munich.

Choi, W.Y. and Park, R.H. (1989) Motion vector coding with conditional transmission. *Signal Processing*, **18**(3), Nov.

Dasarathy, B.V. and Sheela, B.V. (1979) Composite classifier system design: Concepts and methodology. *Proceedings of the IEEE*, **67**(5), 708–713.

Gabriel, K.R. (1971) The biplot graphic display of matrices with application to principal component analysis. *Biometrika*, **58**, 453–467.

Hastie, T., Tibshirani, R. and Friedman, J.H. (2001) *The Elements of Statistical Learning*, Springer.

Hauske, G., Stockhammer, T. and Hofmaier, R. (2003) Subjective Image Quality of Low-rate and Low-resolution Video Sequences. In *Proceedings of International Workshop on Mobile Multimedia Communication*, Munich, Germany, Oct. 5–8.

Heeger, D. (1992) Normalization of cell responses in cat visual cortex. *Visual Neuroscience*, **9**, 181–197.

Igel, C. and Husken, M. (2000) Improving the Rprop learning algorithm. In *Proceedings of the Second International Symposium on Neural Computation*, pp. 115–121, Berlin, ICSC Academic Press.

ITU-T H.264 – Series H (2005) Audiovisual and Multimedia Systems Infrastructure of Audiovisual Services–Coding of Moving Video. International Telecommunication Union, Mar. 2005.

ITU-T Rec. P.910 (2000) Subjective video quality assessment methods for multimedia applications.

ITU-T Rec. P.930 (1996) Principles of a reference impairment system for video.

JVT – Test and Video Group 2003 (2003) Report of The Formal Verification Tests on AVC (ISO/IEC 14496-10 – ITU-T Rec. H.264), Waikoloa, Dec.

Koumaras, H., Kourtis, A. and Martakos, D. (2005) Evaluation of Video Quality Based on Objectively Estimated Metric. *Journal of Communications and Networking*, Korean Institute of Communications Sciences (KICS), **7**(3), Sep.

Kozamernik, F., Sunna, P., Wyckens, E. and Pettersen, D.I. (2005) Subjective quality of Internet video codecs – Phase 2 evaluations using SAMVIQ. *Technical Review*, The European Broadcasting Union, Jan.

Krzanowski, W.J. (1998) *Principles of Multivariate Analysis*, Clarendon Press, Oxford.

Krogh, A. and Vedelsby, J. (1995) Neural Network Ensembles, Cross Validation and Active Learning. *Advances in Neural Information Processing Systems*, **7**, MIT Press.

Kuncheva, L.I. (2005) *Combining Pattern Classifiers, Methods and Algorithms,* Wiley Interscience, New York.

Kusuma, T.M., Zepernick, H.J. and Caldera, M. (2005) On the Development of a Reduced-reference Perceptual Image Quality Metric. In *Proceedings of Conference on Systems Communications (ICW'05)*, pp. 178–184, Montreal, Canada, Aug.

Marziliano, P., Dufaux, F., Winkler, S. and Ebrahimi, T. (2002) A No-Reference Perceptual Blur Metric. In *Proceedings of IEEE International Conference on Image Processing*, pp. 57–60, Sep.

Nemethova, O., Ries, M., Siffel, E. and Rupp, M. (2004) Quality Assessment for H.264 Coded Low-rate and Low-resolution Video Sequences. In *Proceedings of Conference on Internet and Information Technologies (CIIT)*, St Thomas, US Virgin Islands, pp. 136–140, Nov.

Nemethova, O., Ries, M., Dantcheva, A., Fikar, S. and Rupp, M. (2005) Test Equipment of Time-variant Subjective Perceptual Video Quality in Mobile Terminals. In *Proceedings of International Conference on Human Computer Interaction (HCI)*, pp. 1–6, Phoenix, USA, Nov.

Ong, E.P., Lin, W., Lu, Z., Yao, S., Yang, X. and Moschetti, F. (2003) Low Bit Rate Quality Assessment based on Perceptual Characteristics. In *Proceedings of International Conference on Image Processing*, vol. 3, pp. 182–192, Sep.

Pinson, M.H. and Wolf, S. (2004) A new standardized method for objectively measuring video quality. *IEEE Transactions on Broadcasting*, **50**(3), 312–322, Sep.

Richardson, I.E.G. and Kannangara, C.S. (2004) Fast subjective video quality measurements with user feedback. *IEE Electronic Letters,* **40**(13), 799–800, June.

Ries, M. (2008) Video Quality Estimation for Mobile Video Streaming. *Doctoral thesis Vienna University of Technology*, available at http://publik.tuwien.ac.at/files/PubDat_170043.pdf.

Ries, M., Nemethova, O. and Rupp, M. (2005) Reference-free Video Quality Metric for Mobile Streaming Applications. In *Proceedings of DSPCS 05 & WITSP 05*, pp. 98–103, Australia, Dec.

Ries, M., Kubanek, J. and Rupp, M. (2006) Video Quality Estimation for Mobile Streaming Video Quality Estimation for Mobile Streaming. In *Proceedings of Measurement of Speech and Audio Quality Networks*, Prague, Czech Republic.

Ries, M., Nemethova, O. and Rupp, M. (2007a) Motion Based Reference-free Quality Estimation for H.264/AVC Video Streaming. In *Proceedings of IEEE International Symposium on Wireless Pervasive Computing (ISWPC)*, San Juan, Puerto Rico, USA, Feb.

Ries, M., Nemethova, O., Crespi, C. and Rupp, M. (2007b) Content Based Video Quality Estimation for H.264/AVC Video Streaming. In *Proceedings of IEEE Wireless Communications and Networking Conference (WCNC)*, Hong Kong, pp. 2668–2673, Mar.

Ries, M., Nemethova, O. and Rupp, M. (2007c) Performance Evaluation of Mobile Video Quality Estimators. In *Proceedings of the 15th European Signal Processing Conference (EUSIPCO)*, Poland, Sep.

Ries, M., Nemethova, O. and Rupp, M. (2008) Video Quality Estimation for Mobile H.264/AVC Video Streaming. *Journal of Communications*, **3**(1), 41–50.

Rix, A.W., Bourret, A. and Hollier, M.P. (1999) Models of Human Perception. *Journal of BT Tech.*, **17**(1), 24–34, Jan.

Saha, S. and Vemuri, R. (2000) An analysis on the effect of image features on lossy coding performance. *IEEE Signal Processing Letters*, **7**(5), 104-107, May.

Teo, P. and Heeger, D. (1994) Perceptual Image Distortion. In *Proceedings of SID International Symposium Digest of Technical Papers*, pp. 209–212, June.

VQEG 2000 (2000) Final report from the Video Quality Experts Group on the validation of objective models of video quality assessment, available at http://www.vqeg.org/.

Wang, Z., Lu, L. and Bovik, A.C. (2004) Video quality assessment based on structural distortion measurement. *Journal of Signal Processing: Image Communication, special issue on "Objective Video Quality Metrics"*, **19**(2), 121–132, Feb.

Westen, S., Lagendijk, R.L. and Biemond, J. (1995) Perceptual Image Quality Based on a Multiple Channel HVS Model. In *Proceedings of ICASSP*, vol. 4, pp. 2351–2354.

Winkler, S. and Dufaux, F. (2003) Video Quality Evaluation for Mobile Applications. In *Proceedings of SPIE Conference on Visual Communications and Image Processing*, vol. 5150, pp. 593–603, Jul.

Zetzsche, C. and Hauske, G. (1989) Multiple Channel Model for the Prediction of Subjective Image Quality. In *Proceedings SPIE*, vol. 1077, pp. 209–216.

Part VI

Packet Switched Traffic – Evolution and Modelling

Introduction

Part IV consists of four chapters: 'Traffic Description', 'Traffic Flows', 'Traffic Models for High Delay Services' and 'Traffic Models for Specific Services', all written by Philipp Svoboda. The chapters are based on traces taken from the Gi and Gn interfaces of a live 3G network operator in Austria between 2005 and 2007 using the measurement system from the project METAWIN as described previously in Chapters 9 and 10 of this book.

Since 2005 we have realized how the service mix changes over time during the introduction of UMTS in Austria. Similarly to energy usage, traffic and service demand vary largely during the day with some periodic daily, weekly and seasonal behaviour. We cover the application layer and 3rd Generation Partnership Project (3GPP) related parameters such as the Packet Data Protocol (PDP) context in Chapter 12.

In Chapter 13 the concept of traffic flows is introduced. Due to today's advanced services, a single PDP context is no longer sufficient to be assigned to a user or a service. It is very interesting that few users cause most of the traffic while most users cause only a small fraction of the traffic. Such behaviour can be described as so-called 'heavy tailed' distributions, as explained in this chapter. Experimental data of such heavy tailed distributions are difficult to handle as they require huge amounts of observation and thus storage space. The parametric fitting is particularly difficult as such distributions often do not have a mean and/or variance while for a limited set of data such values can always be computed.

The last two chapters describe models for more recent services that were not present at the introduction phase of the Internet, such as email, HTTP, FTP, POP3, gaming and push-to-talk services. Original models from the 1990s are adopted in Chapter 14, further improved and parametrically fitted to our observations. Malicious traffic needs to be treated with care since the large amount of scanning viruses and worms makes the Internet traffic look other than it is.

Particular focus is on new services, 'wireless gaming' and 'push-to-talk', in Chapter 15 as they are expected to become the services with the largest growth. Wireless gaming is a worldwide rapidly growing market, in which already several million people participate. Next to strategy games that have only mild demands on wireless links, action games with quasi real-time requirements are being played. For such services no models existed before and thus first proposals are presented here.

The work reported here by Philipp Svoboda can be found in more detail in his thesis *Measurement and Modelling of Internet Traffic over 2.5 and 3G Cellular Core Networks*, which is available at http://publik.tuwien.ac.at/files/PubDat_170755.pdf as well as in numerous publications cited within.

12

Traffic Description

Philipp Svoboda

12.1 Introduction

First we introduce the traces on which this and Chapter 13 are based. In section 12.2, 'Volume and User Population', we present and discuss basic network values such as: volume share between Universal Mobile Telecommunications Systems (UMTSs) and General Packet Radio Services (GPRSs), volume per user, user population, penetration of handset types and so on. The results are then compared with other recent publications on this sector (Dahmouni *et al.* 2005; Kalden *et al.* 2003; Svoboda *et al.* 2006). The section ends by discussing service mixes, for UMTS and GPRS, respectively. The information gained here is then used to determine which services we go on to look at in detail. In addition to this we analyse the time of day effect for traffic and services and their long-term evolution. These are important parameters for operators to plan link utilization and upgrade.

The equivalent to a dial-in session of a customer attached to its Internet Service Provider (ISP) via a modem is, in the mobile world of 3rd Generation Partnership Project (3GPP), the Packet Data Protocol (PDP) context. Section 12.3 deals with the types of context, their duration and volume and the usage patterns generated over different days (Svoboda *et al.* 2006; Varga *et al.* 2004).

Recall that, as explained earlier in Chapter 9, absolute volumes cannot be disclosed due to our NDA with the network operator. Therefore, we will only present truncated empirical cumulative distribution functions (ecdfs) and relative changes of shares.

12.1.1 Analysed Traces

We present an overview of the traces on which this part of the book is based. We use traces recorded over three years from December 2004 until April 2007. As the traffic load increased

Video and Multimedia Transmissions over Cellular Networks   Edited by Markus Rupp
© 2009 John Wiley & Sons, Ltd

over the years the first traces spanned over several days up to one week, while the most recent traces in April 2007 are four hours around the busiest time on the network.

We recorded most of our traces on the Gn interface. See Chapter 1 for more details about the core network. At this interface we are able to distinguish an individual subscriber via the GPRS Tunnelling Protocol (GTP) tunnel Identification (ID). In our network each Serving GPRS Supporting Node (SGSN) covered a specific radio technology, for example SGSN1 for GPRS traffic and SGSN15 for UMTS traffic. Therefore, it is possible to split the traffic into UMTS and GPRS traffic via the information to which SGSN the user is attached.

With the introduction of Combo-SGSN, an SGSN type was capable of connecting with GPRS and UMTS Radio Access Networks (RANs) simultaneously; this approach is no longer useful. Future tracing will process the RAN-Type flag, which is transmitted in the GTP tunnel and was introduced in Rel. 6, in order to assign traffic according to its RAN type. There were no Combo-SGSNs active until the end of 2007; the RAN type was evaluated using the IP addresses of the SGSNs. Therefore, this does not restrict our trace analysis.

Trace TR_1 is one full week in December 2004 including GPRS and UMTS traces. At this time only a small amount of subscribers had a UMTS UE and even fewer a data-card. Therefore, in this trace the majority of the subscribers were web browsing with mobile terminals.

Traces TR_2 and TR_3 are from one week in September 2005 and from one week in May 2006. The data-cards were already widely deployed and Enhanced Data Rates for GSM Evolution (EDGE) technology was installed in the network. The typical UMTS subscriber is no longer a data-card only user.

Trace TR_4 was recorded in October 2006. We recorded three consecutive days at the end of October. These traces already include High Speed Downlink Packet Access (HSDPA) User Equipments (UEs) but not High Speed Uplink Packet Access (HSUPA).

Trace TR_5 contains several non-consecutive days from April 2007. By this time HSDPA was already introduced widely. These traces served as the basis for the email traffic model.

Trace TR_6 contains two consecutive days in September 2006. This dataset is evaluated only on the flow level in Chapter 13.

12.1.2   Daily Usage Profile for UMTS and GPRS

The daily profile of the network load is an important factor in dimensioning network capacity. The information concerning the busy hour is particularly valuable. This is the time during which the maximum load is present in the network. For network dimensioning the designer can focus on this time span to evaluate the resulting Quality of Service (QoS) parameters. We first show qualitative results for the live network separately for GPRS and UMTS.

Figure 12.1 depicts the load on a Monday in May 2006, part of TR_3. As expected the load for UMTS and GPRS follows a typical daily profile. The peak hour is around 8.00 p.m. This is an important result, as it indicates that the main load in the traffic is due to private customers and not due to business subscribers only. Between UMTS and GPRS there is a factor of 2:1 in terms of generated data rate at the Gn interfaces. As we reveal later, the gap between UMTS and GPRS data rates was wider at the start of the mobile Internet service. However, over the years and with the introduction of EDGE the gap has started to close.

Figure 12.2 presents the number of active or open PDP contexts for 24 hours on a Monday in May 2006. The first interesting point is the fact that even during the night this curve does

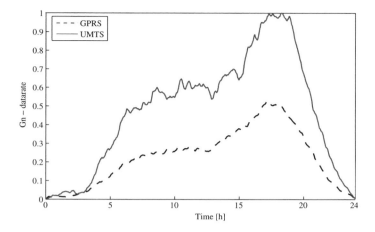

Figure 12.1 Daily profile of Gn data rate for UMTS and GPRS (TR_3, normalized).

not drop to zero, because there are subscribers with an always-on context behaviour. The factor between UMTS and GPRS subscribers is 1:5, which is exactly the opposite of what we found in the data rate case. The smaller UMTS population generates twice the data rate in the core network. The activation rate is the first derivative of the plotted curves. The steep slope of the curves between 5.00 a.m. and 9.00 a.m. indicates a sudden increase in the activation rate, which is assumed to be linked to the subscribers' waking up. Most of the active contexts are torn down during the late evening after 9.00 p.m., one hour after the peak load with respect to the data rate in the network. We assume that this is due to timeouts in the mobile terminals.

12.2 Volume and User Population

At first we obtained results as to how the total traffic and user population is split between different services. The comparison between TR_1 and TR_2 provides an insight into the historical changes in the macroscopic traffic composition during almost one year.

Attached to the 3rd Generation (3G) Core Network there are two different Radio Access Networks (RANs), namely GPRS and UMTS, delivering different link data rate; for further details, see, for example, Holma and Toskala (2004, p. 23). Since UMTS radio deployment is more recent, the number of UMTS base stations is smaller compared to GPRS and is limited to urban and suburban areas. Therefore, GPRS traffic may be characterized by rural areas with a very small amount of traffic compared to urban city centres.

12.2.1 Volumes and User Population in GPRS and UMTS

Table 12.1 shows the relative fractions of users and volumes found in the network for GPRS and UMTS. Users are identified by their IMSI. Note that only 'active' users are accounted for

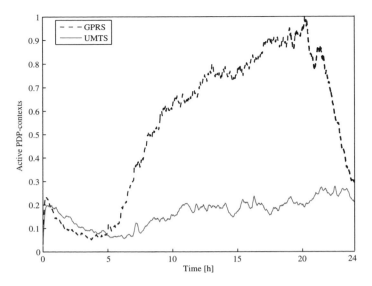

Figure 12.2   Daily profile of number of active PDP contexts for UMTS and GPRS (TR_3, normalized).

here, that is, with at least one PDP-context activation in the measurement window. Attached but inactive users are not accounted for. Note that in general UMTS capable terminals can also access GPRS RANs outside the UMTS coverage. We label as 'GPRS users' those seen exclusively on the GPRS section during the measurement period, while those who accessed UMTS at least once are classified as 'UMTS user'.

A caveat here is that a 'UMTS user' can also generate GPRS traffic. While this traffic is correctly accounted for as GPRS, the user count in GPRS only considers GPRS-only users. This leads to some overestimation of the average per-user traffic for GPRS. In order to avoid such overestimation, a more complex classification of users and traffic is required, considering the GPRS traffic produced by UMTS capable MSs in a separate class. For the sake of simplicity we omit such refinement in this work. A study of TR_1 reveals that fewer than 1% of the subscribers change cell while a packet-switched connection is active. Out of this subset fewer than 1% had changed the RAN technology. This validates our assumption from above.

From Table 12.1 it can be deduced that in TR_1 the number of active UMTS users was approximately two orders of magnitude less than GPRS, but the two groups generated comparable traffic volumes. In fact at that time a large fraction of UMTS mobile stations were 3G data-cards mounted on laptops coupled with flat-rate contracts, while UMTS handheld devices were just starting to spread. In TR_2 the fraction of UMTS users had increased, as the combined effect of additional UMTS subscribers plus legacy GPRS users upgraded to UMTS handheld terminals.

The fact that for UMTS the growth factor in the volume share is substantially less than for the user share suggests that most of the new additional terminals were handheld devices rather than 3G data-cards for laptops, under the assumption that hand-held devices generate less

Table 12.1 Fraction of users and volume in GPRS and UMTS (TR_1, TR_2).

TR_1 (Dec. 04)				TR_2 (Sep. 05)		
RAN	Users (%)	Volume (%)		RAN	Users (%)	Volume (%)
GPRS	98.8	58.3		GPRS	90.0	40.4
UMTS	1.2	41.7		UMTS	10.0	59.6

traffic than laptops owing to a combination of differences in terminal capabilities and billing schemes. However, laptop data-cards are still an important subset in the UMTS terminal population. The comparison between the ratios of volume to user shares shows that the average per-user traffic is larger for UMTS than for GPRS. This was expected, given that UMTS delivers a higher data rate and hence a better user experience than GPRS, despite the fact that they share the same flat rates.

As a next step we directly evaluated the weekly volume generated by individual users. The per-user volume distributions for GPRS and UMTS in TR_2 are reported in Figure 12.3. All values are in Mbyte/week. Figure 12.3(a) reports the complementary cumulative distribution function (ccdf). It can be seen that the median is around 8.2 Mbyte/week for UMTS and only 102 kbyte/week for GPRS. The distribution spans about five orders of magnitude, denoting a large disparity in the user behaviour.

This can be better seen from Figure 12.3(b). There we plot the fraction y of total volume generated by the fraction x of top users in a double log scale (loglog). We conclude that the top 1% of UMTS users generated around 16% of total traffic, while the top 1% of GPRS users generated 47% of total traffic. For the 10% of top users, the cumulated fraction of volume jumps to 59% for UMTS and 92% for GPRS. The GPRS figures reveal that the vast majority of users 'seen' on GPRS are spurious users, generating sporadic or very low traffic volumes. At the other extreme of the range are a small population of heavy users, who transfer massive volumes of traffic.

This suggests the possibility of classifying the users into three basic groups: sporadic, heavy and intermediate users. In general, such classification would be service-dependent. However, as far as the total network load is concerned we can ignore service-specific metrics and consider just the total transferred volume in the measurement period. Hence, we classified the users into three groups based upon arbitrarily chosen threshold values on the weekly volume v: low ($v \leq 1$ Mbyte), medium (1 Mbyte $< v < 100$ Mbyte) and high ($v \geq 100$ Mbyte). We used the same boundary values for UMTS and GPRS. Table 12.2 shows the results for both datasets.

From Table 12.2 we observe that the vast majority of users (70–85%) are in the 'low' group, but they generate only a small fraction of the total traffic (1–4%). The largest change in GPRS from TR_1 to TR_2 is found in the 'high' group: the fraction of heavy users in GPRS increased and they account now for 70% of the total traffic.

In UMTS, we see that the 'low' group was almost non-existent in TR_1 (less than 4% of the users), while in TR_2 the fraction of low users rose to almost 30.5%. This is probably due to the introduction of a large number of hand-held devices, while in TR_1 the predominant terminal type was the laptop data-card.

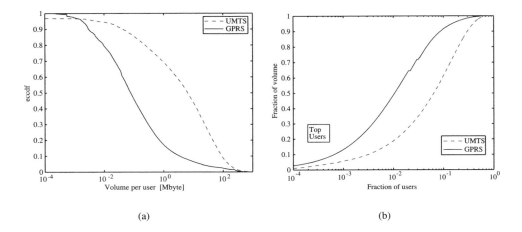

(a)   (b)

Figure 12.3 Distribution of volume per user in Mbyte/week in September 2005 (TR_2). The x-axis had to be truncated to 10^{-4} in order not to disclose the total number of users: (a) volume per user in one week (cdf); (b) fraction of volume cumulated by the fraction of top users.

Table 12.2 Grouping per transferred volume (TR_1 (Dec. 04) and TR_2 (Sep. 05), values in fraction of total).

	GPRS (UMTS)		
TR_1	High (%)	Medium (%)	Low (%)
Users	0.1 (13.0)	29.3 (83.3)	70.6 (3.7)
Volume	7.0 (61.4)%	89.3 (38.6)	3.7 (< 0.1)
TR_2			
Users	2.9 (7.3)	14.1 (62.2)	83.0 (30.5)
Volume	70.2 (91.2)	28.5 (8.8)	1.3 (< 0.1)

12.2.2 Fraction of Volume per Service

Next, we analysed the volume share for each service. The following figures do not include custom services implemented for specific customers, which was pre-filtered based on the APN value on a per-PDP-context basis. Also, we filtered traffic on ports tcp:135 and tcp:445, which are used by several scanning worms; see discussion in section 12.2.5.

The results are presented in Figure 12.4 for UMTS and GPRS, separately. UMTS shows only minor changes in the per-service volume distribution. In GPRS the main changes occur for the Wireless Application Protocol version 1.0 (WAP 1.0) share, which halved from 30% to 14%. This was only partially compensated for by the increase of WAP 2.0.

In both GPRS and UMTS the largest volume share is on WEB. In Figure 12.4, we compare relative shares. Table 12.3 shows the growth factors of each individual service during the last

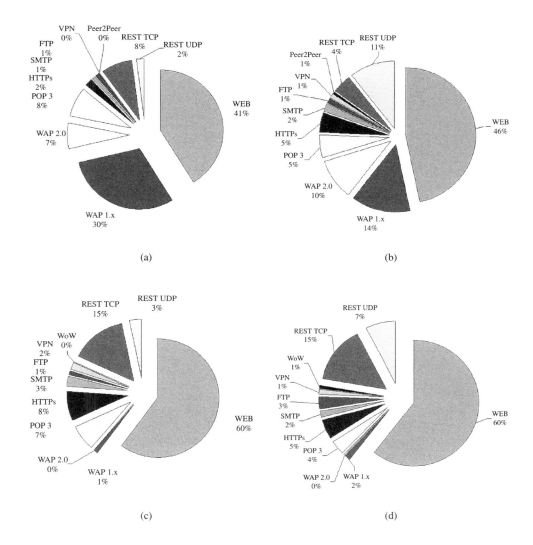

Figure 12.4   Service shares for (a) GPRS TR_1; (b) GPRS TR_2; (c) UMTS TR_1; and (d) UMTS TR_2.

year (recall that the absolute volume cannot be disclosed). From Table 12.3 we learn that in GPRS the web traffic has grown considerably more than WAP. Curiously, it appears that the opposite applies to UMTS, with a WAP growth rate of 800%, double that of the web. This is due to the fact that WAP was only marginally present in UMTS in TR_1 and, as noted above, UMTS capable handsets started to spread later.

Table 12.3 Annual growth factor of per-service volumes ($TR_1 - TR_2$).

Service	WAP 1.x and 2.0	HTTP	POP3, SMTP IMAP	HTTPs	VPN (port1000, etc.)
GPRS (%)	+132	+437	+122	+850	+113
UMTS (%)	+852	+401	+178	+241	+397

Table 12.4 Daily service shares – GPRS and UMTS (TR_1).

	Fraction of volume						
Service	Mon	Tue	Wed	Thu	Fri	Sat	Sun
GPRS							
HTTP	39.4	40.1	39.1	38.8	40.1	40.3	41.8
WAP	35.8	35.4	36.0	34.7	35.4	39.2	32.1
POP3	8.0	8.4	8.2	8.5	8.9	7.0	7.3
SMTP	1.0	1.2	1.1	1.0	0.8	1.0	1.0
HTTPS	1.8	2.2	2.0	2.0	2.4	2.2	2.1
Rest TCP	10.3	9.7	10.9	11.1	8.9	7.8	10.7
Rest UDP	3.7	3.0	2.7	3.9	3.5	2.5	5.0
UMTS							
HTTP	60.7	60.6	62.8	61.0	60.5	60.3	61.0
WAP	2.1	1.9	1.8	2.1	1.7	1.9	2.0
POP3	3.9	4.3	4.4	4.5	3.8	2.0	1.9
SMTP	2.0	2.1	2.1	1.9	1.9	1.5	1.7
HTTPS	5.3	5.3	5.5	5.5	5.2	5.8	5.7
Rest TCP	15.7	14.3	14.6	14.9	14.7	16.3	16.2
Rest UDP	10.3	11.5	8.8	10.1	12.2	12.2	11.5

12.2.3   Service Mix Diurnal Profile

Next, we analysed the stability of service shares on a day-to-day basis, limited to the top services. Table 12.4 shows the fraction of volume per service per RAN accumulated on a day-to-day basis. We found that they have nearly constant day-to-day shares, with only the exception of email traffic which displays a significantly lower share at weekends. We assume that this is due to the fact that business people tend not to use their mobile access on the weekends. An interesting detail is that GPRS users tend to download mainly emails (POP3/SMTP = 8/1), while the same ratio for UMTS users is around two. We can state that the weekly service share is rather representative of single days.

Network planning departments often refer to the busy hour to dimension their networks. To answer the question of whether the weekly traffic share can also be used to adjust traffic shares in a busy hour simulation, we split the traffic further into bins of 10 000 s. The result is presented in Figure 12.5 separately for GPRS and UMTS. The graphs have been rescaled

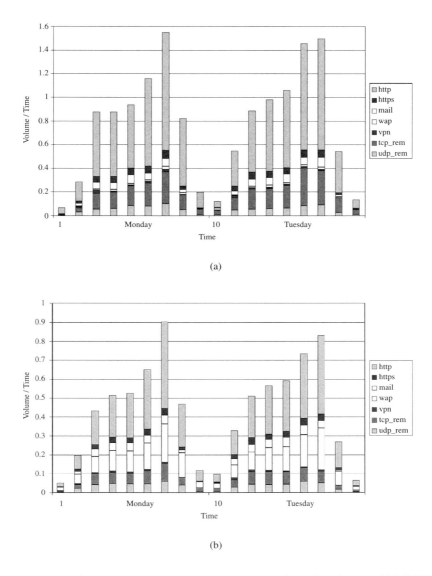

(a)

(b)

Figure 12.5 Time of day effect in TR_2, volume normalized, (a) GPRS TR_2; (b) UMTS TR_2.

by a factor in order not to disclose absolute volumes. From the shape of the curves we see that the busy hour is found between 7 p.m. and 9 p.m. The actual patterns are stable on a day-to-day basis. It can be seen that web is the major player all day long. The UMTS and GPRS patterns are similar to each other if we neglect WAP traffic. The service volume share in the busy hour is shown in Table 12.5. The values are averaged over three busy hours in TR_2.

Table 12.5 Average fraction of volume per service in the busy hour (TR_1).

Service	HTTP (%)	HTTPS (%)	POP3 and SMTP (%)	WAP (%)	VPN (%)	TCP-Rest (%)	UDP-Rest (%)
GPRS	45.1	3.7	5.3	22.4	0.6	16.4	6.5
UTMS	64.2	5.5	8.9	1.7	2.8	10.8	6.1

Table 12.6 Grouping customers per service (TR_1 (Dec. 04) and TR_2 (Sep. 05), values in fraction of total).

Usage	WAP only (%)	WAP and email (%)	Internet (%)
GPRS			
TR_1	89.3	9.3	1.4
TR_2	78.1	5.1	16.8
UMTS			
TR_1	0.1	1.4	98.5
TR_2	11.4	0.9	87.7

12.2.4   Grouping Subscribers per Service Access

We defined three user groups according to their services, as in Kalden *et al.* (2003):

1. WAP only;

2. WAP and email;

3. Internet services (no WAP).

Table 12.6 shows the relative numbers of users in each of the three groups for GPRS and UMTS in December 2004 (TR_1) and September 2005 (TR_2). The first group represents the vast majority of users who use only a mobile phone to browse WAP pages or download ring tones, send MMS via WAP and so on. In TR_1 about 89.3% of all GPRS users are part of this group, compared to fewer than 0.1% of the UMTS users. As noted above in TR_1 most users active on UMTS were data-card users. In TR_2 the share of WAP UMTS has already grown to 11.4%. The second group is able to browse WAP and transfer emails but does not have web traffic. This group of users is small in UMTS, 1.4%, but populated with at least 5% in GPRS. In the last group there is no WAP traffic and mainly web volumes. This includes the mobile offices, users who access the Internet outside their company by their notebooks or smart-phones. In the GPRS part of TR_2 fewer than 17% of the users in this group consume all of the web volume share, which is ≈50% of the total traffic. In the older traces on GPRS fewer than 2% generated all of the web traffic. This is consistent with a result presented in Kalden *et al.* (2003) by Vodafone and Ericsson, a study carried out by a mobile provider in Hungary.

12.2.5   Filtering in the Port Analysis

During the exploitative analysis we detected the presence of a large number of packets directed to ports tcp:135 and tcp:445, mainly Transmission Control Protocol (TCP) SYN in the uplink direction. This is due to several self-propagating worms attached to some infected 3G terminals. The presence of such unwanted traffic is to be expected as laptops with 3G data-cards – often equipped with popular operating systems – coexist nowadays with handsets and smart-phones in the 3G network. For years it has been well known that unwanted traffic is a steady component of the traffic in the wired networks (Pang *et al.* 2004). It is important to understand that such traffic is not the expression of user preference. Therefore, we took it out of our analysis. The filtering rules were set to:

$$\text{TCP: } \sum(\texttt{Uplink-Packets}) \geq 2 \,\&\, \sum(\texttt{Downlink-Packets}) \geq 2$$

and

$$\text{UDP} \sum(\texttt{Uplink-Packets}) \geq 1 \,\&\, \sum(\texttt{Downlink-Packets}) \geq 1.$$

Please note that the filtering was applied mainly to avoid an inflation of the flow table in the analysis; for example, each probing packet of a virus would have been counted as a new flow and in the measurement system each flow holds a set of state information which has to be maintained. This would put additional load on the system, making some of the analysis impossible. The aim of this filtering rule is not to exclude all malicious traffic but to reduce the flows created by the probing of infected terminals. Although we exclude so-called background radiation from the analysis, a network operator should always understand the presence of these activities. Systems such as a stateful packet inspection tool could become overloaded due to the tree-like structure of the mobile core network, in which all the traffic has to be routed over one single interface. A failure of this system will logically disconnect the whole core network from the Internet.

On TCP this rule is set according to the TCP handshake procedure, which generates three packets (pkts), plus a single data packet. In UDP there is no connection setup; therefore, we set the threshold to two packets: one sent and one received packet. These rules filtered 3.4% of the TCP and 2.1% of the UDP traffic in TR_2.

Finally, note that EDGE was introduced in August 2005; therefore, the dataset taken in September 2005 includes data traffic from EDGE terminals, which we refer to simply as 'GPRS'.

12.3   Analysis of the PDP-context Activity

In this section we investigate the activity of each user at the level of PDP-contexts. The PDP-context in a GPRS/UMTS network is comparable to a dial-up process in wired networks: a user has to establish a PDP-context to transfer data via a mobile network (3GPP TS230.60 2003). The dataset sample for this analysis was TR_2 (September 2005), divided into GPRS and UMTS PDP-contexts. To reduce the processing time we took the first three days of TR_2.

12.3.1   Per-user Activity

For each PDP-context j generated by MS i we extracted the total volume v_{ji} and the duration d_{ji}. For each MS i, we considered the following attributes: total number n_i of PDP-contexts, total transferred volume $s_i = \sum_{j=1}^{n_i} v_{ji}$ and total on-time $t_i = \sum_{j=1}^{n_i} d_{ji}$. We show scatter plots of these three attributes with logarithmic binning, split to UMTS and GPRS in Figures 12.6 and 12.7. Each point represents the number of users i within the loglog binning.[1] In all plots there are boundaries due to admissible regions.

The empty region R_1 in Figures 12.6(a) and 12.7(a) relates to the presence of a minimum value for the duration of a PDP-context, say d_{min}, which forces a user with n_i PDP-context to stay online for at least $s_i \geq n_i \times d_{min}$. We obtain that in this case GPRS and UMTS have a similar footprint. In GPRS (Figure 12.7(a)), for $n_i > 100$ there is a region R_2 with correlations. The average PDP-context duration is 12.1 s for this region. We observed a d_{min} of 32 ms.

In Figures 12.6(b) and 12.7(b) the lower limit relates to the maximum available data rate (cumulated uplink and downlink) UMTS: 320.3 kbit/s; GPRS 81.2 kbit/s. The maximum achievable data rate is 384 kbit/s in UMTS and 85.6 kbit/s in GPRS. This result is only a validation of the measurement, as there will always be idle times within a PDP-context. We observe that Figure 12.6(b) is shifted to the right compared to 12.7(b); the mean values are indicated by the black **x** in the scatter plots. This indicates a higher net data rate in UMTS. The spreading is also smaller in UMTS. The upper limit is due to the presence of a timeout for long inactive PDP-contexts.

In Figure 12.7(c) we can spot two different regions, not present in Figure 12.6(c). For $n_i \leq 100$ there is no significant link between the number of connections and the volume transferred. The cluster of highly active users, with many PDP-contexts (>100) shows a correlation between these two values. This indicates service messages of constant size. The slope of the cluster cloud yields an average volume per PDP-context of around 1.8 kbyte as shown in Figure 12.10. More detailed analysis of the size of the PDP-context (in up- and downlink) can be found in the following section on the PDP-context size.

In Figure 12.8(a) there are four curves, two per RAN. The 'IMSI – [RAN]' (there are two RANs – either UMTS or GPRS) curves are the cdf of the variable n_i: the point at $x = X$ shows the fraction of users with $n_i \leq X$. The 'PDP – [RAN]' curves show instead the total 'fraction of total PDP-contexts' generated by all the users with $n_i \leq X$. It is interesting to note that 48.3% of the UMTS users and 31% of the GPRS users had only one PDP-context within the observation period. Considering 100 PDP-contexts it is apparent that more than 99.9% of the users are covered by this limit; however, we obtain only at most 57.4% for UMTS, respectively at most 88.4% for GPRS of all PDP-contexts for these users. From this we learn that the last 0.01% of the population is generating a huge amount of the PDP-contexts (difference from 100%). Further analysis will reveal that these contexts characterize the epdfs of the population.

12.3.2   Distribution of PDP-context Duration

In this section we investigate the distribution of PDP-context duration. Based on the high dispersion of n_i revealed by Figure 12.8(a), a small fraction of users (0.01%) generated

---

[1]The scale has again been multiplied by an arbitrary value so as not to reveal any absolute number.

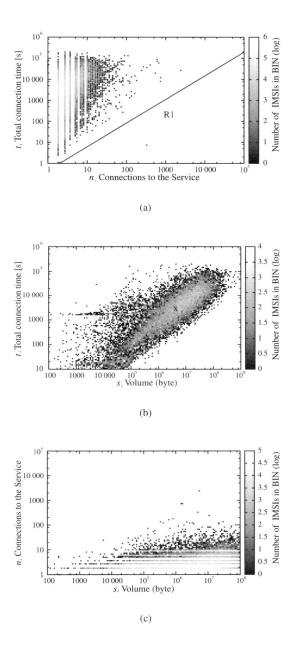

Figure 12.6 Scatter plots of 'Volume', 'Number of Contexts' and 'Total Duration per IMSI' [TR$_2$, UMTS] (the figures have been truncated in order not to show the detailed numbers): (a) t_i versus n_i; (b) t_i versus s_i; and (c) n_i versus s_i.

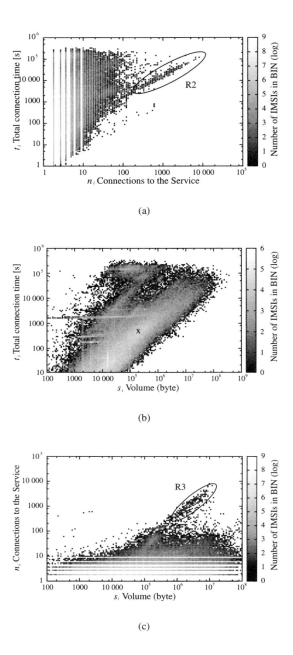

Figure 12.7 Scatter plots of 'Volume', 'Number of Contexts' and 'Total Duration per IMSI' [TR_2, GPRS] (the figures have been truncated in order not to show the detailed numbers): (a) t_i versus n_i; (b) t_i versus s_i; and (c) n_i versus s_i.

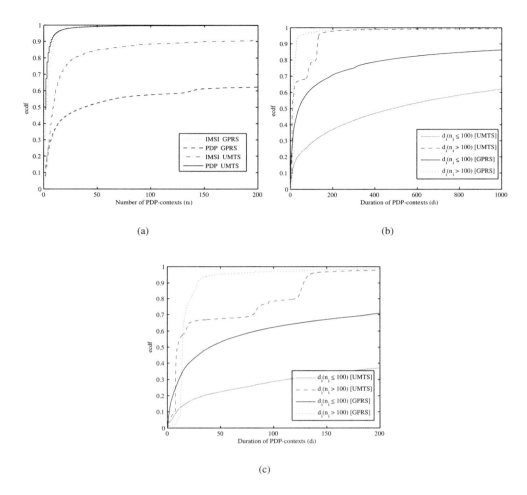

(a)

(b)

(c)

Figure 12.8 (a) Number of PDP-contexts; (b) duration of PDP-contexts; and (c) duration of PDP-contexts – zoom (TR$_2$).

values of n_i one to two orders of magnitude larger than the rest of the population. We know that a simple probability density function (pdf) of the duration d_{ji} would be mainly characterized by those few users with a very high number of PDP-contexts. These are assumed to originate from periodic polling of custom services with an intrinsically different behaviour from other users. An off-line inspection of the payload would confirm this assumption. However, as mentioned in the introduction this is not allowed due to data privacy issues. We therefore considered the distributions of the variable d_{ji} 'conditioned' to a particular value of n_i. This is represented in a compact way by the plots in Figures 12.9(a) and 12.9(b), where each discrete value of x represents the empirical distribution of the variables d_{ji} conditioned to the value of n_i given in the abscissa. Note that the values of the

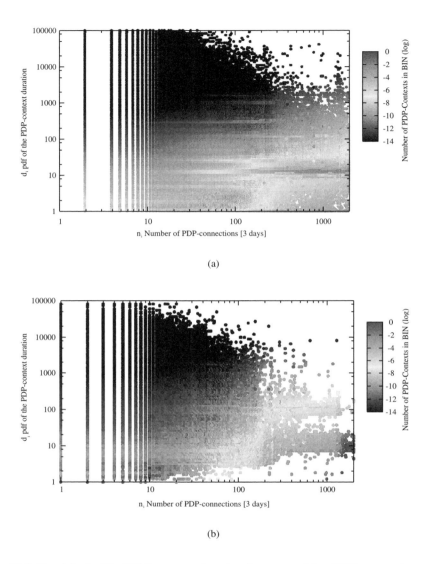

(a)

(b)

Figure 12.9 Empirical pdf of PDP-context duration for users with x PDP connections within three days (weighted, see text) (TR_2): (a) d_i versus n_i (GPRS); (b) d_i versus n_i (UMTS).

empirical histogram (in log binning) across each column have been normalized by the counts per bin multiplied by the size of the corresponding bin. The plot allows a comparison of the distribution of PDP-context duration for users with different activities. The plot reveals that low-activity users (say ≤100 PDP/week) generate PDP-contexts with a different distribution of the duration from that of high-activity users. The low-activity group exposes a distribution with a plateau ranging from 1–100 s. For the high-activity users the PDP-context duration

Table 12.7 Model parameter for PDP-context on-time (TR_2).

Group	Weight		Distribution		
GPRS					
$n_i \leq 100$	1		Weibull	$\lambda = 187.82$	$k = 0.368$
$n_i > 100 + 0.179$		log-normal		$\mu = 5.13$	$\sigma = 2.44$
	+0.821		step	11.3 s	–
UMTS					
$n_i \leq 100$	1		Weibull	$\lambda = 935.88$	$k = 0.48$
$n_i > 100 + 0.36$		log-normal		$\mu = 4.91$	$\sigma = 2.13$
	+0.421		step	8.2 s	–
	+0.053		step	88.2 s	–
	+0.167		step	131.2 s	–

tends to be concentrated on smaller values. In UMTS there are two peaks around 8 and 80 s. In GPRS the high-activity group has a peak at 11 s and a high plateau below 5 s duration. For more details, see the results in Table 12.7. Based on such figures, we decided to split the users into two main groups based on the value of n_i and an arbitrary set boundary of 100 PDP-contexts/week. The first group, with $n_i \leq 100$, gathers 99.6% of the users as depicted in Figure 12.8(a) and at most 57.4% (UMTS), respectively at most 88.4% (GPRS) of the PDP-contexts, while in the second group 0.4% of users generate 40% of all contexts. Assuming a constant arrival rate of the second group, the resulting fraction is a function of the observation period, for example six days would double the number of PDP-contexts in the second group. It is likely that the second group is dominated by automatically accessed services, for example polling, remote-control. The ecdf of the PDP-context duration for each group and RAN are shown in Figure 12.8(b). The curves for IMSIs with at most 100 PDP-contexts fit a log-normal function, with parameters provided in Table 12.7. We observe that the duration of a PDP-context is larger for UMTS users. For IMSIs with a larger number of PDP-contexts, the distribution has discrete steps. In GPRS we see one major step at 11.3 s for users with $n_i > 100$, in UMTS the first step is already at 8.2 s. In UMTS this group has two additional steps at 88.2 s and at 131.2 s. This indicates that there are additional services active on UMTS. We split the fit for this cdf into two different components. The first component is a log-normal distribution, as used for the first group. The second component is a step function, which probably originated from the automatic services. Table 12.7 shows the parameters for the two groups per RAN. The second group is split into several lines, which have to be added to gain the total cdf.

12.3.3 The Volume of a PDP-context

In this section we investigate the size of the PDP-contexts in terms of total volume. Recall that the volume is defined as the sum of the IP layer bytes (GTP payload) of user traffic, without considering the volume of signalling messages. In this section we distinguish between uplink and downlink volumes. We used the same grouping as above ($n_i \leq 100$ and $n_i > 100$). In Figure 12.10 we plot the cdf of downlink and uplink volume per PDP-context split into the two user groups. Figure 12.10(a) shows the results for GPRS and 12.10(b) for UMTS. There

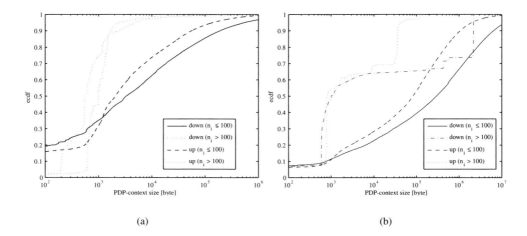

(a)   (b)

Figure 12.10 Size of PDP-contexts divided into two user groups (TR_2): (a) GPRS (b) UMTS.

Table 12.8 Number of bytes per PDP-context (TR_2).

Direction	Distribution	
$n_i \leq 100$		GPRS
Uplink/bytes	Weibull	$\lambda = 169.27$ k, $k = 0.39$
Downlink/bytes	Weibull	$\lambda = 633.19$ k, $k = 0.34$
$n_i \leq 100$		UMTS
Uplink/bytes	log-normal	$\mu = 8.22, \sigma = 2.23$
Downlink/bytes	Weibull	$\lambda = 10.52$ k, $k = 0.30$

is obviously a lower limit for the size of a PDP-context. This is due to the fact that we pre-filtered all PDP-contexts with no data packets, that is spurious PDP-contexts and erroneous procedures (they accounted for 2% of all PDP-contexts in our dataset). Note that in GPRS ≈20% and in UMTS ≈8% of the PDP-contexts in each group have fewer than 100 bytes. Table 12.8 shows the fitting for the up- and downlink bytes per PDP-context.

In GPRS more than 70% of all contexts are accumulated in the step near 1 kbyte. In UMTS only ≈50% of the contexts are covered by this step. However, there are two additional steps. Together these steps add up to more than 85% of the PDP-contexts in UMTS. This indicates that the services with many PDP-contexts not only have a step in the duration of their PDP-context, but also in the size of the transferred volume.

12.3.4   Total Volume and Number of PDP-contexts per Group

In conclusion we summed up the number of contexts and volume generated per group and compared the results. Using the same grouping as above we accumulated the volume and the

Table 12.9 Fraction of 'Users', 'Number of PDP-contexts' and 'Volume' divided into two groups: $n_i \leq 100, n_i > 100$ (TR$_2$).

Technology group	UMTS		GPRS	
	$n_i \leq 100$	$n_i > 100$	$n_i \leq 100$	$n_i > 100$
Number of users (%)	$\geq$99.9	<0.1	99.9	0.1
Generated PDP-contexts (%)	88.6	11.4	57.6	42.4
Total volume (%)	97.6	2.4	97.9	2.1

number of PDP-contexts for $n_i \leq 100, n_i > 100$. Table 12.9 shows how the total volume maps to the different groups and how many users are inside. It is interesting to see that for GPRS less than 0.1% of the users generate $\approx$42% of all PDP-context activity in the network but only 2.1% of the total volume. In UMTS the effect is smaller, however; more than 10% of the traffic is generated by less than 0.1%. The conclusion we have drawn from the numbers is that the group with $n_i \leq 100$ generates the main amount of the payload volume.

12.4 Detecting and Filtering of Malicious Traffic

The dataset TR$_2$ for this analysis was recorded in September 2005. Starting with the raw dataset, in Figure 12.11(a), we observed several problems. The first problem we detected was a high number of TCP connections with only one SYN packet. This is an indication for port scans and/or virus activities. A (computer) virus is a self-propagating program which reproduces itself by infecting other programs. Like a biological worm the program harms the resources of its host system. In contrast to a (computer) worm, a virus will not spread to other computers on its own. The so-called (computer) worm is a computer program, which is capable of spreading via networks from one infected host to many other computers, for example by sending out spam emails.

A worm program will try to use exploitations on other systems in the network in order to gain access to the system. Therefore, an infected computer will intensively scan its network to find other computers that can be infected. A worm will target typical services hosted on most computers (such as Server Message Block (SMB) protocol on Windows machines), therefore the number of connection setup attempts on such ports infers worm activity.

The fact that the TCP port for SMB is seen in the top ten in Figure 12.11(a) supports the hypothesis of a self-spreading worm performing port scans. More information on this topic can be found in Staniford *et al.* (2002). SMB is used by Windows-driven operating systems for sharing files, printers and communication interfaces. Some SMB implementations of Microsoft suffer from buffer overflows, which can be exploited by various worms and viruses. Taking a look at Figure 12.11(b) we find that SMB takes the major share of the connections. We therefore conclude that this is due to a virus spread and we set filter rules to eliminate this spurious traffic. At this point we should also mention that filtering undesired traffic can have a strong impact on the reliability of traffic prediction. For example, it can happen that the huge number of ignored TCP connections takes down some WEB-caching device located in the core network. As soon as we have filtered all TCP-connections with fewer than three

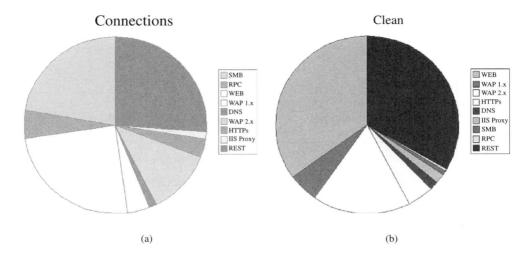

Figure 12.11 Number of connections in UMTS per service (TR_2): (a) unfiltered; (b) filtered result.

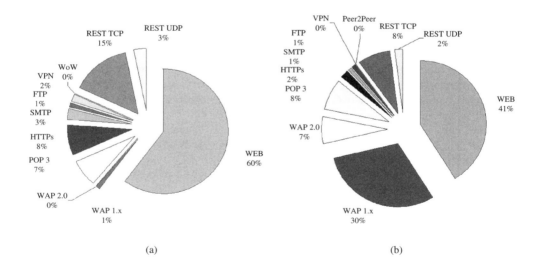

Figure 12.12 Volume service share (TR_2, filtered): (a) UMTS; (b) GPRS.

packets, this is equal to the number of packets for one TCP-handshake (RFC-793 1981), nearly 99.53% of the SMB traffic disappeared as depicted in Figure 12.11(b).

Table 12.10 Number of connections in UMTS sorted in terms of SYN packets (TR$_2$).

Protocol	Port	Description	Total number	Clean number	Clean/total %
TCP	445	MS-SMB	105 966 578	497 273	0.47
TCP	135	MS-RPC	25 068 717	256 533	1.02
TCP	80	HTTP	7 226 108	6 499 828	89.9
UDP	53	DNS	1 624 423	587 311	36.15
TCP	8001	WAP 2.0	1 663 524	1 593 378	95.78
TCP	443	HTTPs	799 929	707 273	88.41
TCP	4662	edonkey	292 737	204 920	70.0
ALL			154 234 535	14 650 613	9.50

The second port used by worms is the Remote Procedure Call (RPC) protocol. This port became infamous by the outbreak of the BLASTER-worm (Staniford *et al.* 2002). The filtering rules also had considerable impact on this port. Table 12.10 shows the impact of the filtering rules on the different services. We filtered for TCP $\geq$ 4pkts, and for UDP $\geq$ 2pkts.

We learned that valid DNS connections are affected by this rule, too. This indicates problems with the DNS resolution for some customers. A valid connection for an address resolution has to have at least two packets (RFC-2929 2000). Results for the volume per service can be found in Figure 12.12.

References

Dahmouni, H., Rosse, D., Morin, B. and Vaton, S. (2005) Impact of Traffic Composition on GPRS Performance. In *Proceedings of the 19th International Teletraffic Congress; Specialist Seminar Network Usage and Traffic (ICNP 96)*, pp. 1405–1409, Berlin, Germany.

Holma, H. and Toskala, A. (2004) *WCDMA for UMTS, Radio Access For Third Generation Mobile Communications,* John Wiley & Sons, Ltd.

IETF RFC 793 1981 (1981) Transmission Control Protocol (TCP).

IETF RFC 2929 2000 (2000) Domain Name System (DNS).

Kalden, R., Varga, T., Wouters, B. and Sanders, B. (2003) Wireless Service Usage and Traffic Characteristics in GPRS Networks, In *Proceedings of the 18th International Teletraffic Congress,* pp. 981–990, Berlin, Germany.

Pang, R., Yegneswaran, V., Barford, P., Paxon, V. and Peterson, L. (2004) Characteristics of Internet Background Radiation. In *Proceedings of the 4th ACM SIGCOMM Conference on Internet measurement*, pp. 27–40, Sicily, Italy.

3GPP TS 23.060 2003 (2003) General Packet Radio Service (GPRS), Service description, v.4.1.0.

Staniford, S., Paxson, V. and Weaver, N. (2002) How to Own the Internet in Your Spare Time. In *Proceedings of the 11th USENIX Security Symposium*, pp. 149–167, Berkeley, CA, USA.

Svoboda, P., Ricciato, F., Hasenleithner, E. and Pilz, R. (2006) Composition of GPRS/UMTS Traffic: Snapshots from a Live Network. In *Proceedings of the Fourth International Workshop on Internet Performance, Simulation, Monitoring and Measurement*, pp. 113–124, Salzburg, Austria.

Varga, T., Bert, H. and Sanders, B. (2004) Analysis and Modelling of WAP Traffic in GPRS Networks. In *Proceedings of the 16th ITC Specialist Seminar on Quality of Experience*, pp. 1–9, Antwerpen, Belgium.

13

Traffic Flows

Philipp Svoboda

In this chapter we analyse the network traffic of the Transmission Control Protocol (TCP) and User Datagram Protocol (UDP) flow level at the Gn interface of a mobile core network. We analyse, for each flow duration, size in up- and in downlink directions. A TCP or UDP flow can be viewed as the next smaller entity when compared with the PDP-context introduced in Chapter 12. A flow is, in the perspective of a network engineer, a set of packets transmitted by an application between two nodes.

Internet measurements carried out in Wide Area Network (WAN) environments showed heavy tail behaviour in the size of flows, especially of TCP flows (Giambene and Tommasi 2005; Leland *et al.* 1993; Morin and Neilson 1995; Park *et al.* 1996). Heavy tailed distributions show a relatively high probability in regions far from the mean or median. Section 13.1 introduces several basics such as the definition of heavy tails and TCP or UDP flows in the network. In the case of a UDP stream the definition of a flow is not unique, therefore it is important to clarify the usage of these terms (RFC 5103 2008). Section 13.2 introduces a set of methods to fit parameters. We apply these methods in the follow-up sections and fit numerous parameters from empirical data to known density functions. In section 13.3 we first depict the length of the flows. We then analyse the tails, far from the mean or median, and the body, close to the mean or median, of the empirical distributions. We repeat this analysis for several datasets recorded over three years.

13.1 Introduction to Flow Analysis

In teletraffic engineering the answer to the question of whether a flow has heavy tail properties is very important (Leland *et al.* 1993) since heavy tail distributions in queuing systems lead to unexpectedly high peaks, basically everywhere where means or variances appear in performance measures. However, for the performance of TCP the body of the distributions is

Video and Multimedia Transmissions over Cellular Networks   Edited by Markus Rupp
© 2009 John Wiley & Sons, Ltd

also an important input parameter for simulation setups. We therefore focus on the tails and analyse separately the body of the various distributions found in the network.

Short TCP flows have an impact on the performance of TCP as many retransmission strategies of this protocol require a certain amount of outstanding packets to work properly (RFC 2018 1996; RFC 3517 2003). In applications with flows of fewer than ten packets (for example WAP 2.0), retransmission events especially in a high RTT environment decrease the Quality of Service (QoS).

In wired networks short flows make up for a very small percentage of the traffic (Claffy *et al.* 1995). This is due to the fact that optimizing an application for wireless channels is equal to transmitting the data efficiently, for example by compressing the sent data. In a wireline network, applications are not optimized in this way, therefore the flows are much larger. Heavy tails are important for network performance of network elements, but short tails may be important for user-perceived performance in several applications.

13.1.1   Heavy Tailed

Heavy tailed distributions are closely linked to self-similar processes (Feldman and Whitt 1997). The 'heavy tail' refers to a property of some statistical distributions, for example Power law, Pareto and so on. It describes distributions that consist of two parts: a highly populated region and a low-populated one that converges asymptotically. Although the events in the tail have a very low probability they have a very large value and therefore impact the statistical moments of the distribution. In other words the tail converges slower than $1/x^2$, sometimes even slower than $1/x$. A distribution is called 'heavy tailed' if:

$$P[\mathbf{X} > x] \sim x^{-\alpha}, \quad \text{as } x \to \infty, \ 0 < \alpha < 2. \tag{13.1}$$

From this definition we learn that a heavy tailed distribution has an infinite variance and may even have an infinite mean value. We consider two cases: first, $0 < \alpha \leq 1$ and secondly, $1 < \alpha < 2$. In the second case the variance does not exist and in the first case also the mean value is infinite. To explain this fact we directly use the definitions of these two values. If the probability distribution of $\mathbf{X}$ allows for a probability density function (pdf) $f(x)$, then the expected value is computed as:

$$E(\mathbf{X}) = \int_0^a x f(x) \mathrm{d}x + \int_a^\infty x f(x) \mathrm{d}x = \mu; \quad \alpha > 0,$$

where the value of a is selected to split the body from the tail. If the integral for this part is infinite, its mean is also infinite. From equation 13.1 we can derive $f(x) \sim x^{-1-\alpha}$; the argument within the integral is $x \cdot f(x) \sim x^{-\alpha}$. Note that the integrals are starting at zero as we are working with positively distributed values only. In the first case the argument in the integral stays larger than x^{-1}. The result for the tail part of the integral will therefore be infinite. Since the variance depends on the mean, it cannot exist in case the mean is already infinite. In the second case the mean does exist. Its variance is given by:

$$Var(\mathbf{X}) = E[(\mathbf{X} - \mu)^2].$$

Inserted into the definition of the expectation value we obtain:

$$Var(\mathbf{X}) = \int_0^a f(x) \cdot (x - \mu)^2 \mathrm{d}x + \int_a^\infty f(x) \cdot (x - \mu)^2 \mathrm{d}x.$$

Again we only focus on the tail part which is for $a > 0$ the right integral. We are in the same situation as for the mean in the first case. Note that a measured sample will always have a mean and a variance; the tail parameter can only be estimated on a part of the distribution. This fact makes it very difficult to extract heavy tailed behaviour from measurement samples.

13.1.2 The Flow

In this paragraph we introduce the definition of a flow as we implemented it in our tracing modules. A flow in a network is the set of all packets exchanged between two nodes in the Internet which belong to one application. It does not have to be bi-directional, for example a push service will generate a uni-directional flow. The flow is described by three numbers: *size in bytes, number of packets, duration*. For each flow, the size and the number of packets is split into up- and downlink direction.

In a flow the uplink direction is defined from the client to the server. This is not always equal to the network definition of uplink which is defined from the User Equipment (UE) to the Gateway General Packet Radio Service (GPRS) Supporting Node (GGSN) node. Therefore, when the server is at the UE side the flow direction differs from the network direction.

The TCP Flow

The TCP protocol is connection oriented. Therefore, we can directly implement the definition of the TCP standard to define a connection in the tracing system. A TCP connection is defined via the quadruple: IP address and port for each source and destination.

The flow starts with a three-way handshake. The source sends a packet with the *TCP SYN* flag set. The destination replies with the *TCP SYN ACK* flag set. The handshake is finalized with a *TCP ACK* flag granting the receive of the last packet. Now the connection is established and the two nodes may transfer data. The flow ends with a similar procedure where the *TCP FIN* flag is set. In case of an out-of-order tear down, one node signals a *TCP RST* flag, closing the connection immediately. The analysis module in use has a small state machine following these procedures. The TCP protocol allows bi-directional data transfers. However, often the data is nearly uni-directional, only ACK packets flow in the downlink direction. To identify the flow direction we analyse the up- and downlink flow size and order it according to the larger value.

The UDP Flow

For UDP traffic we define a connection as the union of all packets featuring the same quadruple (source/destination addresses and corresponding ports) within the same PDP-context and with a maximum inter packet spacing of less than T where we followed Corvella and Taqqu (1999) which proposes $T = 10$ min.

This choice might be critical when comparing the two implementations of the same service adopting different transport protocols, for example WAP 1.x (over UDP) and WAP 2.0 (over TCP). We verified that the average number of WAP connections per user is approximately the same for both implementations: 1.3 and 1.4 for WAP 1.x and WAP 2.0, respectively. This supports our UDP connection definition. Note that in the literature there exist several other definitions for the UDP timeout (Claffy *et al.* 1995; Zhang *et al.* 2002).

The Application Flow

The definition of an application flow is important for the following analysis. Based on the knowledge of the destination port we can use the well-known ports, for example ports below 1024, to identify the involved service. For example, a TCP flow with a destination port 80 or 8080 is accounted as an HTTP flow.

This rule set works well where there is no software tunnelling its data over well-known ports to bypass firewall systems in place, as for example 'Skype' does (see Bonfiglio *et al.* 2007; Chen *et al.* 2006; Ronald and Dodge 2008; Rossi *et al.* 2008, for more details on the methods implemented in Skype).

An analysis of the flows, for the top ten services, showed that less than 0.01% of the flow was affected by software tunnelling through well-known ports in our recorded traces. However, this number may change in case the number of mobile Internet subscribers increases.

Hypertext Transfer Protocol (HTTP)   Flows from any port to the destination ports 80 and 8080 are accounted as HTTP traffic. In a first attempt we tried to run `tcptrace`[1] over our traces in order to obtain a more detailed picture of whether these flows really carry HTTP traffic. However, `tcptrace` was not able to cope with the huge amount of traffic numbers with which we were dealing. The actual load numbers in September 2006 were in the order of hundreds of gigabytes of traffic on the Universal Mobile Telecommunications System (UMTS) link for one day; this is far above the limit that `tcptrace` can process. Therefore, we stayed with the port based classification rule.

File Transfer Protocol (FTP)   The rule set for FTP is the same as for HTTP: all TCP flows with a destination port 20 are accounted as FTP traffic. The signalling port of FTP was ignored for our analysis. This is not a strong limitation as only a small amount of traffic is generated by the signalling part of this application.

Wireless Application Protocol (WAP) 1.X   The assignment for WAP 1.x flows was different. WAP 1.x runs on UDP traffic and therefore we have to use the flow definition from section 13.1.2 to estimate a UDP flow. In a WAP session the terminal connects with the WAP gateway, which then connects to the page requested by the browser of the terminal. Monitoring on the Gn link allows us to track the destination IP address of a request. If we find the address of the WAP 1.x gateway in the destination address field we account the traffic as WAP 1.x traffic. Since the WAP traffic can only work with this gateway we were able to identify 100% of the WAP 1.x traffic.

Wireless Application Protocol (WAP) 2.0   WAP 2.0 runs on top of a TCP connection and was treated similarly to WAP 1.x. Again the terminal connects with a WAP gateway, which then fetches and processes the data for the terminal. In our network there are different gateways for WAP 2.0 and WAP 1.x. Therefore, we are able to track these types of traffic via the destination IP addresses at the Gn link trace. The assignment is precise again.

---

[1] Tcptrace 6.6.1, available at http://www.tcptrace.org.

13.1.3 Protocol Shares

In this paragraph we analyse the number of connections sorted out by the specific protocols. We split the graphs into UMTS and GPRS to gain a better overview on the traffic. In section 12.2 we learned that the service mix, derived from the volume of each connection, on GPRS and UMTS is evolving towards a service share measured in wireline networks. This analysis focuses on the number of connections. Again, we filtered out all connections with fewer than three packets and also those with no packet in either up- or downlink direction.

Table 13.1 depicts the TCP/UDP share for UMTS and GPRS for different datasets. The share of TCP connections is around 90%. The value is stable over the last three years. Nevertheless there is a difference compared to traces from wireline networks (Thompson *et al.* 1997). The TCP/UDP connection share is nearly identical for UMTS and GPRS. This fact again indicates that subscribers use mobile Internet services.

Table 13.1 TCP/IP flows from 2005 to 2007.

| Date | Nov. 05 | Sep. 06 | Sep. 07 |
Averaged over	7 days	2 days	2 days
UMTS			
TCP (%)	86.2	92.4	88.0
UDP (%)	13.8	7.6	12.0
GPRS			
TCP (%)	87.4	91.0	87.2
UDP (%)	12.6	9.0	12.8

13.2 Fitting of Distributions to Empirical Data

In this chapter we require several techniques to fit our empirical data to analytical distributions. This section introduces the necessary steps. To fit an empirical dataset to an analytical distribution four basic steps are necessary. First the data has to be independent of a sample-by-sample basis and stationary without a trend. Secondly, we have to choose a class of analytical distributions which we think best fits our data. Thirdly, the parameters for the distribution are estimated. Finally, testing the goodness of the fit tells us if our choice at step two was correct. In case step four fails we have to restart at step two. For the following paragraphs the empirical sample is given by $X = \{x_1, x_2, \ldots, x_n\}$.

13.2.1 Pre-evaluation of the Dataset

Stationarity and Ergodicity

A stationary process has the same pdf for all times or positions. A stochastic process is called ergodic if its ensemble averages equal appropriate time averages. In our case we investigated for weak stationarity, considering only the first- and second-order moment of our samples.

The mean was calculated in a sliding window of size k:

$$X'(k) = \frac{1}{k} \sum_{i=k'}^{k'+(k-1)} x_i. \tag{13.2}$$

In order that the sample averages $X'(k)$ converge to a constant mean μ', equation 13.2 requires the process to be ergodic. We assume weak ergodicity from the nature of the underlying network (many subscribers, random radio and network conditions). Now we can define a variance σ' within the sliding window:

$$\sigma'^2 = \frac{1}{k-1} \sum_{i=k'}^{k'+(k-1)} (x_i - \mu')^2. \tag{13.3}$$

The mean was then visually inspected for trend components. None of our traces showed a trend component. We expected such a result, as our largest traces span only one week. From first results we learned that the heavy tailed regions start at a probability of 10^{-5}. Therefore, we need more than 10^5 samples. This is valid for all sets.

Serial Correlation

In a time series, where observations have a natural order in time, there may be a correlation between two successive values. The autocorrelation function (ACF) calculates the correlation for all possible time lags of the time series. It is the correlation between two values of the same sequence x_i and x_{i+k}. The normalized autocorrelation of a WSS sequence is defined as

$$r_k = \frac{E\{(x_i - \mu')(x_{i+k} - \mu')\}}{\sigma'^2}. \tag{13.4}$$

High values of the ACF indicate that values with this specific time lag are correlated strongly whereas for time lags with less correlation the ACF is also low. For white noise it can be shown that 95% of the sample autocorrelation values should lie between $\pm\sqrt{n}$ since the sample autocorrelation values are asymptotically normally distributed with zero mean and variance $1/n$. If the autocorrelations are below this boundary, the sequence is likely to be random, that is, white noise. For further reading see Schlittgen (1995) and Davis and Brockwell (2002). There are different notations to indicate independence; one way is

$$|r_k| \le \frac{\Phi^{-1}(1 - \delta/2)}{\sqrt{n}}, \quad k = 1, 2, \ldots, \tag{13.5}$$

where $\Phi(x)$ is the percent point function (Papoulis and Unnikrishna 2002), the inverse of the cumulative distribution function (cdf) of the standard normal distribution and δ is the significance level to assume independence.

13.2.2   Parameter Estimation

The parameter estimation is to extract an estimator. In the optimal case this estimator extracts all information from the input and is implementable with reasonable effort. The estimator takes measured data as an input and produces an estimate of the parameters. In the general case we used the Maximum Likelihood Estimator (MLE) to estimate the parameters of the distributions. If heavy tails were present in the input data we split the distributions into a body and a tail part. The body part was fit to several analytic distributions and can be used for large-scale simulations, for example to detect bottlenecks (Svoboda *et al.* 2007). The parameters for the tail part, which we define to be dominated by a heavy tail effect, were extracted with the scaling method (Corvella and Taqqu 1999).

Maximum Likelihood Estimation

The MLE is a method of calculating the best way of fitting a mathematical model to some measured data. The method was originally introduced by Sir R.A. Fisher in 1912 (Alderich 1997). Loosely speaking, given a set of data and a chosen underlying probability model, the MLE selects the values for the model parameters that make the data more likely than other values would do. For the normal distribution the MLE has a unique solution. However, in more complicated scenarios this may not be the case. An MLE estimator is defined in the following way: given a family of probability distributions D_Θ with the parameter Θ and a pdf f_Θ, we compute the probability density associated with our observed data, $f_\Theta(X(x_1, x_2, \ldots, x_n | \Theta))$. We want to find a fit for the parameter Θ. The likelihood function of Θ with given $x_1, x_2, \ldots, x_n$ is

$$L(\Theta) = f_\Theta(x_1, x_2, \ldots, x_n | \Theta). \tag{13.6}$$

The value of Θ is now estimated by maximizing the function $L(\Theta)$. This is the MLE of Θ, defined as

$$\hat{\Theta}_{\text{MLE}} = \arg\ \max_\Theta L(\Theta). \tag{13.7}$$

The problem simplifies considerably if we now assume that our dataset drawn from a particular distribution is independent and identically distributed. In this case the likelihood is a product of n univariate probability densities:

$$L(\Theta) = \prod_{i=1}^{N} f_\Theta(x_i | \Theta). \tag{13.8}$$

A monotone transformation does not affect the position of an extreme value. Therefore we only have to search the maximum of the logarithmic sum:

$$\log L(\Theta) = \sum_{i=1}^{N} \log f_\Theta(x_i | \Theta). \tag{13.9}$$

The maximum of this expression can be found by solving the partial differential equations

$$\frac{\partial \log L(\Theta)}{\partial \Theta} = \mathbf{0}. \tag{13.10}$$

Phase Type Approximation of an Empirical Distribution

The phase type distribution was introduced in Feldman and Whitt (1997). This method delivers an analytical tractable approximation of a general distribution. The paragraph will focus on hyper-exponential phase type distribution in order to approximate an empirical distribution derived from measurements. There are several different methods to estimate the parameters (Feldman and Whitt 1997). Here we introduce the EM-algorithm for PH approximation from Khayari *et al.* (2003). It is an iterative method which needs only the empirical data and has a complexity that grows in a linear way with the number of phases present in the model.

The following paragraph describes the EM-algorithm for phase type distributions (Khayari *et al.* 2003). Let the number of phases of a hyper-exponential distribution be I. There are three parameters in the algorithm: $c_{1,...,I}$, $\lambda_{1,...,I}$ and the precision ϵ defining the terminating condition of the iterative algorithm. For every phase let

$$p(x_j|\lambda_i) = \lambda_i \exp^{-\lambda_i x_j}, \tag{13.11}$$

$$p[x_j|(c_i, \lambda_i)] = \sum_{i=1}^{I} c_i p(x_j|\lambda_i). \tag{13.12}$$

Each iteration starts with $c_i = c_i'$, $\lambda_i = \lambda_i'$ and then computes

$$p(i|x_j, \lambda_i) = \frac{c_i p(x_j|\lambda_i)}{p[x_j|(c_i, \lambda_i)]}, \tag{13.13}$$

$$c_i' = 1 \bigg/ N \sum_{j=1}^{N} p(i|x_j, \lambda_i), \tag{13.14}$$

$$\lambda_i' = \frac{\sum_{j=1}^{N} p(i|x_j, \lambda_i)}{\sum_{j=1}^{N} x_j p(i|x_j, \lambda_i)}, \tag{13.15}$$

until $|c_i - c_i'| < \epsilon$ and $|\lambda_i - \lambda_i'| < \epsilon$. The resulting values for c_i and λ_i fully describe the phase type distribution.

The Scaling Method

An empirical distribution derived from Internet measurements typically consists of two parts: a body and a (heavy) tail. The Hill estimator (Hill 1975) offers a simple way to estimate the tail parameters. However, this method needs a starting point well away from the body of the distribution.

Another method of finding the slope α of the tail parameters was introduced by Corvella and Taqqu (Corvella and Taqqu 1999). It utilizes a special behaviour of variables with a heavy tail distribution. Let $\mathbf{X} = \{x_1, x_2, \ldots, x_n\}$ be a dataset and x_i^m the aggregated sums of non-overlapping blocks of observations of size m. If $\mathbf{X}$ has a heavy tail property, the distributional properties of x_i^m will also show heavy tail behaviour. This method aggregates the dataset of several different steps and then estimates for each step the slope parameter α. The heavy tail region is chosen on a comparison of the different α estimates at the different aggregation levels. In our analysis we applied the method of Corvella and Taqqu (1999), which was coded into a tool called AEST. This tool outputs an estimate for the slope α of the tail. Take, for example, a Pareto distribution given in the following equation:

$$f(x;\ k,\ x_m) = \begin{cases} \dfrac{k x_m^k}{x^{k+1}}, & \text{for } x \geq x_m > 0 \\[2mm] 0, & \text{else.} \end{cases} \tag{13.16}$$

The slope α equals $k + 1$ in this case.

13.2.3   Goodness of Fit

The last step is the verification of the estimated parameters. A goodness of fit describes how well a statistical model fits the observed data. A typical measure of goodness of fit summarizes the discrepancy between observed and expected values from the chosen model. We looked at two methods: the chi-square and the Kolmogorov–Smirnov (KS) test.

The chi-square test is more general in the application. It compares a sum of differences between observed and expected outcome frequencies. The goodness of fit is found by comparing the resulting value to the chi-square distribution. However, this test needs an input parameter set in advance, which has an impact on the result. If the data follows arbitrary distributions, setting this parameter is not trivial. Therefore, we use the KS test in this chapter.

The KS test determines whether two underlying one-dimensional probability distributions differ. We assume $F_n(x)$ to be an empirical distribution function of n independent and identically distributed observations of x_i. Next, we choose $F(x)$ to be the underlying distribution the dataset follows. Then the KS test is

$$D = \sup_x |F_n(x) - F(x)|. \tag{13.17}$$

A critical value for D can be defined. Note sup Z is the supremum of the set Z. Loosely speaking if Z is a subset of T then sup Z is the smallest element of T which is greater than or equal to all elements of Z. Based on the value of D the assumption that $F_n(x)$ is a member of $F(x)$ will then either be accepted or rejected. The value depends on the distribution and the level of significance chosen. For empirical measurements D is often above the critical value as the data does not originate from a single analytical distribution. Therefore, the value of D is used to find the distribution that best fits the measurement data.

13.3   Flows Statistics

In section 13.3.1 we compare the ecdfs of different datasets recorded between the autumn of 2005 and that of 2007. Section 13.3.2 validates the dataset with respect to independence and stationarity. After its validation, in section 13.3.3 we check for heavy tail behaviour in the dataset. After preprocessing we fit the datasets to analytic distributions.

13.3.1   Evolution of the TCP/UDP and Application Flow Lengths from 2005 to 2007

In this section we provide a first overview to our dataset. Figure 13.1 depicts the ecdf for all flows separately split into UMTS (left) and GPRS (right). Figures 13.1(c) and (d) show how the TCP flow size grew over three years, again for UMTS and GPRS, separately. Although the traffic amount did increase more than one order of magnitude between 2005 and 2007, we learned from the figures that the shape of the distribution did not change at all. In the GPRS case the distribution was stable between 2005 and 2006 and grew only slightly in 2007. Therefore we decided to analyse only one dataset in more detail. We chose the dataset TR_6 from September 2006 as it is right in the middle of our observation time period. The UMTS flows (Figure 13.1(a)) for WAP have a dedicated step at around 200 bytes. We assume that these steps originate from WAP chat applications which are very popular for younger subscribers. Often such steps in an empirical measurement cannot be modelled with a simple analytic distribution alone. We thus use the phase type approach instead. The UDP flows in UMTS are one order of magnitude smaller than the TCP flows. This result fits our numbers for the volume transported via UDP and TCP. It is interesting to note that the HTTP curve is nearly identical to the TCP curve. We assume that most of the TCP flows are in fact generated from HTTP or similar protocols such as HTTP Secure (HTTPS).

The results for the GPRS flows (Figure 13.1(b)) are similar to UMTS. Again there is a step of $\approx 13\%$ in the WAP distributions. The step in GPRS is smaller ($\approx 13\%$) than in the UMTS case ($\approx 30\%$). Note that the overall number of transferred bytes over WAP is one order of magnitude larger in GPRS; therefore, the smaller step includes more traffic than in the UMTS case. In GPRS there is a second step in the WAP distributions at around 20 kbyte. Based on manual inspection of a part of these samples, we assume that the step is caused by MMS services. The HTTP flows are again very similar to the TCP flows, with a slightly larger gap this time. Then, at first glance we assume that fitting TCP, UDP, HTTP is straightforward.

13.3.2   Example Validation of the Datasets

In this section we tested our dataset for stationarity and independence. We inspected all flows from the September 2006 dataset TR_6. The lack of correlation was tested with the ACF and an FFT analysis. For all flows the ACF stayed below a 1% significance level and the FFT plot did not show any dominating frequency. To test for a trend we analysed moving average plots with different levels of aggregation. The different moving average plots did not show any trend. Note that for the FFT we reduced the datasets to the peak hours of 7 p.m. to 11 p.m.; a full dataset of several days has a daily periodicity due to the lack of traffic in the night. None of the FFT analysis showed a dominant frequency.

In order to test the long time stationarity we split the trace into bins of 1000 s and calculated the ecdf for each bin. A Kullback–Leibler distance (Kullback and Leibler 1951) was used to evaluate the distance between the subsamples. It is a non-commutative measure to compare the difference of two probability distributions, for example $Q(x)$ and $P(x)$. The value of the metric is zero, if and only if $P(x)$ equals $Q(x)$ and is larger than zero in any other case. Our results are all below 0.1 indicating a very small difference between the curves. Therefore, we conclude that the flow distributions are valid over all the day. Note that in some traces only very low traffic occurs during the night hours so we excluded these bins from the investigation as they did not have enough events to produce a meaningful pdf.

Figure 13.2 depicts a sample of this analysis in the form of a scatter plot. The x-axis represents the time of day, the y-axis is the size of the flow. Both axes are binned and each bin is shaded according to the number of events that fell into it. The intensity bar is normalized

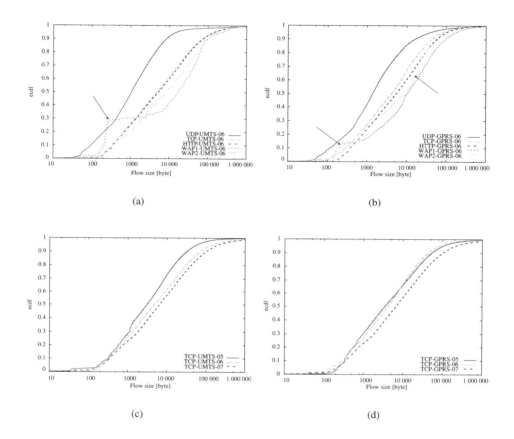

Figure 13.1 Flow length of all services and growth over the years 2005–2007: (a) All flows UMTS; (b) All flows GPRS; (c) TCP evolution UMTS; (d) TCP evolution GPRS.

for every 1000 s subsample in order to produce an empirical pdf. In other words, for each $x =$ constant the intensity information along the y-axis represents the pdf of the dataset.

13.3.3 Scaling Analysis of the Heavy Tail Parameter

In this paragraph we investigate the flows for heavy tail behaviour. We analysed TCP, UDP, HTTP, WAP 1.x and WAP 2.0 flows from the dataset TR_6 recorded in September 2006. In the following figure we depict on the right side the GPRS traces and on the left side the UMTS traces. The first group of Figures 13.3 depicts the observations for TCP and UDP flows. The highlighted parts of each curve (see arrow in Figure 13.3) are estimated heavy tail regions. A heavy tail distribution would be entirely marked. If only a part of the distribution is marked, this indicates the fact that the underlying distribution has two parts: a body and a (heavy) tail. A very small marked region should not be considered as a presence of heavy tail behaviour (Corvella and Taqqu 1999). For TCP the estimator delivers a α close to one,

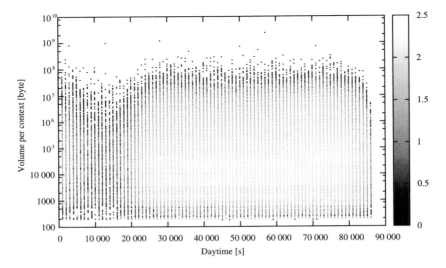

Figure 13.2 TCP flow size in 1000 s time bins plotted for one day (UMTS, September 2006 (TR$_6$)).

regardless of the technology. Both flows show clear indications of heavy tailed behaviour, while the UDP flows for GPRS and UMTS did not show heavy tailed regions.

The next group of Figures 13.4 depicts the scaling analysis for HTTP flow length. A large part of the distribution, especially the tail, follows the heavy tail assumption. The parameter α is estimated with $\alpha_{UMTS} = 1.04$, respectively $\alpha_{GPRS} = 1.13$. This is consistent with measurements obtained in the wired network. It is interesting to see that the results for GPRS and UMTS are similar, although the access technology of UMTS provides a higher data rate.

The last service tested was WAP. In Figure 13.5 we depict one example for WAP 1.x over UMTS. The regions are scattered along the curve. In such cases Corvella and Taqqu (1999) suggested that the distribution may follow a log-normal or Weibull but is not heavy tailed. The scaling method serves only as an indication for heavy tails but not as a rigorous proof. Visual inspection is used to decide from case to case.

The estimated values for α, estimated with the Hill estimator, can be found in the title of the graphs. As in the UDP case we consider these distributions free of heavy tails. From Figure 13.5 we learn that in 2006 WAP access originated mainly from GPRS terminals. Note that due to the small number of WAP terminals (Corvella and Taqqu 1999) the UMTS dataset contains fewer than 4000 events.

13.3.4 Fitting Flow Size and Duration

In this paragraph we search for an analytical distribution which best fits our measured data. First we applied fits for single analytical distributions to keep the modelling simple. In the previous section we identified heavy tails in the TCP and the HTTP flows. Therefore, we split the data for these flows into a body and a tail part. In the literature of wired Internet

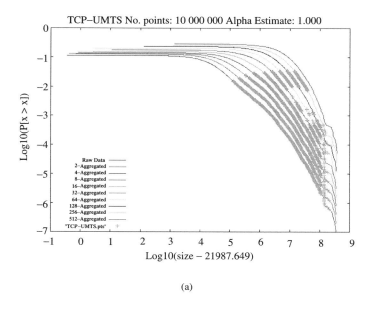

(a)

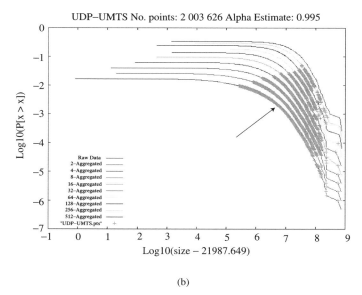

(b)

Figure 13.3 Scaling analysis for heavy tail regions in TCP flows (September 2006, TR_6): (a) TCP, UMTS; (b) UDP, UMTS.

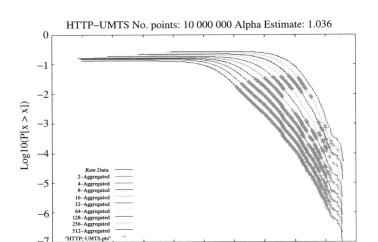

(a)

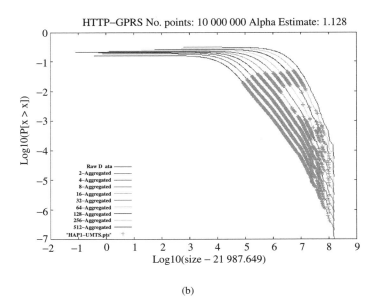

(b)

Figure 13.4 Scaling analysis for heavy tail regions in HTTP flows (September 2006, TR_6):
(a) HTTP, UMTS; (b) HTTP, GPRS.

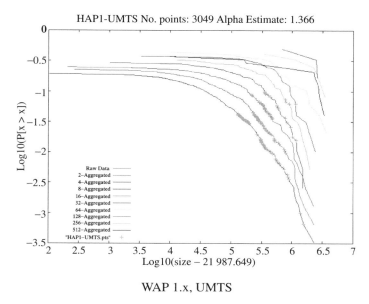

WAP 1.x, UMTS

Figure 13.5 Scaling analysis for heavy tail regions in WAP flows (September 2006, TR$_6$).

measurements (Corvella *et al.* 1998) a split at 10 kbyte is chosen. However, in our trace the heavy tail region starts at values that are one order of magnitude larger. We assume that this comes from the fact that nearly ten years passed between the measurements in the paper and our traces.

We used the MLE to fit the parameters of the analytical distributions. The MLE solution can often be calculated in one run; this was an important decision criterion as our sample sizes were very large, for example one day in September 2006 had more than 50×10^7 flows. In case one distribution did not fit the ecdf we fitted a phase type distribution. We implemented the EM algorithm in C++ to achieve a better performance figure.

Table 13.2 lists the best fits for the flows. First of all we have to mention that all analytical fits failed the KS test. None of the MLE solutions could fulfil the critical value of the KS statistics. Hence, we only present the analytical distribution with the best KS score. Note that we found that the MLE solution could be easily improved by manual adjustments to the parameters. The reason that the method performed so poorly is that the large range which the input values span forces the algorithm to fit the tail rather than the body. The results in Table 13.2 display the log-normal distribution which fitted best the non-heavy tail datasets, in respect to the body parts. This again is consistent with previous results from Internet measurements.

HTTP makes up for more than 60% of the TCP traffic; the graphs for HTTP and TCP are nearly identical. Therefore, we investigate solely the HTTP flows. The following block of figures presents the empirical complementary cdf (eccdf) of the HTTP flows with the log-normal fit added to the graph.

The four plots in Figure 13.6 display the eccdfs for HTTP flows. For each technology we plotted two graphs, one in loglinear, which shows the body and one in loglog, which

Table 13.2 Best fits of the empirical distributions in downlink (Sept. 2006, TR_6).

Flow	Distribution	Parameters		KS-stat
UMTS				
TCP tail	Pareto	$x_m = 100000$	$k = 1.1070$	0.0366
TCP body	log-normal	$\mu = 8.5085$	$\sigma = 2.0565$	0.0913
UDP	log-normal	$\mu = 6.8780$	$\sigma = 1.8541$	0.0832
HTTP tail	Pareto	$x_m = 100000$	$k = 1.0890$	0.0331
HTTP body	log-normal	$\mu = 8.5739$	$\sigma = 2.0200$	0.0933
WAP 1.x	PH-2	$\lambda_0 = 3.12 \times 10^{-4}$	$c_0 = 6.12 \times 10^{-1}$	—
		$\lambda_1 = 6.89 \times 10^{-5}$	$c_1 = 3.77 \times 10^{-1}$	—
WAP 2.x	PH-2	$\lambda_0 = 4.33 \times 10^{-4}$	$c_0 = 7.12 \times 10^{-1}$	—
		$\lambda_1 = 9.72 \times 10^{-5}$	$c_1 = 4.01 \times 10^{-1}$	—
GPRS				
TCP tail	Pareto	$x_m = 100,000$	$k = 1.2070$	0.0955
TCP body	log-normal	$\mu = 8.2979$	$\sigma = 1.9436$	0.1132
UDP	log-normal	$\mu = 7.1892$	$\sigma = 1.9011$	0.0832
HTTP tail	Pareto	$x_m = 100000$	$k = 1.2130$	0.0310
HTTP body	log-normal	$\mu = 8.5830$	$\sigma = 1.8763$	0.1132
WAP 1.x	PH-2	$\lambda_0 = 1.33 \times 10^{-4}$	$c_0 = 5.34 \times 10^{-1}$	—
		$\lambda_1 = 7.68 \times 10^{-5}$	$c_1 = 2.12 \times 10^{-1}$	—
WAP 2.x	PH-2	$\lambda_0 = 3.44 \times 10^{-4}$	$c_0 = 6.89 \times 10^{-1}$	—
		$\lambda_1 = 7.41 \times 10^{-5}$	$c_1 = 2.99 \times 10^{-1}$	—

focuses on the tail. The loglinear plots (Figures 13.6(a) and (c)) show that the log-normal distribution follows the body part well, if we neglect to take into consideration the lower end of the empirical data. From the loglog plots we can clearly identify that the tail cannot be fitted with a log-normal, which starts to decay much faster than the empirical data at around 10^5. This result is consistent with the estimation of the heavy tail region we derived with the scaling method of the AEST (Corvella and Taqqu 1999) tool. An interesting finding is the fact that the body parts of UMTS and GPRS, Figures 13.6(a) and (c), respectively, are very similar. We conclude that the body part of the distribution is linked to the protocol type rather than to the access speed of the user terminal.

13.3.5   Mice and Elephants in Traffic Flows

In section 13.3.3 we showed that the TCP flows in our trace are heavy tailed. A side effect of heavy tailed distribution is the presence of so-called mice (small flows) and elephants (very large flows). Many papers discuss this effect; see Brownlee (2002) for further references.

The mice generate the large part of the connections. However, the elephants generate the traffic in the network. The presence of elephant flows allows a strong simplification in simulation setups. First, only a small number of TCP connections have to be simulated and, secondly, these connections are long and therefore boundary effects (for example, slow start) can be omitted. Figure 13.7 displays the effect for the UMTS and the GPRS networks. Both plots show two cdfs. The curve labelled 'Size' is the ecdf of the flow sizes of the

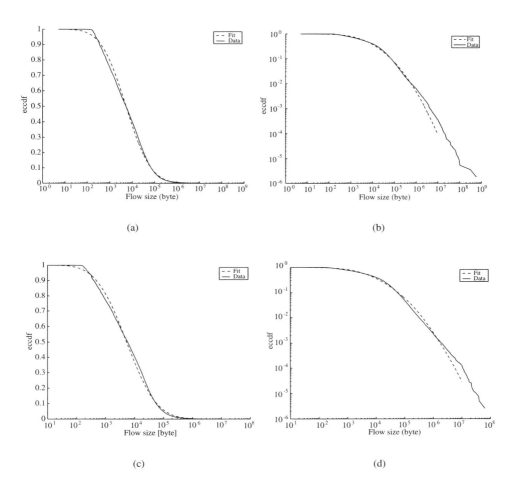

Figure 13.6 Fitting body and tail of the HTTP flows: (a) HTTP, linlog, UMTS; (b) HTTP, loglog, UMTS; (c) HTTP, linlog, GPRS; (d) HTTP, loglog, GPRS.

TCP connections. The curve labelled 'Total Volume' depicts the volume generated by all flows less than or equal to x. The UMTS cdf curve of the flows reveals that 99% of them are below 94.62 kbyte. However, the remaining 1% generates 63.2% of the total volume. In GPRS the 99% limit is 76.11 kbyte. Here, the remaining part accounts for 48% of the traffic. UMTS and GPRS have about the same upper bound for mice flows. Due to the higher data rate the elephants are larger in UMTS. In the WAP 1.x and 2.0 flows the elephants have a smaller impact of 22% and 19%, respectively. In this application the limited capability of the terminals does not allow for very large flows. As the distribution of the flow sizes of HTTP is very similar to those of TCP the results are also valid for HTTP flows. In a high RTT environment we cannot neglect the mice flows. A simulator will have to consider effects such as the slow start.

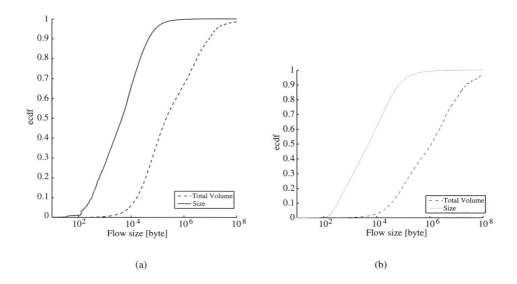

(a)   (b)

Figure 13.7 TCP flow size and originating volume: (a) UMTS; (b) GPRS.

References

Alderich, J. (1997) A. Fisher and the making of maximum likelihood 1912–1922. *Statistical Science*, **12**(3), 162–176.

Bonfiglio, D., Mellia, M., Meo, M., Rossi, D. and Tofanelli, P. (2007) Revealing Skype traffic: When randomness plays with you. *SIGCOMM Computer Communications*, **37**(4), 37–48.

Brownlee, N. (2002) Understanding Internet traffic streams: Dragonflies and tortoises. *IEEE Communications Magazine*, **40**, 110–117.

Chen, K.T., Huang, C.Y., Huang, P. and Lei, C.L. (2006) Quantifying Skype user satisfaction. *SIGCOMM Computer Communications*, **36**(4), 399–410.

Claffy, K.C., Braun, H.W. and Polyzos, G.C. (1995) A parameterizable methodology for Internet traffic flow profiling. *IEEE Journal of Selected Areas in Communications*, **13**(8), 1481–1494.

Corvella, M.E., Taqqu, M.S. and Bestavros, A. (1998) Heavy-tailed probability distributions in the World Wide Web, a practical guide to heavy tails. *Statistical Techniques and Applications*, 3–25, 1998.

Corvella, M.E. and Taqqu, M.S. (1999) Estimating the heavy tail index from scaling properties. *Methodology and Computing in Applied Probability*, **1**, 55–79.

Davis, R. and Brockwell, P. (2002) *Introduction to Time Series and Forecasting,* Springer, New York.

IETF RFC 2018 1996 (1996) TCP Selective Acknowledgment Options.

IETF RFC 3517 2003 (2003) A Conservative Selective Acknowledgment (SACK)-based Loss Recovery Algorithm for TCP.

IETF RFC 5103 2008 (2008) Bidirectional Flow Export Using IP Flow Information Export (IPFIX)

Feldmann, A. and Whitt, W. (1997) Fitting Mixtures of Exponentials to Long-tail Distributions to Analyze Network Performance Models. In *Proceedings of 16th Annual Joint Conference of the IEEE Computer and Communications Societies. 1997 (INFOCOM'97)*, vol. 3, pp. 1096–1104.

Giambene, G. and Tommasi, C. (2005) 3G Wireless Core Network Dimensioning in the Presence of Self-Similar Traffics. In *Proceedings of the Second International Symposium on Wireless Communication Systems*, pp. 223–227.

Hill, B.M. (1975) A simple general approach to inference about the tail of a distribution. *The Annals of Statistics*, **3**, 1163–1174.

Khayari, R.E.A., Sadre, R. and Haverkort, B.R. (2003) Fitting World-Wide Web request traces with the EM-algorithm. *Special Issue: Internet Performance and Control of Network Systems: Performance Evaluation*, **52**(2–3), 175–191.

Kullback, S. and Leibler, R.A. (1951) On information and sufficiency. *The Annals of Mathematical Statistics,* **22**(1), 79–86.

Leland, W.E., Taqqu, M.S., Willinger, W. and Wilson, D.V. (1993) On the Self-similar Nature of Ethernet Traffic. In *Proceedings of SIGCOMM 1993*, vol. 1, pp. 183–193, San Francisco, California.

Morin, P.R. and Neilson, J. (1995) The Impact of Self-similarity on Network Performance Analysis. *Technical Report*, Carleton University.

Papoulis, A. and Unnikrishna, P.S. (2002) *Probability, Random Variables, and Stochastic Processes,* McGraw-Hill.

Park, K., Kim, G. and Corvella, M. (1996) On the Relationship between File Sizes, Transport Protocols, and Self-Similar Network Traffic. In *Proceedings of the 1996 International Conference on Network Protocols (ICNP '96)*, IEEE Computer Society, Washington, DC, USA.

Ronald, C. and Dodge, J. (2008) Skype Fingerprint. In *Proceedings of the 41st Annual Hawaii International Conference on System Sciences (HICSS '08)*, pp. 484–491, Hawaii, USA.

Rossi, D., Valenti, S., Vglia, P., Bonfiglio, D., Mellia, M. and Meo, M. (2008) Pictures from the Skype. In *Proceedings of the ACM SIGMETRICS Demo Competition*.

Schlittgen, R. (1995) *Zeitreihenanalyse,* Oldenbourg.

Svoboda, P., Ricciato, F. and Rupp, M. (2007) Bottleneck footprints in TCP over mobile internet accesses. *IEEE Letters of Communications*, **11**(11), 839–841.

Thompson, K., Miller, G. and Wilder, R. (1997) Wide-area Internet traffic patterns and characteristics. *IEEE Network*, **11**(6), 10–23.

Zhang, Y., Breslau, L., Pacson, V. and Shenker, S. (2002) On the Characteristics and Origins of Internet Flow Rates. In *Proceedings of the Conference on Applications, Technologies, Architectures, and Protocols for Computer Communications (SIGCOMM '02)*, vol. 32(4), pp. 309–322, ACM, New York, NY, USA.

14

Adapting Traffic Models for High-delay Networks

Philipp Svoboda

In this chapter we present application-specific traffic models. Network traffic is obviously driven by users interacting with applications. Therefore, these models emulate the user behaviour at the application layer of the network model. The first section 14.1 gives a motivation as to why we try to model user behaviour on the application level. Section 14.2 presents new parameters for HyperText Transfer Protocol (HTTP) models. We use a model from Choi and Limb (1999) and fit the parameters according to our measurements. The same approach is followed in section 14.3 for the FTP model. Section 14.4 presents two different models for email traffic. For POP3 we propose a new model that extends older models with a login process as an important add-on for high-delay networks. Secondly, we present a Simple Mail Transfer Protocol (SMTP) model similar to FTP.

14.1 Motivation

In teletraffic engineering we can choose different layers of the protocol stack in which we want to model our traffic. Moving to lower layers reduces the parameters we have to model, for example the Medium Access Control (MAC) layer can be modelled by the packet size and rate. However, such models cannot answer questions that are related to the application layer of the network, for example how does the traffic change if a bottleneck is present?

Due to the small number of parameters such models are favoured in simulation scenarios with many nodes. We presented such an approach in Chapter 13 where we investigated the distribution of the TCP/UDP flows. Simulating the traffic on top of TCP/UDP flows allows simulation scenarios that investigate interaction at layer three of the OSI model, for example TCP retransmissions or TCP RTT change in the presence of a bottleneck

Video and Multimedia Transmissions over Cellular Networks   Edited by Markus Rupp
© 2009 John Wiley & Sons, Ltd

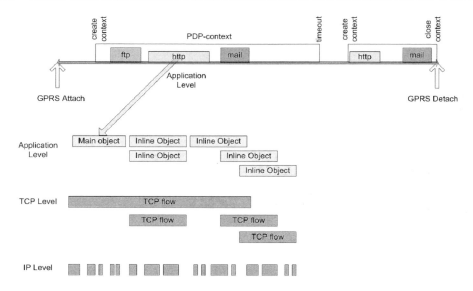

Figure 14.1 Traffic modelling on different layers of the network.

(Ricciato *et al.* 2007). However, interaction at the user level can only partly be mapped onto TCP/UDP properties. For example, it is not possible to simulate changes in the application such as an increase of the mean webpage size. These events can only be evaluated with a model which acts at the application level. Figure 14.1 displays three different layers to simulate the traffic.

In this chapter we present models for the top services observed in our network. For applications where appropriate models exist, for example HTTP and FTP, we fitted the parameters according to our mobile network. Where the models did not meet our needs, we improved them, for example POP3, or for new applications such as online gaming we introduced new traffic models.

Advantage of Traces from a Cellular Mobile Network

Constructing a traffic model on the application level is often difficult. The main drawback to such models is the fact that some parameters of the models have to be extracted on a per user basis. For example, we know from the service mix that there are 10 Mbyte of FTP traffic per hour and that there are 10 000 subscribers in the network. If we had no knowledge of the number of subscribers accessing FTP services, the resulting average FTP traffic per user would be 1 kbyte/h. Although this example exaggerates the problem, a good application model needs additional information on how the service usage is distributed among the user population. Figure 14.2(a) illustrates the idea of an average user. In section 12.2 it was shown that the service usage per user can be divided into three large groups: Wireless Application Protocol (WAP) only, WAP and email, and Internet services (no WAP). Figure 14.2(b) visualizes this approach. There are two ways to collect these parameters: a modified application software logging the data or an Internet access where each user has to

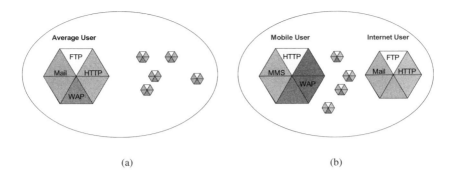

Figure 14.2 HTTP model measurements: (a) identical service mix for all users; (b) individual service mix for each user.

authenticate himself. A measurement based approach on a modified application software is limited due to the limited number of participants. However, in our case every IP packet, in fact every GTP packet, can be assigned to the subscriber. Therefore, we are able to track and extract the parameters on a user base. We therefore also investigate the distribution of the service call rate for the different traffic sources.

14.2 Modelling HTTP Browsing Sessions for the Mobile Internet Access

The HyperText Transfer Protocol (HTTP) is a protocol to transmit data at the application layer between hosts. The protocol was designed in 1989 by Tim Berners-Lee at CERN in combination with the Uniform Resource Locator (URL) and the HyperText Markup Language (HTML). It is a communication scheme to transmit data units which are parts of websites in the World Wide Web (WWW) and defined in the RFC 2616 (RFC 2616 1997). HTTP is a stateless protocol allowing asynchronous connections between client and server. It needs a reliable connection to transmit data. Mainly TCP is used for this purpose although it can also run on other reliable protocols. The data on the server side is encoded using a markup text format called HTML or eXtended Markup Language (XML). The URL interconnects the objects embedded within one webpage. The basic operation between two hosts follows a request/response scheme as shown in Figure 14.3. The client establishes a connection with the server and sends a request to the server (Get) including additional information about protocol version, client capabilities and modifiers. The server responds with a message (Reply) about his status including server capabilities, possible error codes, page content and the main object, which is the main webpage including all URLs necessary to fetch all embedded objects on this page. If there are embedded URLs in the main object, the client starts another Get message subsequently for all objects on the page. This process stops when the last embedded object has been downloaded by the client and the page transfer is now complete.

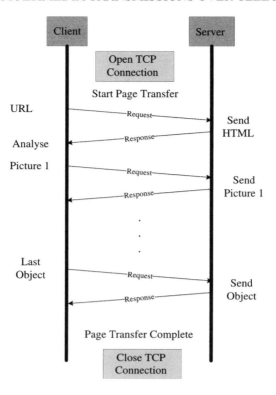

Figure 14.3 HTTP transfer of a webpage (persistent connection, HTTP 1.1).

Up to now there have existed two different versions of HTTP 1.0 and HTTP 1.1; the latter is a down-compatible extension of the previous version. In HTTP 1.0 the client/server relation can only set up separate connections for every request. In this case a client creates a new TCP connection for each object request separately, for example images on the page, closing it right after the response is received. The TCP protocol is not optimized for this kind of data transfer. In fact the slow start mechanism will harm the performance of such protocols. As one of the main advantages HTTP 1.1 offers so-called persistent connections. In such a connection the client can request multiple objects at once. Hence, there is only one open TCP connection for the whole webpage. The drawback to this method is the fact that there are more open connections that must be handled by the servers.

The second performance improvement in HTTP 1.1. is pipelining. This feature enables the client to request multiple objects without waiting for the response from the server. In combination with the persistent connection this feature fills the available resources much more efficiently. In general the HTTP header may hold optional information not standardized, which allows special applications to implement modified data communications.

14.2.1 HTTP Traffic Model

We expected the HTTP model parameters to be different for UMTS and GPRS. Therefore, we collected the values separately. To extract the application level data we used the `tcptrace` tool (Osterman 2005) with the add-on HTTP-parser. Although the huge size of our traces sometimes led to program crashes, we were able to extract a useful number of user data parsed. A shell script parsed the HTTP data and dumped the session parameters: reading time, objects per page, pages per session, object size, page size, download time and so on. The results were then fitted and plotted. In the following sections we show the results for UMTS and GPRS separately to distinguish the RAN settings.

Publications by Stuckmann (Stuckmann 2003; Stuckmann and Hoymann 2002) on this area reuse the typical traffic models, with the standard parameters, from Barford and Corvella (1998), Choi and Limb (1999), Mah (1997) and Villy *et al.* (2000). Therefore, we compare our measurements directly with these models.

In this section we applied a filter rule at the user level. Users that generated fewer than 100 kbyte over the whole trace and the top 1% were removed from the dataset. Figures 14.4(a) and (b) show the ecdfs for the page and the object parameters extracted from the traces. They show a solid curve for GPRS and a dashed line for UMTS and the dotted lines are the fits obtained for the parameters.

Comparing these figures we can reveal two facts concerning the different access technologies. First, UMTS users tend to visit larger HTTP pages (mean in UMTS $\approx$ 6.2 kbyte versus mean in GPRS $\approx$ 2.1 kbyte) – see Figure 14.4 (a.2) – than GPRS users. This can also be seen in the ecdf of the object sizes in one page, Figure 14.4 (b.2). Secondly, the download time of the UMTS HTTP pages is the same as for the GPRS pages. This point is important because it could indicate a hard limit on the user side, for example a maximum tolerance time a user will wait for a page. The Radio Access Network (RAN) in UMTS is approximately three times faster than in GPRS; this manifests itself in a measured difference of the mean HTTP page size.

Figure 14.4(a) shows the number of objects per page as an ecdf. It is interesting to see that there are over 50% of requests resulting in a page with only one object. In the literature (Reyes *et al.* 1999) a gamma distribution with mean 2.5 is proposed for the objects per page, which is higher than we have measured. To clarify this misfit we analysed the raw traces. We found many automatic program updates. For example, anti-virus or Windows updates were using the HTTP protocol. These programs accessed only one large object – the update file. In addition to this, many HTTP requests originated from cache updates in the network. We can only guess on this topic as, in order to protect the privacy of the subscribers, we are not allowed to investigate the data at full length.

Figure 14.5 displays the parameter reading or thinking time. We define the 'reading time' as the time span between the arrival of the last packet of the previous page and the first packet of the new page. Comparing UMTS and GPRS we found that this parameter does not depend on the RAN. The values in Figure 14.5, which were fitted with a Maximum Likelihood Estimator (MLE) and some manual fine tuning, depict similar results as found in Mah (1997). The curve has two regions: a steep start indicating fast hopping from page to page and a long tail for the largest 20%. Only every fifth reading time is above 18 s, therefore we conclude that users read fewer than 20% of the pages to the full extent.

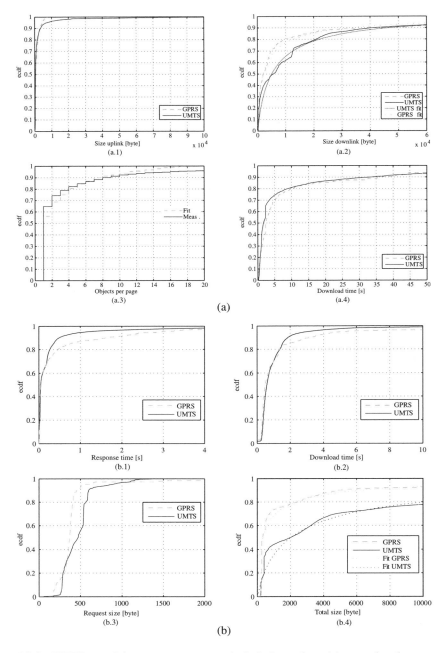

Figure 14.4   HTTP model measurements and their best fits: (a) page level parameters; (b) object level parameters.

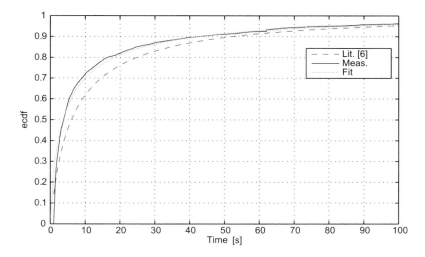

Figure 14.5 HTTP reading time.

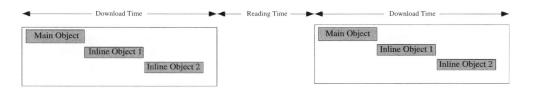

Figure 14.6 HTTP browsing session.

Fitting the Model Parameters

In this paragraph we present the fittings for our dataset and compare it to the distributions found in the literature (Mah 1997; Paxson 1995). Figure 14.6 depicts the model of an HTTP application. The page is constructed from one main object and several inline objects, for example images. After the transmission the user will read the page indicated by the reading time between two pages.

For this evaluation we analysed the HTTP traces of about 1000 users, part of the TR_1. Due to the small size of our dataset there would certainly be no heavy tailed behaviour visible. Note that the observation of a heavy tailed behaviour needs a large dataset, for example if we want to detect events with a probability of larger than 10^{-6}, we would need at least 10^6 samples. Therefore, we fitted the parameters to different distributions and chose the best fit based on the KS-statistic as we did in Chapter 13.

Table 14.1 displays the best fits for the parameters of the HTTP model for UMTS and GPRS. Here, in contrast to the results in section 13.3.4, we see a clear difference between the fitted parameters for UMTS, respectively GPRS. As this special analysis is based on data from autumn 2004, a time when UMTS was deployed for only a few months, we conjecture,

Table 14.1 HTTP parameter fittings.

		Parameters	
Model parameter	Distribution	μ	σ
UMTS			
Reading time	log-normal	1.620	1.501
Size main object	log-normal	8.530	1.743
Object size	log-normal	5.830	2.270
Number of objects	gamma	5.614	213.4
GPRS			
Reading time	log-normal	1.631	1.544
Size main object	log-normal	7.600	2.194
Object size	log-normal	7.620	1.503
Number of objects	gamma	5.321	246.2

based on the results from section 13.3.4, that for actual results the parameters for UMTS and GPRS are identical.

The fit for the reading time parameter delivers smaller results than found in Choi and Limb (1999) in both technologies. We assume that multi-tabbed browsing, one browser transfers several pages in parallel, affects this parameter toward lower values. Although tabbed browsing was not widely in use at that time, we have to note the fact that this analysis was performed for the top users in the network, not for an average or sporadic web browsing customer. To verify the numbers we extended the metric 'Gn-duration' so that it reports page requests starting in parallel to active page downloads. Visual inspection showed a similar result from this data source. For both technologies the best fit was achieved with a log-normal distribution. The size of the main object and the object size also followed a log-normal distribution. The GPRS and UMTS fits for the object size differed widely. We assume that UMTS subscribers tend to download larger objects within the HTTP pages visited, for example embedded streams or images. In the following section we focus on the difference between our measured distributions as compared to those from the literature.

Comparing the different parameters

We now investigate how the parameters of the various models differ in detail. In addition we add some information taken from the top 1000 sites in the Netbooster log-file extracted with the IE_Control tools (Svoboda 2007; Svoboda *et al.* 2006). The Netbooster is an HTTP caching device located at the Gi interface. Modifying the IE_Control tool delivered the following information: number of inline objects and size of main object. Figure 14.7 shows the reading time (a), the size of the main object (b) and the number of inline objects (c) for the different models.

From Figure 14.7(a) we can conclude that the reading time in the UMTS network behaves like the SURGE model. The HTTP model from Choi and the model in the 802.20 standard do not fit the dataset. If only the size of the main object, depicted in Figure 14.7(b), is considered, the measurements follow the model of Choi. The best fit for the parameter of the SURGE

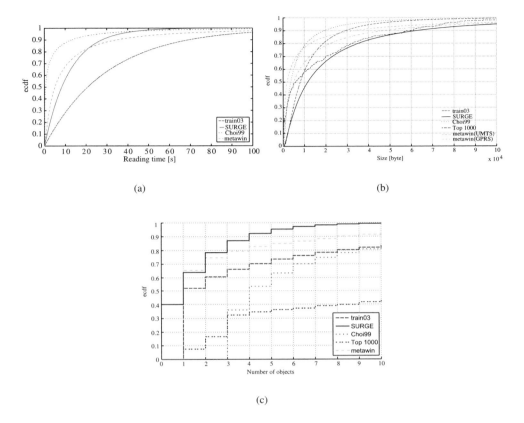

Figure 14.7 Comparison of HTTP-model parameters: (a) reading time; (b) size of the main object; (c) number of inline objects.

model still showed a big error when compared to the empirical measured data. We assume that the distribution does not fit our measured dataset.

14.3 Modelling FTP Sessions in a Mobile Network

The File Transfer Protocol (FTP) allows bulk transfers of large files through the Internet. FTP is a network protocol defined in the RFC 959 (RFC 959 1985). The RFC states the following objectives for the use of FTP: sharing of files, indirect or implicit use of remote computers, hiding of variations in file storage systems from the user, transferring data reliably and efficiently. It supports exchange of files over any TCP/IP based network as well as manipulations of files at the server, for example deleting or listing files. It uses exclusively TCP. An FTP server will by default listen at port 21 for incoming clients' connection attempts. The client is granted access to the file system of the remote host in an operating

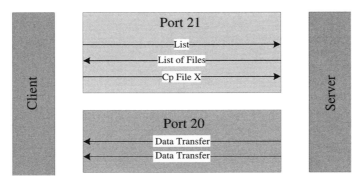

Figure 14.8   FTP file transaction in active mode.

system style manner. It is possible to download and upload, respectively, several files within one session.

FTP implements out-of-band control, thus data and signalling is transmitted over different connections as shown in Figure 14.8. An FTP client will access the server at port 21 to browse the file system. The download itself will take place in a second TCP connection. In active mode the FTP client opens a port, randomly chosen from above 1023, and sends the FTP server this number waiting for the server to initiate the data transfer. The server will open port 20 and connect to the client port starting the data transfer.

FTP allows for two different data formats: American Standard Code for Information Interchange (ASCII) and binary. In the binary mode a transaction takes place bit by bit. This mode is used for program binaries or data that is not plain text. For plain text files the FTP application supports ASCII transfers. In this mode the ASCII code of a symbol is transmitted between server and client. This enables the client to translate text to the local ASCII settings, for example replacing an end-of-line code from a Unix server with the symbol for Windows computers.

The original FTP definition has some drawbacks such as clear text passwords, high latency and no integrity checking. In 2007 FTP transfers are largely replaced by HTTP, for example web mail applications contain parts of HTTP and FTP services.

14.3.1   Modelling FTP Sessions

The data transfer of an FTP session can be interpreted as a single large TCP flow. Therefore, we propose a very simple model for FTP transfers which is depicted in Figure 14.9. The application has two states corresponding to 'on' and 'off'. In the 'off' state the application is inactive and remains until a timeout expires. After the timeout the application enters the 'on' state. In this state a file transfer via TCP is initiated, the size of the file is derived from a distribution fitted to the empirical data we extracted from our traces. Such a model can be found in many papers (Graja *et al.* 2001; Staehle *et al.* 2000).

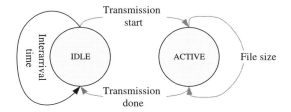

Figure 14.9 State diagram for an FTP traffic model.

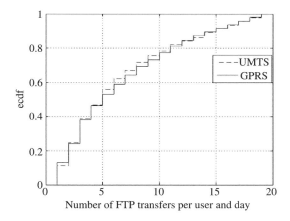

Figure 14.10 FTP service calls per user and day.

14.3.2   Fitting the Parameters

The FTP model we propose has the following parameters: file size, service calls per user and number of users. We used the traces from September 2006 to evaluate the parameters. The best MLE fit for the file size was log-normal ($\mu = 8.5534, \sigma = 2.112$). This is a similar result to that found for the TCP connection in section 13.3.4.

A scaling analysis revealed no clear signs of a heavy tailed behaviour. This result is in contrast to the literature (Stuckmann 2003). However, the FTP share in the network is very small and therefore the sample size is not significant enough. Note that the resolution of an empirically generated cdf cannot exceed the reciprocal of the events in the sample; for example given 10 000 measured events the smallest step, equal to a unique event x_{unique} at a specific value, is $P(x_{unique} = 10^{-4})$. The share of FTP users among the subscribers was 2.43%. The service call pattern per user is depicted in Figure 14.10. Again the service call distribution is independent from the access technology. The best MLE fit for both curves is an exponential distribution with $\mu = 6.542$.

Summarizing the FTP model we have found out that only a small part of the subscribers accesses this service and that the file sizes we obtained did not show signs of heavy tails.

14.4   Email Traffic Model: An Extension to High-delay Networks

In this section we introduce a traffic model for the email service POP3 in a mobile environment. As shown in Chapter 12, this service is in the top three of the service share. Although it generates smaller traffic loads than web browsing does, most business customers rely upon a proper function of this service. In contrast to wire line connections 3rd Generation (3G) networks impose a high delay in their links. Therefore, a proper modelling of this service needs a different approach than in the low-delay case. In a low-delay scenario a login process can be omitted and the email transfer is reduced to some kind of push service only.

Trace TR_5 used in this section consists of four different one-day periods in April 2007 that were recorded on the Gn interface. Measurements in a live network are likely to encounter an impact generated from some heavy hitters. So-called heavy hitters are users generating the main amount of traffic. Normally, this user group holds only a small part of the population. However, these users still characterize many measured pdfs. Also, in this case we found that more than 20% of all email requests originated from fewer than 0.25% of the users. We filtered out this group in a preprocessing step. The detection was implemented by marking all IMSIs as 'bad' which had more than 500 requests in six hours in any of the traces. This limit is according to public rules for DNS blacklists. More than 99% of the users had fewer than 80 requests in the full week. This rule excluded fewer than only 0.32% of the users from the traces, but astonishingly more than 25% of the emails.

14.4.1   Email Protocols of the Internet

In the Internet the following three protocols are used to transmit most of the email traffic: Simple Mail Transfer Protocol (SMTP) (RFC 2821 2001), Post Office Protocol v.3 (POP3) (RFC 1939 1996) and Internet Message Access Protocol (IMAP) (RFC 2060 1996). SMTP is a protocol to send emails between local systems. The protocols POP3 and IMAP were designed for clients who only sporadically connect to the email server. Emails are stored on the server and transferred to the client on demand. All protocols rely on the TCP protocol for an error-free end-to-end delivery.

Simple Mail Transfer Protocol

SMTP is a text-based, simple protocol which can send a message and other objects to one or more recipients. The user's client will look up his SMTP server in the local configuration and deliver the message to this server. An SMTP server, by default, listens on port 25 for incoming connections. TCP is used underneath as a transport protocol. The first SMTP server will then look up the eMail eXchange (MX) Domain Name Service (DNS) entry of each recipient's domain name and will then forward the message to the SMTP server of the recipient. SMTP only features push deliveries of email. To pull messages from a server the recipient has to use protocols such as POP3 or IMAP. Figure 14.11 depicts this process.

The original SMTP has no authentication functions for the sender as every user could use any SMTP server to send his messages. Therefore, spam emails are a massive problem for this server. There are some extensions to the original SMTP standard, such as STMP-AUTH. However, they are impractical and therefore not implemented. Some operators only allow

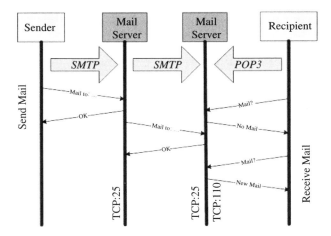

Figure 14.11 Mail transport between two users.

their users to access the SMTP server to minimize the problem of spam attacks. Although there are strong shortcomings in this protocol there is no real competitor for delivering email through the Internet.

Post Office Protocol v.3

The first version of POP was defined in reference RFC 918 (1984). In 2007, clients implemented the third version of POP, called POP3, which is defined in the reference RFC 1939 (1996). The protocol allows for end users with dial-up connections to retrieve emails from servers; see Figure 14.11. POP3 definition intends that the user downloads all available messages at once and deletes them from the server thereafter. There is an option to leave messages stored on the server. However, most POP3 commands identify the messages by their number only. That causes problems if someone deletes an email manually at the server. To avoid this, the Unique Identification Listing (UID) command is used. This command adds a unique ID to the message at the server side. The client can now distinguish new from old emails. In contrast to SMTP, the POP3 application needs authentication. A standard server listens on TCP port 110 for incoming connections. After a session the TCP connection is closed; several emails may be downloaded with the same connection.

Internet Message Access Protocol

The basic operation of the Internet Message Access Protocol (IMAP) is similar to POP3. It is an application layer protocol. The server operates on TCP port 143. The current version of IMAP is IMAPv4r1; it is defined in references RFC 2060 (1996) and RFC 3591 (1999).

IMAP offers more sophisticated functions to manipulate emails. While POP3 was designed to download emails from the server and read them off-line, IMAP also supports an online mode, which leaves messages on the server until the user deletes them. IMAP supports connected and disconnected operations, offering more flexibility. In POP3 only one client at a time can be connected to a mailbox, while IMAP supports simultaneous interaction of multiple clients to the same mailbox. In addition to this, IMAP provides message state information. Within POP3 this information is locally generated by client interaction. Some email manipulation can be executed directly on the server side, for example searching for keywords.

In spite of the advantages of IMAP there are also several drawbacks. IMAP is much more complex than POP3 and users may consume more resources on the server side, for example via a message search.

14.4.2   A POP3 Email Model for High RTT Networks

Email models in the literature cover only the email message size (Brasche and Walke 1997; Paxson 1995). However, in order to receive his email, a user has to authenticate his account with his password. This login process transfers only a small amount of data and is neglected in common models. We decided to use our measurement data to build up a model for email traffic better suited for mobile Internet access technologies. Figure 14.12(a) explains the principle idea behind these thoughts. The aim is to separate the login procedure from the email download itself. This has two advantages: first, we can model the different payloads on the IP-level separately (login = small payload, download = large payload) and, secondly, we can make use of the model to estimate a Mean Opinion Score (MOS) value for this service. To estimate the MOS it is important that the login process is a well-defined procedure that can be mapped to a number of small packets travelling in up and down directions. A high RTT, as in GPRS, will then result in a lower service experience.

Extracted Parameters   For the first step of the extraction process we evaluated the number of users connecting to the email services. Due to the NDA restrictions we cannot disclose the absolute numbers. However, we are allowed to present normalized values. In the measurement period 22.1% of the customers used email. This is consistent with the results we obtained in section 12.2. Note that in this context we calculated the fraction of active subscribers to email users. An active subscriber has to have at least one PDP-context, meaning that only data active users were counted. Figure 14.12(b) presents the state model for the email model using the introduced parameters.

Service Request Rate   Next, we evaluated the number of service calls generated per subscriber per day. Figure 14.13 shows the ecdf of service calls per user. As in the previous sections we see that both technologies, GPRS and UMTS, represented by the solid and the dashed lines, respectively, are very similar to each other. This indicates that the email service usage is decoupled from the technology. Further investigation in this topic showed that other parameters such as the email size do not differ between the technologies. Therefore, we did not extract the following parameters for GPRS and UMTS separately. Considering a request-generation process combined for all users in the network we extracted the inter-arrival time between two consecutive service calls. The parameter for 1000 active users in the busy hour follows an exponential distribution with the rate $\lambda_r^{-1} = 0.01102$. We implemented the service generation using an exponential distributed rate.

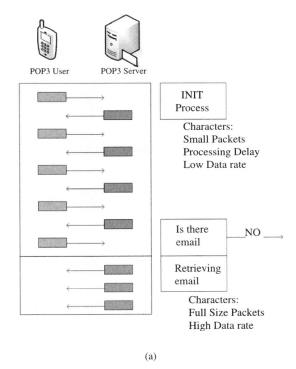

(a)

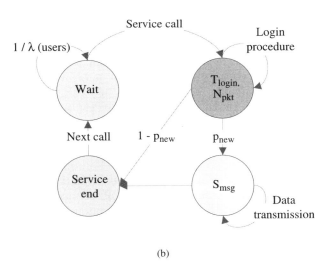

(b)

Figure 14.12 Simulation model for POP3 service: (a) service call of a single user; (b) model states for traffic generation.

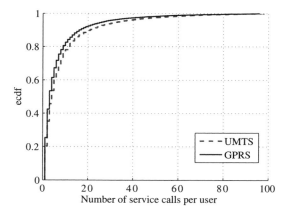

Figure 14.13 Number of POP3 service calls per user and day for UMTS and GPRS.

Probability of Login-only Session   According to our email model we had to discriminate between a simple login with no email download from a login with an email download. Without a parser at the application layer we had to find a compromise in detecting a service call with only a login process. An analysis of the POP3 protocol (RFC 1334 1992) revealed that a normal login process should have fewer than 13 packets, including the TCP handshake. We set an arbitrary limit of 15 packets in up- and downlink directions for connections we considered to be login only. Further, we also set a lower limit to five packets in up- and downlink directions, as the lower limit for a valid connection. This rule was set to filter all port scans and attacks against any email server. The final numbers were 22.4% of the logins with an email transfer and 77.6% with no email transfer. In other words the probability that a user has a new email in his mailbox, called p_{new}, equals 22.4%. Therefore, simulating the login process separately is an important part of the model. The volume for all detected login connections is below 300 byte.

Size of an Email Message   The email size, S_{msg} is the next parameter evaluated. In our case the name 'email size' is a bit misleading, as the POP3 service can transfer many emails in one connection. The parameter is the volume a user will transfer if he is connected to the service once there is a new email in his mailbox.

   A scaling analysis of the email sizes did not show any heavy tailed behaviour. As in Chapter 13 we applied an MLE fit to a set of distributions and chose the best fit based KS statistic. Figure 14.14 presents the fitted values for the login and the email size part. We also included values from the Finnish University NETwork (FUNET) (Brasche and Walke 1997) and the values suggested by Paxson (Paxson 1995; Stuckmann 2003). The FUNET model was derived from traces collected from the Finnish university network and its research network. It uses a Cauchy distribution with a cut-off, S_{max}, for the maximum message size

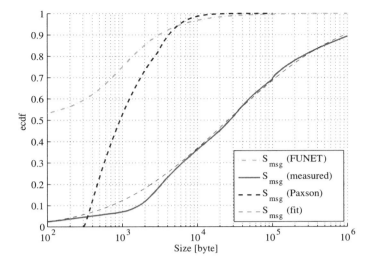

Figure 14.14 Volume transferred per service call.

of 10 kbyte:

$$f_{\text{Cauchy}}(x) = \begin{cases} \dfrac{1}{\pi \ F_{\text{Cauchy}}(S_{\text{max}})} \dfrac{1}{1 + (x - 0.8)^2}, & 0 \leq x \leq S_{\text{max}}, \ x \in \mathbb{N} \\ 0, & \text{else.} \end{cases} \tag{14.1}$$

Paxson (Paxson 1995) characterizes the mail size using two $\log_2$ distributions, similar to a log-normal but with a base of two, one for small and one for large emails. A fixed quota of 300 byte is added to both distributions in order to model the application overhead, S_{head}, within the login process of every email download. The paper already mentions that it is problematic to model the overhead as it is a function of the number of servers the email has to pass, similar to a hop count, in order to reach its destination. Every server adds some lines to the header of the email, thereby increasing the size of the header. It is assumed that 20% of all emails follow the large distribution while the remaining 80% originate from the small distribution. The crossover between the two distributions was set to 2 kbyte. The email size is limited to 100 kbyte.

The values for both models already include the login process. Therefore, the cdf starts at higher values than our email size. The median value from Paxson for the email size is more than one order of magnitude smaller than the actual values measured in the 3G mobile network. The FUNET estimation for the email size is even smaller than in the Paxson model. We assume that this was due to the age of the models. We fitted the parameter mail size with a log-normal distribution. The login process consists of a number of packets, N_{pkt}, with a fixed packet size of 30 bytes payload each. A uniform distribution was used to model the varying number of packets per service call. The processing time at the server side, T_{login}, is also implemented as a uniform distributed variable.

In this paragraph we summarize the model. Table 14.2 compares the fittings of the parameters for the different models. Note that the Paxson values are based on a log-normal

Table 14.2 Model parameters for the email model for both literature and our model. (Note that the $\log_2$-normal distribution is similar to the log-normal distribution, with a different base parameter of 2.)

Model parameter	Distribution	Parameter	
Paxson			
S_{msg} small (byte)	$\log_2$-normal	$\mu = 10.0$	$\sigma = 2.75$
S_{msg} large (byte)	$\log_2$-normal	$\mu = 9.5$	$\sigma = 12.8$
S_{head} (byte)	Constant	$x = 300$	—
S_{max} (byte)	Constant	$x = 100\,k$	—
FUNET			
S_{msg} (byte)	Cauchy	$a = 0.8$	$b = 1.0$
S_{max} (byte)	Constant	$10\,k$	—
New model			
S_{msg} (byte)	log-normal	$\mu = 10.151$	$\sigma = 2.8096$
N_{pkt} (byte)	Uniform	$min = 10$	$max = 12$
T_{login} (ms)	Uniform	$min = 100$	$max = 200$
λ_r^{-1} (s)	Exponential	$\mu = 0.011$	—
p_{new}	Constant	$\mu = 0.224$	—

distribution with base two. The parameter λ_r^{-1} given in Table 14.2 is based on 100 active users within the busy hour, which lasts from 7.45 to 8.45 p.m., averaged for four days. The parameters of the model are depicted in Figure 14.12(b). If a service request is generated, the model enters the login phase. This phase is executed for every user. At the end of the login process the service call will start downloading an email with the probability p_{new} and terminates the session after the successful transfer of the email. If the model decides not to download an email message the service call will terminate right after the login phase. The number of users in the network can be changed by increasing the request rate.

14.4.3   Simulation Setup

In this section we describe the simulation setup that we used to verify our new model. We used ns-2 to verify our email model. A basic setup with two nodes was used to simulate the underlying network conditions. The link parameters were adopted to parameters similar to the RAN we have been monitoring. The data rate was set to discrete values: 40, 64, 128 and 384 kbit/s. These values reflect the setting in the network for UMTS-DCH and GPRS. We simulated the RTT as a uniform distribution from 200 to 500 ms, according to values we have measured for the TCP-RTT in Ricciato *et al.* (2007). Into this artificial scenario the new email agent was implemented. The agent supports email transfers with and without login. It is therefore possible to extract the impact of the login session separated from the email download process. The application uses the email size according to our measurements in the network. In addition to this we implemented the email sizes as suggested by Stuckmann (2003).

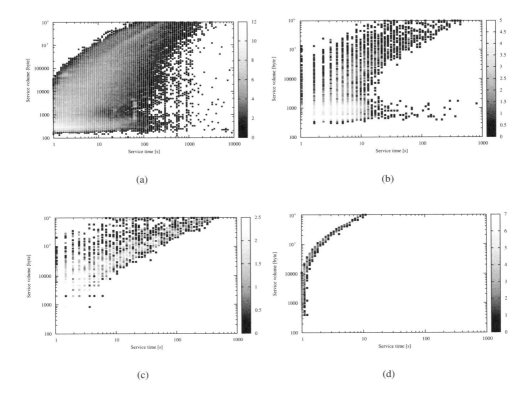

Figure 14.15   Scatter plot of the email service: service time versus service volume (log binning): (a) measured result; (b) simulated: with login; (c) simulated: no login; (d) simulated: with login, ADSL.

Measured Service Footprint   For the first result we extracted a scatter plot of the service time over the data volume transferred. Each point in the scatter plot is equal to one POP3 service call of a client in the network. Hence a single user can be represented with several points in these scatter plots. In this scatter plot both axes were binned into logarithmic bins. The grey scale of each dot indicates the number of occurrences at this specific $[x, y]$ position in the plot. Figure 14.15(a) gives the service footprint recorded in the mobile core network at the Gn interface of the provider.

There are two main clusters visible in Figure 14.15(a). The lower cluster is due to users who only access the system and do not find any new email in their account. They complete the login procedure but do not transmit any emails. This is visible from the fact that they only have a very limited traffic volume per service call. The wide spread in time reveals the large variation in the round trip time for the different terminal connections. This is consistent with our finding in Ricciato *et al.* (2007). The second cluster above 1 kbyte is generated by terminals which have new emails waiting at the server for delivery. This cluster is modelled by the email size parameter we extracted in the previous section. The data rate of these terminals is limited to the parameters of the RAN. Therefore, there is an upper boundary at a net data rate of approximately 1 Mbit/s.

14.4.4   Simulation Results

We now setup a simulation according to section 14.4.3 in ns-2. In this section we present the results of this simulation. To compare the output of our model with the measured results presented in Figure 14.15(a) we ran three different setups: with login and without a login process, which is similar to the Paxson model, where in fact only the size of the emails is distributed differently. The first two scenarios should analyse the impact of a separate login process emulation. The third setup is used to show possible gains over common models.

The outcome of the first scenario is depicted in Figure 14.15(b). The figure is very similar to the measured result in Figure 14.15(a). The 2D correlation coefficient of this result with the measured data is 0.91, which indicates a high similarity. However, the variation in the direction of the x-axis is smaller than for the recorded results. This is due to the fact that we use only a simple emulation of the underlying RAN system, introducing only small variation.

The result of the second scenario in Figure 14.15(c) differs strongly from the recorded trace in Figure 14.15(a). Here the correlation is only 0.3, which indicates the large difference between the results. As the login emulation is not in place, there is no cluster and no additional delay from the network. The data is pushed directly to the mobile terminal.

Finally, we simulate the first scenario with new link parameters set according to an ADSL line: 1 Mbit/s of data rate and 10 ms of RTT. In Figure 14.15(d) the results for this setup are presented. The link delay is one order of magnitude smaller than in the mobile RAN case. Therefore, the result is similar to a scenario where the login process was neglected. We conclude that this is also the reason why common email models, designed for fast LAN connections, skipped the login process.

In concluding the POP3 model we learned that only 20% of subscribers access this application. As in most of the previous sections the parameter distributions are similar for UMTS and GPRS. The email size did not show heavy tailed properties. A comparison based on TCP footprints showed that a simulation with an explicit implementation of a login process delivers better results.

References

Barford, P. and Corvella, M. (1998) Generating representative Web workloads for network and server performance evaluation. *Proceedings of Measurement and Modeling of Computer Systems*, vol. 1, pp. 151–160, New York, USA.

Brasche, G. and Walke, B. (1997) Concepts, services and protocols of the new GSM Phase 2+ General Packet Radio Service. *IEEE Communications Magazine*, **35**, 94–104.

Choi, H. and Limb, O. (1999) A Behavioral Model of Web Traffic. In *Proceedings of the Seventh Annual International Conference on Network Protocols (ICNP '99)*, vol. 1, pp. 327–338, Washington, DC, USA.

IETF RFC 918 1984 (1984) Post Office Protocol.

IETF RFC 959 1985 (1985) File Transfer Protocol (FTP).

IETF RFC 1334 1992 (1992) PPP Authentication Protocols.

IETF RFC 1939 1996 (1996) Post Office Protocol Version 3.

IETF RFC 1945 1996 (1996) Hypertext Transfer Protocol – HTTP/1.0.

IETF RFC 2060 1996 (1996) Internet Message Access Protocol.

IETF RFC 2616 1997 (1997) Hypertext Transfer Protocol – HTTP/1.1.

IETF RFC 2821 2001 (2001) Simple Mail Transfer Protocol.

IETF RFC 3591 1999 (1999) Internet Message Access Protocol Version 4.

IETF RFC 2929 2000 (2000) Domain Name System (DNS).

Graja, H., Perry, P. and Murphy, J. (2001) Development of a Data Source Model for a GPRS Network Simulator. In *Proceedings of the First Joint IEEI/IEEE Symposium on Telecommunications Systems Research*, vol. 1, pp. 1–10, Dublin, Ireland.

Mah, B. (1997) An Empirical Model of HTTP Network Traffic. In *Proceedings of Conference on Computer Communications (INFOCOM'97)*, vol. 2, pp. 592–600, Kobe, Japan.

Osterman, S. (2005) `tcptrace`, available at: http://www.tcptrace.org.

Paxson, V. (1995) Empirically-derived analytic models of wired-area TCP connections. *ACM Transactions on Networking*, **6**(5), 226–244.

Reyes, G., Gonzalez-Parada, A. and Casilari, F. (1999) Internet Dial-up Traffic Modelling. In *Proceedings of the Specialist Seminar on Performance Evaluation of Wireless and Mobile Systems*, vol. 1, pp. 1271–1280, Edinburgh, Scotland.

Ricciato, F., Vacirca, F. and Svoboda, P. (2007) Diagnosis of capacity bottlenecks via passive monitoring in 3G networks: an empirical analysis. *Computer Networks*, **51**(4), 1205–1231, March.

Staehle, D., Leibnitz, K. and Tran-Gia, P. (2000) Source Traffic Modelling of Wireless Applications. In *Technical Report, Universität Würzburg*.

Stuckmann, P. (2003) *Traffic Engineering Concepts for Cellular Packet Radio Networks with Quality of Service Support*. PhD Thesis, Aachen, Germany.

Stuckmann, P. and Hoymann, C. (2002) Performance Evaluation of WAP-based Applications over GPRS. In *Proceedings of IEEE International Conference on Communications (ICC 2002)*, vol. 5, pp. 3356–3360, New York, USA.

Svoboda, P. (2007) `IEControl`, available at: www.nt.tuwien.ac.at/fileadmin/users/psvoboda/ie_control.zip.

Svoboda, P., Karner, W. and Rupp, M. (2007) Modeling E-mail Traffic for 3G Mobile Networks. In *Proceedings of the 18th Annual IEEE International Symposium on Personal, Indoor and Mobile Radio Communications (PIMRC'07)*, vol. 1, pp. 344–349, Athens, Greece.

Svoboda, P., Keim, W., Ricciato, F. and Rupp, M. (2006) Measured Web Performance in GPRS, EDGE, UMTS and HSDPA With and Without Caching. In *Proceedings of the Eighth IEEE International Symposium on a World of Wireless, Mobile and Multimedia Networks*, vol. 1, pp. 56–62, Helsinki, Finland.

Villy, B., Arne, J. and Rasmussen, J. (2000) Internet Dial-up Traffic Modelling. In *Proceedings of the 15th Nordic Teletraffic Seminar*, vol. 1, pp. 11–23, Lund, Sweden.

15

Traffic Models for Specific Services

Philipp Svoboda

In this chapter we again present service traffic models. In contrast to the previous chapter these models were built from scratch and are related to the recently emerging and fast growing applications in the sector of online gaming and Push to Talk (PTT). Section 15.1 introduces three new models for online games. Each model covers one genre of online games: first person shooter, real time strategy and massive multiplayer online games. A purely theoretical model is presented for PTT in section 15.2. The model was implemented in MATLAB according to Nokia PTT specifications and ITU 59.9 artificial conversations. Service shares at the Internet backbone links of today (2007) show that new services take over the lead of HTTP. In terms of volume, peer-to-peer networks generate more traffic than all web browsing customers. However, in a mobile cellular network, volume is a costly resource for the customer. Therefore, we did not focus on such applications but, rather, investigated new applications for which the customer may be willing to pay extra in order to stay connected everywhere. We chose online gaming and PTT as new emerging markets; online gaming because today most of the games have a kind of social platform. The users want to stay online even on weekends in the countryside. The fees for the games are charged monthly. Therefore, the customers want to be online as often as possible. Figure 15.1 depicts the growth of the number of subscribers for a popular massive multiplayer game, namely World of Warcraft (WoW). Already in 2006 we found more than 1000 gamers of WoW in the mobile network in Austria (population eight million). In total more then 7% of the *high* group, as defined in Table 12.2, mobile customers played online games. Therefore, online games are already an important service for customers.

PTT is a similar system to Voice over Internet Protocol (VoIP), which we consider to be an important service that may replace classical phone communication. Therefore, we focused on this service for building a traffic model. At the beginning of our work (2004) only a

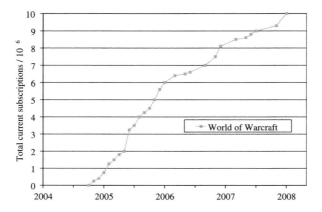

Figure 15.1 Growth of subscriptions to the online game WoW.
(Source: http://www.mmogchart.com/.)

few analyses showed signs of online games existing (Färber 2002, 2004) and PTT was not launched until then. Therefore, we decided to develop new models for these types of service.

15.1   Traffic Models for Online Gaming

Online gaming is an application that evolved from applications called multiplayer games. In multiplayer games two or more players enter the same game either in competition or in cooperation mode. The early adaptations were realized at the same machine; later the players took part on separate computers connected via a network. The first known implementation was coded in 1959 by William A. Hinginbotham (Armitage *et al.* 2006). He used an oscilloscope to show a virtual tennis game at the Brookhaven National Laboratory. The commercial success started with Galaxy Wars in the 1970s (Armitage *et al.* 2006). However, at this point a computer was far too expensive for a normal household; therefore, games did not spread quickly. The introduction of cheap home computers such as the Commodore C64 changed the scene. In the early 1990s id Software released a game called Doom (Armitage *et al.* 2006). It was a first-person shooter; although not the first of its kind, it had an important new function implemented: network gaming. This feature yielded a big success and was from then on a must-have for follow-up games. Today some games can be played online only.

Network Architecture   There are many different network architectures for online games to establish communication between nodes. The simplest realization is depicted in Figure 15.2(a), in which the events of all players are exchanged between the nodes and then executed synchronously. There is no need for a dedicated server node; however, if one computer crashes the communication is broken. In real-time strategy games a setup such as shown in Figure 15.2(a) is often used. Here, this is no shortcoming as such games need all players to stay connected until the end. The architecture does not scale well as every added client increases the information to be transferred between the clients. The server–client architecture, as shown in Figure 15.2(b), scales better in terms of participating users. All clients

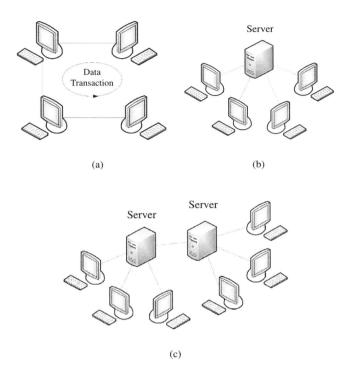

(a)   (b)

(c)

Figure 15.2   Online gaming network architectures: (a) peer-to-peer; (b) client–server; (c) client–server cluster.

report the input updates from their players to a central server node which can physically run on one of the clients. The server processes the inputs and broadcasts the results to the client computers. This setup scales better for a large number of players, as the client information is only reported once to one node. The server then sends only the updates relevant to the specific client, for example only information within the view of the player has to be updated.

For very large user populations, as in Massively Multiplayer Online Games (MMOG) where more than 10 000 users join the same online game, the client load may be split to several servers that run as a loose cluster as depicted in Figure 15.2(c).

Modelling Online Games   In the recent past the popularity of online games has grown quickly (MMOGChart 2008). Even in our observed mobile cellular network the traffic share of online gaming traffic reached a level of around 2% in 2006 (Svoboda *et al.* 2007). The relatively high load combined with the fact that this type of traffic shows significant different characteristics compared to common and well-known traffic sources such as World Wide Web (WWW), email or File Transfer Protocol (FTP) makes it interesting to analyse. In Asia, MMOGs are much more popular than in Europe. Therefore, there are already some publications based on games such as ShenZhou-Online (Färber 2002; Lang *et al.* 2004;

Zander and Armitage 2005). In this chapter we provide new traffic models for the following three different types of online games: First Person Shooter (FPS), Real Time Strategy (RTS) and MMOG.

15.1.1   Traffic Model for a Fast Action Game: Unreal Tournament

FPS games are the ultimate benchmark for mobile systems. These games tolerate only a small amount of delay and jitter (Svoboda and Rupp 2005), which are system inherent in wireless connections. Today's 3G networks cannot provide low-delay connections suitable for FPS games. However, with the deployment of High Speed Downlink Packet Access (HSDPA) and High Speed Uplink Packet Access (HSUPA), delays in mobile networks may be low enough for FPS games. We thus investigated the traffic patterns of an FPS game.

Unreal Tournament 99 (UT 99) is a game quite popular in the world of FPS. It has a dedicated server infrastructure as depicted in Figure 15.2(b). In an FPS the games are viewed in a first person perspective and the goal is: shooting opponents. An important factor in this type of game is timing and accurate movement, making it challenging to set up on wireless systems. UT 99 sends data via the UDP transport protocol, as most real time applications. In the previous section we modelled services running on top of TCP. In contrast to UDP, which offers only error correction codes for the packets, TCP offers reliable transmission of data including retransmission of lost segments. Therefore, data transmission in TCP has to be modelled as data units from an application. In UDP, however, there is no protocol interaction. We can directly model the traffic patterns of an application via the packet size and the *interarrival-times* of the packets.

The following model for UT assumes that the communication between server and client is independent of the packet loss at the interface level. While in wired networks packet loss can often be neglected, this is not possible in a mobile cellular network. However, there are several publications which show that the applications we simulate, for example UT and SC, can operate with up to 30% of packet loss (Armitage and Steward 2004; Armitage *et al.* 2006; Beigbeder 2003). This is achieved by the complex prediction of user movement and actions. The target packet loss in the network under test is set to 1%. We showed in Ricciato *et al.* (2007) that even in the case of a bottleneck in the core network, the packet loss did not exceed 5%. Therefore, we did not consider correlation between packet loss and source rate.

Packet Level Analysis

As mobile access technologies of today are too slow to serve as a transport network for FPS applications, we could not collect our traces from the METAWIN system. Therefore, we recorded eight hours of packet-switched traffic on a PC connected to the Internet via an ADSL modem, game version 1.32.

Figure 15.3 displays the packet-level analysis of a UT 99 trace. As discussed above, we focus on packet size and packet interarrival times for server and client separately. Histograms for the server node are depicted in Figures 15.3(a) and 15.3(c). We can observe that the server is transmitting packets of varying size at a nearly constant rate. We assume that the server sends regular packets to keep all clients updated. The size of the packets changes depending on the amount of events the server has to report.

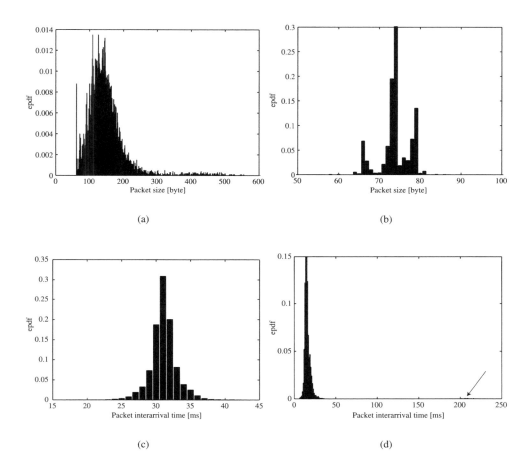

Figure 15.3 Packet level analysis for Unreal Tournament 99: (a) sent packet size (server); (b) sent packet size (client); (c) interarrival times (server); (d) interarrival times (client).

In contrast to this, the client, as we recognize from Figure 15.3(b) and (d), transmits packets at a varying rate with nearly constant size. The client sends packets depending on the user interaction. There are some events visible at 200 ms interarrival time; notice the arrow at the right side in Figure 15.3(d). A visual inspection of the traffic time series in combination with the ingame videos revealed that there are periods of low uplink traffic for the time span when a user is dead and has to wait until he can rejoin the game. Based on these findings we designed a state space to model the different levels of traffic.

Traffic Model for UT 99

The traffic model for UT 99 consists of three states. A player starts the session in the 'Change Level' state. After a timeout the state is automatically changed to 'Active'.

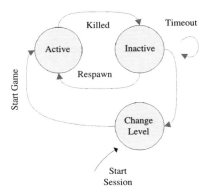

Figure 15.4 State diagram for a UT99 traffic model.

The game starts and server and client begin to transmit packets. The client transmits packets with an interarrival time on the left side of the histogram in Figure 15.3(d). If the player is killed, the interarrival time of the client is around 200 ms corresponding to the second cluster in Figure 15.3(d). The player will stay for some timeout in the killed state until he can rejoin the game again. After the session time the state automatically changes to 'Change Level' and the cycle can restart again. In the 'Change Level' state we assume as a simplification that both server and client do not transmit any packets. The parameters which we needed to implement our model were: packet size and interarrival time for client and server, session time, respawn timeout and the time it takes to change a level. In the following paragraph we extracted these parameters and implemented the model into ns-2.

Parameter Fit and Simulation Results   In this paragraph we present the fit for the parameters which we defined for the traffic model. A UDP packet has an upper limit defined by the Maximum Transfer Unit (MTU); therefore, we are not required to search for heavy tailed behaviour. Table 15.1 shows the best MLE fits for the packet size and interarrival times. The respawn timeout and the time between two consecutive levels were set to discrete values according to the observed mean value from the trace.

Figure 15.5 displays two ecdfs for the uplink data rate, the measured data in solid and the ns-2 simulation result in dashed. For this graph we simulated one gaming session of 15 min. We realize that the model produces an ecdf similar to the empirical data we measured.

Concluding the UT 99 traffic model we learned that this FPS has different states and each of them has different traffic patterns. As the application uses UDP as a transport protocol the traffic can be modelled simply by fitting size and interarrival time of the packets to an analytical distribution. The output from our ns-2 implementation yielded results close to the empirical data. As we did not have access to the source code in order to understand the origin of the correlation in up- and downlink, such correlation was not taken into account.

Table 15.1 Model parameters for UT 99. Note the first variable is log-normal!

Model parameter	Distribution	Parameter
Packet size (server)	log-normal	$\mu = 4.21, \sigma = 0.657$
Packet size (client)	normal	$\mu = 73.53, \sigma = 2.331$
Interarrival time (server)	normal	$\mu = 31.23, \sigma = 1.221$
Interarrival time$_{low}$ (client)	normal	$\mu = 32.70, \sigma = 8.739$
Interarrival time$_{high}$ (client)	normal	$\mu = 218.50, \sigma = 15.130$
Respawn timeout	constant	4.329 s
Session time	constant	15.42 min
Change level timeout	constant	23.61 s

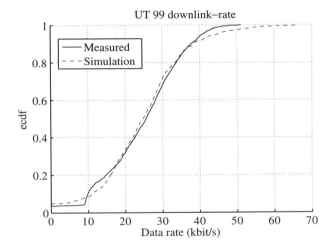

Figure 15.5 Simulation result for ns-2 implementation of the UT99 traffic model.

15.1.2 Traffic Model for a Real Time Strategy Game: Star Craft

Star-Craft (SC) is one of the most popular RTS games. There are three tasks to master in this kind of game: first, collect resources (often two types), secondly, use them wisely to build units and thirdly, finally go into combat. Timing in these games is not that crucial, but higher delays also result in problems for micro management of the built units. The average delay of a 3rd Generation (3G) network is approximately 200 ms. SC could be played well when we ran the game via a UMTS DCH connection (Svoboda and Rupp 2005).

Like UT 99, SC uses UDP packets in the local network. We again recorded packet traces and extracted the size and interarrival time of such packets. SC has a network structure as displayed in Figure 15.2(a). All nodes participating in the local network run synchronously. The failure of a single node results in blocking the others. Again we assume that the

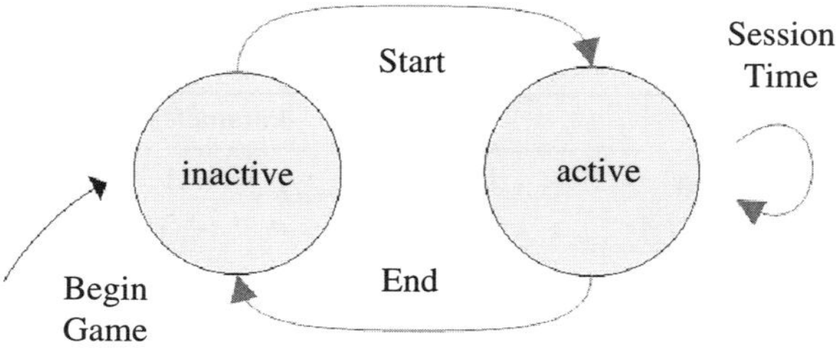

Figure 15.6 State diagram for an SC traffic model.

parameters of the model are independent of the packet loss at the interface. This decision is based on the same arguments we presented for the UT traffic model in section 15.1.1.

Packet Level Analysis   We recorded a packet trace of a one-hour-game session. From this dataset we extracted the parameters' packet size and interarrival time. Both parameters had a nearly constant value. The interarrival time was uniformly distributed between 31 and 33 ms. The packet size sent from each client was between 75 and 78 byte.

Traffic Model   In the last paragraph we showed that the traffic patterns of SC are very simple. A two-state model, as depicted in Figure 15.6, can generate appropriate traffic. The only missing parameter is the session time. We extracted this parameter from 400 games, which were recorded over a timespan of three years. The empirical data for the session time had two peaks, one at around 8 min and a smaller one at around 19 min. We modelled the session time as a combination of two normal distributed variables $t_s = 0.62 \cdot N(8.2, 2.3) + 0.38 \cdot N(18.7, 3.7)$.

15.1.3 Traffic Model for a Massive Multiplayer Online Game: World of Warcraft

Within the last few years, games such as World of Warcraft (WoW) made MMOGs very popular. An MMOG intends to create a virtual environment which is populated simultaneously by thousands of players. The big attraction arises from the fact that only a few characters are simulated by the server, called Non Playing Characters (NPCs), while the rest are humans.

The session times of MMOGs are typically higher than those of round based action games, such as FPSs. FPS games join short matches in the order of 15 min (Svoboda *et al.* 2007). Users stay for several matches in one session. Due to these tournaments the sessions have a periodical behaviour. It is therefore common to model only one tournament and repeat the result (Lang *et al.* 2004; Svoboda and Rupp 2005). MMOG gamers will play one continuous session with varying parameters, impacting the traffic patterns. MMOGs motivate to go online regularly to interact with the virtual environment. Therefore, the timespan between consecutive sessions may show a periodic profile.

From a network traffic engineer's perspective this type of application pushes a new class of service: the real-time interactive service over TCP. The real time restriction is caused by the interactive nature of the application. The gamer is interacting with others via the server node. Therefore, responsiveness of the underlying transport network is crucial for a good subjective gaming experience. The maximum tolerable delays are much larger than those known from other online games. Common real-time systems use UDP because retransmitted packets will have too high a delay and are discarded anyway. In WoW and many other MMOGs, TCP serves as a transport protocol. TCP is connection oriented and offers a reliable connection. This attribute is well suited for MMOGs, preventing error propagation during long sessions. The RTT can be reduced by using small packets. In WoW we are now dealing with an application generating large numbers of very small TCP packets in one long stream. Services such as web browsing generate many small connections. WoW is the first game we could monitor on our 3G live network. In fact we were quite surprised to find WoW in the top ten TCP services in one of our traces. However, this was only true for a single day in May 2006, part of TR_3.

Measurement Setup

In order to obtain the different parameters we used two setups for our measurements, an active and a passive approach, shown in Figure 15.7. The first setting is based on active measurements. This setup is able to monitor user session parameters, IP packet size and interarrival time. The IP parameters were recorded using a monitoring PC running the WoW client, which was connected to the Internet via a 1 Mbit ADSL link. In addition to this an ethereal session running at the PC recorded the packet traces. The datasets consist of several gaming sessions with a sum of about 20 h net active time, recorded in May 2006. The user behaviour was extracted using a shell script recording the session time of two groups with five players each. The script ran as a background job, dumping start and stop time of the wow.exe file. These outputs contain the session duration for each player. To verify these values from the small sample we used a portion of the TR_6 dataset.

The second dataset (TR_6) contains two days of traffic from September of 2006. Analysing this trace we found approximately 1000 different users playing WoW. We were surprised by the fact that these users added up to nearly 1% of the total TCP traffic in the core network. We showed in Chapter 12 that for non-HTTP services a share of 1% is a relatively large number in mobile networks. In fact other common services such as FTP had a smaller volume share.

Considering the two sources we were able to verify the session time of the small sample with the one collected from the large population. Note that the users monitored in the mobile core network most probably used a flat rate contract; therefore, a similar behaviour is expected.

Traffic Characterization

MMOGs focus on the accurate execution of client inputs. This has an impact on the transport protocol. WoW sends its packet through TCP connections, while common FPSs transmit their packets via UDP. The TCP ACK mechanism will protect the delivery of the data packets by the underlying physical network.

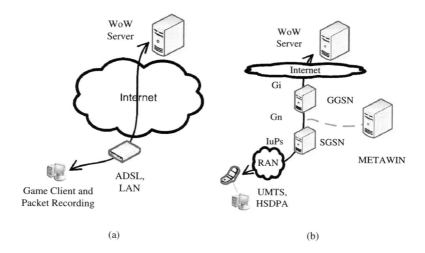

Figure 15.7 Measurement setups: (a) wired networks; (b) mobile networks.

Basic Analysis of the Traffic   Analysing the TCP stream from WoW, we observed that the two flags ACK and PSH are set for most of the packets. In near real-time usage both flags are used, although this is uncommon in normal TCP transmission. A TCP agent has several performance features (IETF RFC 793 1981), one of which is delayed acknowledgment. Instead of sending an ACK packet for each datagram, the receiver will wait for some time and then acknowledge a bunch of packets all together. On low-delay links this will reduce the number of sent ACK packets. However, in near real-time applications which do not tolerate delays, this is a bad choice as it will introduce extra delay in the case of a retransmission.

Another throughput optimization is achieved by the gathering of small chunks of data. The Nagle algorithm in the TCP agent uses a lower threshold in order to send blocks of data (IETF RFC 896 1984). As TCP is connection oriented, the way programs transmit data has fundamental differences compared to UDP. In UDP the program has to take care of packetizing the data, whereas in TCP the program hands over its payload to the agent without preprocessing; for example, there is no problem for the program to hand over more than MTU bytes because the TCP agent will split them later. Due to this behaviour the TCP agent has no detailed information on a time stamp to send a datagram. To avoid the transmission of packets with very small payload that leads to a large overhead, a threshold was established. In an application with relative low data rates this will introduce a high delay. Known applications such as telnet suffer from this fact. Non-varying data rate will add jitter. The PSH flag forces the TCP agent to send the output buffer immediately, bypassing the threshold.

Packet Traces   In this paragraph we analyse packet traces recorded during several sessions. Our datasets contain ≈ 20 hours of gaming. Figures 15.8(a) and (b) give a snapshot of the down- and uplink data rate time series. To plot the data rate, we used a bin size of 1 s and no running average. The median downlink data rate is 6.9 kbit/s. In uplink we observe less traffic with a median of 2.1 kbit/s. From the figure we also learn that there are some high

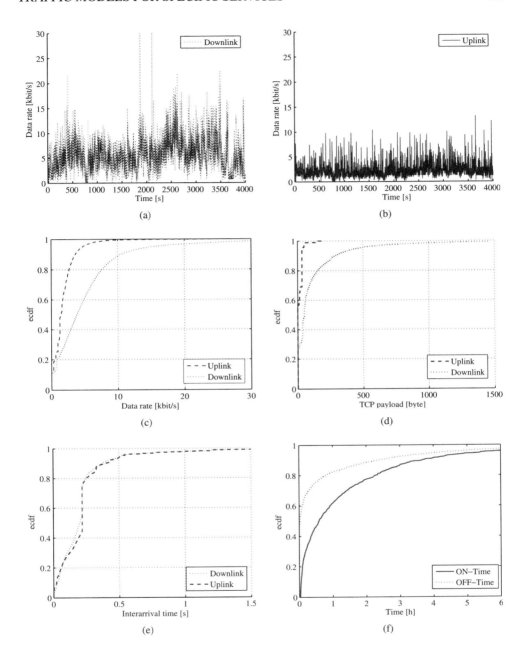

Figure 15.8 Traffic snapshot and summary of the packet-level analysis: (a) snapshot of a downlink data rate time series; (b) snapshot of an uplink data rate time series; (c) data rate; (d) packet sizes; (e) packet interarrival time; (f) packet interarrival time.

peaks ($\geq$ 64 kbit/s) in the downlink direction. Comparing the traces with a recorded video we can correlate the peaks to scenes with high environment interaction, for example many players nearby. We did not use this information in our model.

Figure 15.8(c) presents the cumulative distribution function (cdf) of the up- and downlink data rate. Compared to the snapshot in Figure 15.8(a) we observed some non-negligible part of high downlink bandwidth. A detailed analysis of the packet trace time series showed that these peaks occur mainly at the beginning of the sessions. Figure 15.8(d) illustrates the empirical cumulative distribution function (ecdf) of the packet sizes in the up- and downlink directions. The figure displays that in the up- and downlink, 28% and 57% of the packets, respectively, have no payload. This is due to the fact that these packets are only acknowledgments (ACKs).

Figure 15.8(e) shows the ecdf of the interarrival time of packets in up- and downlink directions. From this we learn that the client has a target update interval of approximately 220 ms indicated by a step in the cdf. The maximum values range up to 1.5 s. Up- and downlink curves are similar.

User Behaviour With the METAWIN system we recorded the session times for a large population ($\approx$ 1000 gamers). Next, we focused on the question of whether or not using a mobile network had a negative impact on the active time (on-time). Therefore, we also extracted session times from two groups of users, each consisting of five gamers connected via fixed line access. Ten persons were monitored by a shell script for one month. We split the users into two groups according to their Internet access speed. Group one used a standard ADSL link with 1 Mbit/s/256 kbit/s and group two had a link speed of 1.5 Mbit/s/512 kbit/s at uplink/downlink.

Figure 15.8(f) visualizes the ecdf for the session times obtained from TR_6. The on-time exceeds the value of 2.75 hours in more than 20% of the samples. It is interesting to see such high values of active service times in a mobile core network. This indicates that the mobile Internet access is used to replace wired Internet access links. The tail of the off-time curve indicates that the players are online on a regular time frame. The stars and crosses in Figure 15.9 show the values obtained by the second dataset extracted from wired Internet accesses. We concluded that the results of all three datasets were similar. Therefore, we fit our model parameters according to the TR_6 dataset.

Simulation Results Based upon our findings in section 15.1.3 we developed an ns-2 script to model a WoW-like client to sever connections. The following paragraph discusses the introduction of two new parameters that adapt the measured values to inputs for the simulation. In contrast to UDP, using a TCP connection between server and client introduces interaction at the protocol level. Therefore, it was necessary to transform the input parameters obtained in the packet traces into suitable parameters to drive a simulation. In other words we could not make direct use of the packet sizes measured, but had to derive a data unit from the view of the application. We then fitted these new parameters and implemented them into the ns-2 simulation environment.

Filtered Packet Trace Analysis In the previous paragraph we showed that there is a high fraction of ACK packets in the trace. The first straightforward approach was to filter all

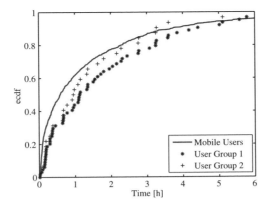

Figure 15.9 Session times: mobile versus wired Internet access.

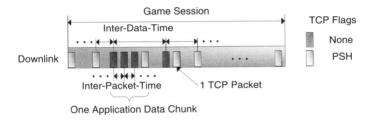

Figure 15.10 Definition of application-data-size and inter-data time.

packets transmitting no payload. Please note that removing all ACK packets would also remove payload packets, as the client can mark a normal payload packet to confirm previous traffic.

Analysing the resulting packet trace, we realized that not all of the packets carry a PSH flag. Extracting these packet sizes we obtained mainly full MTU size. We assume that in this case the application sends datagram units larger than the MTU. The TCP service then splits the units down to MTU in order to transmit them. The last packet of this block is assigned with a PSH flag. Based on this idea we created two artificial values. The inter-data-time replaces the interarrival time for packets and the application-data-size substitutes the packet size. Figure 15.10 illustrates a detailed picture of the parameter mapping.

The following figures visualize the ecdfs for the two parameters. Figure 15.11 displays the ecdfs of the application-data-size in up- and downlink directions. The uplink curve shows several high discrete steps. We conjecture that these values represent simple commands that are often used, such as 'walk', 'attack' or 'open'. More complex commands, represented by a larger data size in the cdf, have a low probability. As we will show later, the update rate is around 220 ms. Using scripts which execute multiple commands may change the distribution.

The downlink curve ranges up to 3000 bytes. In the downlink the application-data size is larger than in the uplink. We applied a continuous distribution for the smooth downlink

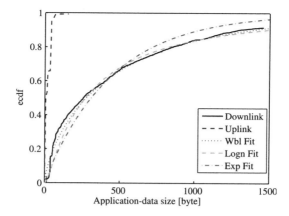

Figure 15.11 Model parameter: application-data size.

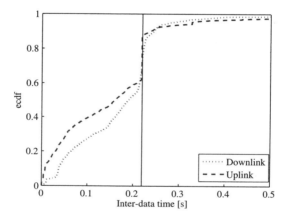

Figure 15.12 Model parameter: inter-data time.

curve. A Weibull distribution fitted best. A comparison between Figures 15.8(d) and 15.11 points out that only the downlink stream is affected by the chosen add-up rule.

Figure 15.12 shows the ecdf of the inter-data time. The filter process reduces the step size at approximately 220 ms. The step occurs in up- and downlink. We assume that this is some threshold coded to the software. Both curves have a relative linear part followed by a step at around 220 ms, indicating a uniform distribution. The tail of the curves shows that the threshold is sometimes exceeded by network delays.

Simulation Model   In ns-2 standard TCP agents only support unidirectional communication. Although client and server were modelled separately from each other, a full featured ACK process needs a bi-directional TCP implementation on the ns-2 level. A bi-directional agent permits a realistic communication between two nodes, for example a payload packet that carries an ACK information. We used the `fulltcp` (Issariyakul and Hossain 2008) agent implemented in ns-2 to model the two-way connection; the agent generates ACK packets according to the TCP algorithms. Therefore, we directly implemented the parameters extracted in the last paragraph to drive our simulation. The following paragraphs discuss the parameter fit. The parameters were fit by MLE and by selecting the distribution with the best KS statistic.

Server → Client We fit a Weibull distribution to the dataset as shown in equation 15.1. The resulting parameters for the distribution were $\lambda = 426$ and $k = 0.8196$. The real distribution has an upper limit $l = 3010$ byte, so we implemented a cutoff to model this limit

$$n = f_{\text{wbl}}(l; k, \lambda)$$

$$f_{\text{wbl}}(x; k, \lambda) = \begin{cases} \dfrac{1}{n} \cdot \dfrac{k}{\lambda} \cdot \left(\dfrac{x}{\lambda}\right)^{k-1} \cdot e^{-(x/\lambda)^k}, & x \geq 0 \\ 0, & \text{else.} \end{cases} \tag{15.1}$$

The inter-data time was harder to synthesize. We decided to use a joint distribution of three random variables. All three processes are modelled uniformly distributed. The resulting description of the pdf is given in equation 15.2. All parameters were fitted by regression to the steps in the dataset ($a = 218.3$, $b = 251.2$, $c = 1\,500$ ms). Hence the curves for inter-data time in up- and downlink directions are similar; we applied equation 15.2 to both.

$$f(x) = \begin{cases} 0.620 \cdot \dfrac{1}{a - 0}, & x = [0 \ldots a) \\[2ex] 0.257 \cdot \dfrac{1}{b - a}, & x = [a \ldots b) \\[2ex] 0.123 \cdot \dfrac{1}{c - b}, & x = [b \ldots c] \\[2ex] 0, & \text{else.} \end{cases} \tag{15.2}$$

Client → Server The application-data size in the uplink had discrete steps, as depicted in Figure 15.11. We implemented a source generating three different sizes of packet (equation 15.3 with $a = 6$, $b = 19$, $c = 43$ byte). The parameters were estimated using the average over the different datasets. The original probability of those steps adds up to 98% in the cdf

$$f(x; a, b, c) = 0.52 \cdot \delta(x - a) + 0.14 \cdot \delta(x - b) + 0.34 \cdot \delta(x - c). \tag{15.3}$$

To model the session times we compared several fittings using Weibull, log-normal and neg-exponential distributions. The best fit for the on-times was obtained using a Weibull distribution with the parameters $\lambda = 4\,321$ and $k = 0.7813$ in equation 15.1. The off-time fitted best a log-normal distribution equation 15.4. The resulting parameters were $\mu = 5.512$ and $\sigma = 2.434$

$$f(x; \mu, \sigma) = \frac{1}{x\sigma\sqrt{2\pi}} \cdot e^{-(\ln x - \mu)^2/2\sigma^2}. \tag{15.4}$$

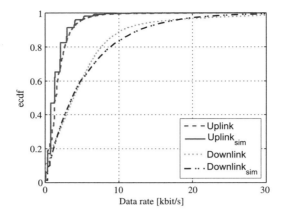

Figure 15.13 Simulated data rate.

Simulation Results

These distributions were implemented in an ns-2 script. The simulation setup consists of a server–client pair using two-way TCP agents. The data rate recorded in 10 h of simulation time is plotted in Figure 15.13. For a direct comparison we presented the cdf and added the original values given in Figure 15.8(c). We noticed that the simulation produces a similar distributed data rate. However, a closer look at the time series revealed the weakness of the simple modelling as there are no activity bursts as in Figure 15.8(a).

Concluding the traffic model for WoW, we learned that this MMOG uses reliable TCP connections to transmit data between server and client. It is the first game we could monitor within our live 3G network and, in September of 2006, TR_6 was, in terms of volume, within the top ten TCP services. In TCP, due to interaction of the protocol with the network, we cannot directly model packet size and interarrival time. We had to extract a pseudo element called the data element by mending together consecutive TCP packets. The ns-2 simulation code based on our model generated a traffic pattern with a cdf nearly identical to the measured traffic.

15.2   A Traffic Model for Push-to-Talk (Nokia)

The push-to-talk service in mobile networks is a new service and therefore most of the information presented here will be of a theoretical nature. The task Push-to-talk over Cellular (PoC) should provide voice services on cheap packet-switched links. To fulfil this requirement, the code rate of the Adaptive Multi Rate (AMR) codec is set to a minimum and relatively high delays will be accepted. The push-to-talk service is a one-way communication using the PoC standard. The voice data is encoded using the AMR codec with a code rate of 5.17 kbit/s. The transport protocol is Real-time Transport Protocol (RTP) on top of UDP. Signalling is transmitted via Session Initiating Protocol (SIP) to a media gateway managing the connections.

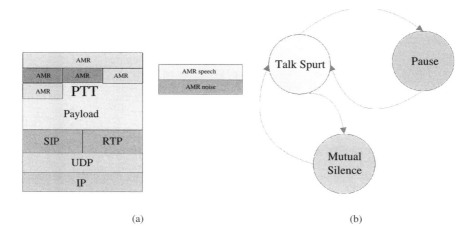

(a)   (b)

Figure 15.14 Push-to-talk over Cellular: protocol stack and state diagram (ITU 59.9). (a) protocol stack for PoC; (b) PoC state diagram.

15.2.1 AMR: Facts from the Data Sheets

The The voice data from one user is coded using AMR. The AMR codec splits the data stream into frames with a timespan of 20 ms (3GPP TS 26071 2004). The output signal consists of three different classes of bit stream (A, B, C), which are coded differently. A holds the important information, and for this reason is protected by the best code. The resulting frame length is 103 bit. The AMR encoder supports activity detection. With this feature it is possible to stop sending data while the user does not speak but listens to the conversation. Absolute silence is not very comfortable for the conversational partner, therefore the so-called 'comfort noise' was introduced in the AMR codec. In the silent period smaller packets containing random noise are transmitted. These packets have 39 bit of payload. An AMR frame is equal to the total payload that has to be transferred for a 20 ms time slot over the PTT bearer. Additionally, we have to consider the overhead of the surrounding protocols. Let us start from the bottom. In the PoC standard the transport protocol is IPv6 but recently there has been an extension so that also IPv4 will be possible. IPv6 has a 40 byte overhead. The next protocol on top of IP is UDP, which consumes 12 bytes of overhead information (IETF RFC 768 1980). UDP supports only connectionless data exchange. The session management has to be processed in the higher layers; see RTP or application level. Finally, the last protocol is RTP, introducing an additional 8 bytes (IETF RFC 1889 1996). In total we now have 60 bytes of overhead and only 15 bytes of information. This problem is solved by adding up 4 to 20 AMR frames in one RTP packet. This increases the payload to about 300 bytes per packet and reduces the overhead to 20%. Figure 15.14 shows the described protocol stack for PTT services.

A further part of the traffic is the heartbeats. PTT services do not have a constant connection. The server has to determine if the user is still there waiting for data. This problem is solved by sending 'empty' packets every 300 s of inactivity, the so-called heartbeats.

Table 15.2 Parameters for artificial conversational speech.

Parameter (s)	Duration (s)	Rate (%)
Talk spurt	1.004	36.53
Pause	1.587	61.47
Double talk	0.228	6.59
Mutual silence	0.508	22.48

15.2.2 Parameters for Artificial Conversational Speech

There are several studies about the talk and silent periods in a human conversation (Deng 1995). The ITU P.59 (ITU P59 1993) standard recommends four different parameters, as depicted in Table 15.2, to model artificial conversational speech. The distribution of the talk duration and talk spurt (the activity period of one user is called spurt in ITU P59) duration is approximated with an exponential function, the pause parameter by an exponential function plus a constant. The activity factor for speech is assumed to be 0.3–0.4. The activity factor states the ratio between talk time and listening time. The likelihood of around 0.5 indicates silent periods of both speakers at the same time.

15.2.3 PTT Model

We used the parameters found in the ITU standard and implemented this to a MATLAB simulation file. There were not enough active users in the network to find an empirical model, and so we decided to build a theoretical model for this type of traffic. The MATLAB script uses the ITU recommendation for the length of a connection and the information of the PoC standard for generating traffic in the activity phase of a user. The overhead of the RTP is also included in the simulation. In a recent publication (James *et al.* 2004) the delay limit for VoIP is given as 200 ms. Setting the tolerated delay to a fixed value is a strong simplification of the problem. In reality this parameter is linked to the packet loss as can be seen in the quality metrics for VoIP, for example E-model (ITU G.107 2000) or Perceptual Evaluation of Speech Quality (PESQ, ITU P.862 2001). With no real measurements to evaluate the packet loss, we ignored this in the simulation setup.

This corresponds to ≈ 10 AMR packets per RTP frame. The MATLAB simulations use these settings to evaluate data rate consumption of various numbers of users per cell. The simulation evaluates x users per run and is re-run until the resulting total average data rate changes by at most 0.1% as compared to the last run.

The results are shown in Figure 15.15. From this figure we can conclude that the overhead depends on the working point of the system. This is due to the fact that small bursts suffer from the large header introduced by RTP. For 100 users, the 99% mark is found at 259.7 kbit/s. The simulation was performed for 100 users over 60 h, using the parameters described above. Figure 15.16 depicts the number of parallel active users ($\equiv$ conversations) and the number of parallel talking users ($\equiv$ acting as data source) in the scenario. This is equal to the probability of parallel used lines (GPRS slots or UMTS bearers) for 100 users. Where there are 100 users in a cell, a blocking probability of less than 1% needs at least six channels. Although GPRS is a packet-switched system, there are some hard limits to

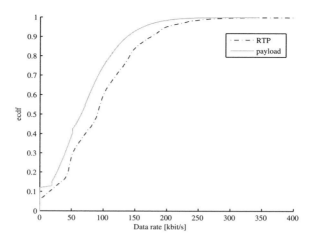

Figure 15.15 Simulated data rate for 100 PTT users.

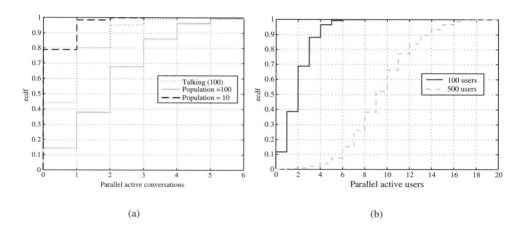

(a)   (b)

Figure 15.16 Simulated number of parallel active users (simulation time: 10^5s): (a) 100 users; (b) 500 users.

the assignment of slots for uplink and downlink. This could result in a blocking probability higher than shown in these results.

There were not yet enough active users to derive a model from the measurements, so we used the ITU recommendations (ITU P59 1993) to build a one-way communication model for PoC services. Using this model we derived the minimum number of parallel channels for different user groups and plotted the result in Figure 15.16. With further monitoring and a growing popularity of this service we can adapt our MATLAB script to the datasets measured.

- 100 users with 1% blocking → 6 slots

- 500 users with 1% blocking → 16 slots

Finally, we want to point out that we did not consider the signalling traffic from the SIP protocol. Therefore, the results are only rough estimations for the minimum number of time slots consumed by this service. A standard GSM cell offers 8 slots per frame. Each slot can carry up to two different voice calls. Therefore, the maximum load for a single cell, using the Erlang-B model, is equal to 8.9 Erlang. Based on the model parameters given in Table 15.2 and a blocking probability of 1% a GSM cell can serve up to 53 customers. The PTT simulation can serve 162 customers within one cell, improving the performance of the cell by a factor larger than three. Note that in the PTT case we only use seven slots for the simulation as the first slot has to broadcast the signalling information in the GPRS cell.

References

3GPP TS 26.071 2004 (2004) AMR Speech Codec: General Description.

Armitage, G. and Stewart, L. (2004) Some Thoughts on Emulating Jitter for User Experience Trials. In *Proceedings of 3rd ACM SIGCOMM Workshop on Network and System Support for Games (NetGames'04)*, vol. 1, pp. 13–18, Portland, Oregon, USA.

Armitage, G., Claypool, M. and Branch, P. (2006) *Networking and Online Games – Understanding and Engineering Multiplayer Internet Games,* John Wiley & Sons, Ltd.

Beigbeder, T. (2003) The Effects of Loss and Latency on User Performance in Unreal Tournament 2003. In *Proceedings of the Third ACM SIGCOMM Workshop on Network and System Support for Games (NetGames'03)*, vol. 1, pp. 144–151, Portland, Oregon, USA.

Deng, S. (1995) Traffic Characteristics of Packet Voice. In *Proceedings of the 1995 IEEE International Conference on Communications (ICC'95)*, vol. 5, pp. 1369–1374, Seattle, WA, USA.

Färber, J. (2002), Network Game Traffic Modelling. In *Proceedings of the First Workshop on Network and System Support for Games (NETGAMES '02)*, vol. 1, pp. 53–57, Braunschweig, Germany.

Färber, J. (2004) Traffic Modelling for fast action network games. *Proceedings of the Multimedia Tools Application*, **23**(1), 31–46.

IETF RFC 768 1980 (1980) User Datagram Protocol (UDP).

IETF RFC 793 1981 (1981) Transmission Control Protocol (TCP).

IETF RFC 896 1984 (1984) Congestion Control in IP/TCP Internetworks.

IETF RFC 1889 1996 (1996) A Transport Protocol for Real-Time Applications (RTP).

ITU P59 1993 (1993) Telephone transmission quality objective measuring apparatus: Artificial conversational speech.

ITU G.107 2000 (2000) The E-model, a computational model for use in transmission planning.

ITU P.862 2001 (2001) Perceptual evaluation of speech quality (PESQ): An objective method for end-to-end speech quality assessment of narrow-band telephone networks and speech codecs.

Issariyakul, T. and Hossain, E. (2008) *Introduction to Network Simulator* ns-2, Springer, Berlin.

James, J.H., Bing, C. and Garrison, L. (2004) Implementing VoIP: A voice transmission performance progress report. *IEEE Communications Magazine*, **42**(7), 36–41.

Lang, T., Branch, P. and Armitage, G. (2004) A Synthetic Traffic Model for Quake. In *Proceedings of the 2004 ACM SIGCHI International Conference on Advances in Computer Entertainment Technology (ACE'04)*, vol. 1, pp. 233–238, Singapore.

MMOGChart (2008) Available at: http://www.mmogchart.com/.

Ricciato, F., Vacirca, F. and Svoboda, P. (2007) Diagnosis of capacity bottlenecks via passive monitoring in 3G networks: an empirical analysis. *Computer Networks*, **51**(4), 1205–1231, Mar.

Svoboda, P. and Rupp, M. (2005) Online Gaming Models for Wireless Networks. In *Proceedings of the Ninth IASTED International Conference Internet and Multimedia Systems and Applications*, vol. 1, pp. 417–422, Grindelwald, Switzerland.

Svoboda, P., Karner, W. and Rupp, M. (2007) Traffic Analysis and Modeling for World of Warcraft a MMOG. In *Proceedings of the 18th Annual IEEE International Communication Conference (ICC'07)*, vol. 1, pp. 1612–1617, Glasgow, UK.

Zander, S. and Armitage, G. (2005) A Traffic Model for the Xbox Game Halo 2. In *Proceedings of the International Workshop on Network and Operating Systems Support for Digital Audio and Video (NOSSDAV '05)*, vol. 1, pp. 13–18, Stevenson, Washington, USA.

Index